AF361015

Research on Submarine Hydrothermal Activity and Its Material Circulation, Magmatic Setting, and Seawater, Sedimentary, Biologic Effects

Research on Submarine Hydrothermal Activity and Its Material Circulation, Magmatic Setting, and Seawater, Sedimentary, Biologic Effects

Editors

Zhigang Zeng
Yuxiang Zhang
Zuxing Chen

Basel • Beijing • Wuhan • Barcelona • Belgrade • Novi Sad • Cluj • Manchester

Editors
Zhigang Zeng
Institute of Oceanology,
Chinese Academy of Sciences
Qingdao, China

Yuxiang Zhang
Institute of Oceanology,
Chinese Academy of Sciences
Qingdao, China

Zuxing Chen
Institute of Oceanology,
Chinese Academy of Sciences
Qingdao, China

Editorial Office
MDPI
St. Alban-Anlage 66
4052 Basel, Switzerland

This is a reprint of articles from the Special Issue published online in the open access journal *Journal of Marine Science and Engineering* (ISSN 2077-1312) (available at: https://www.mdpi.com/journal/jmse/special_issues/Submarine_Hydrothermal).

For citation purposes, cite each article independently as indicated on the article page online and as indicated below:

Lastname, A.A.; Lastname, B.B. Article Title. *Journal Name* **Year**, *Volume Number*, Page Range.

ISBN 978-3-0365-8978-7 (Hbk)
ISBN 978-3-0365-8979-4 (PDF)
doi.org/10.3390/books978-3-0365-8979-4

Cover image courtesy of Zhigang Zeng

Contents

About the Editors

Zhigang Zeng

Zhigang Zeng is a Professor of the Institute of Oceanology, Chinese Academy of Sciences, China. He has more than 20 years of experience in research on submarine hydrothermal activities and the synergetic metallogenic mechanism of magma, fluid, rock, sediment, seawater, and organisms in hydrothermal fields. He established a 'two-stage, six-process' model for investigating hydrothermal activities, proposed a new hypothesis of the same source and different sinks of hydrothermal activity, cold-seep and gas hydrate, and took the lead in proposing the framework of submarine hydrothermal geology. He has published 145 research papers with collaborators, and published the book "Submarine Hydrothermal Geology" and "Submarine Hydrothermal Geology of the East Pacific Rise".

Yuxiang Zhang

Yuxiang Zhang is an Associate Research Fellow of the Institute of Oceanology, Chinese Academy of Sciences, China. His research interests include subduction zone material recycling, and the magmatic process and its relations with hydrothermal systems. He has published 13 1st-author research papers, including 10 SCI papers, which are published in Science Advances, Journal of Geophysical Research: Solid Earth, Journal of Petrology, Lithos, etc.

Zuxing Chen

Zuxing Chen is an Associate Research Fellow of the Institute of Oceanology, Chinese Academy of Sciences, China. His research primarily focuses on utilizing stable isotope compositions of redox-sensitive metal elements, such as Cu and Fe, in arc basalts to evaluate the redox state and transport of slab-derived fluids in subduction zones. He is the first author of 13 research papers, including 11 SCI papers, which are published in Geochimica et Cosmochimica Acta, Journal of Geophysical Research: Solid Earth, Lithos, Journal of Asian Earth Sciences, etc.

Preface

Submarine hydrothermal activity has been a focus of research since its discovery in the 1970s. It is widely believed that submarine hydrothermal activity offers significant prospects for the development of mineral and biological resources, and also that it has a considerable impact on the immediate seawater environment. The study of submarine hydrothermal activity is a multi-disciplinary topic that includes tectonics, petrology, mineralogy, sedimentology, chemistry and oceanography. The development of hydrothermal circulation is controlled in tectonic and magmatic settings, including extensional tectonics where shallow magma reservoirs exist. The chemical compositions of hydrothermal fluids, which are affected by fluid–rock interactions beneath the seafloor and potentially by magma degassing are recorded in hydrothermal minerals, sediments, and seawater. Naturally, such a broad area of research cannot be advanced without the development of exploratory and analytical techniques.

Recent improvements in science and technology greatly assist the development of research on these topics. This Special Issue aims to advance the understanding of the latest progress in research on submarine hydrothermal activity and its material circulation, magmatic setting, and seawater, sedimentary, and biologic effects. This issue includes 13 research works, which cover the chemical and isotopic compositions of seafloor hydrothermal sulfides and fluids, the magmatic process in shallow magma reservoirs and its relations with the hydrothermal system, subduction chemical recycling, and progress in the exploratory and analytical technologies for seafloor hydrothermal activity. We hope these works can help readers better understand submarine hydrothermal activity.

Zhigang Zeng, Yuxiang Zhang, and Zuxing Chen
Editors

Journal of
Marine Science and Engineering

Article

Chemical and Isotopic Composition of Sulfide Minerals from the Noho Hydrothermal Field in the Okinawa Trough

Zhigang Zeng [1,2,3,*], Zuxing Chen [1,4], Haiyan Qi [1,4] and Bowen Zhu [1,3]

1   Seafloor Hydrothermal Activity Laboratory, CAS Key Laboratory of Marine Geology and Environment, Institute of Oceanology, Chinese Academy of Sciences, Qingdao 266071, China; chenzuxing@qdio.ac.cn (Z.C.); qihaiyan@qdio.ac.cn (H.Q.); zhubowen@qdio.ac.cn (B.Z.)
2   Laboratory for Marine Mineral Resources, Qingdao National Laboratory for Marine Science and Technology, Qingdao 266071, China
3   University of Chinese Academy of Sciences, Beijing 100049, China
4   Center for Ocean Mega-Science, Chinese Academy of Sciences, 7 Nanhai Road, Qingdao 266071, China
*   Correspondence: zgzeng@qdio.ac.cn; Tel.: +86-532-82898525

Abstract: Studies of the element contents and isotopic characteristics of sulfide minerals from seafloor hydrothermal sulfide deposits are a significant method of investigating seawater-fluid mixing and fluid-rock and/or sediment interactions in hydrothermal systems. The seafloor hydrothermal sulfide ores from the Noho hydrothermal field (NHF) in the Okinawa Trough (OT) consist of pyrrhotite, isocubanite, sphalerite, galena, and amorphous silica. The Rh, Ag, Sb, and Tl contents mostly increase in galena as the fluid temperature decreases in the late ore-forming stage. In the sulfide minerals, the rare earth elements are mainly derived from the hydrothermal fluids, while the volcanic rocks and/or sediments are the sources of the sulfur and lead in the sulfide minerals. After the precipitation of galena, the redox state becomes oxidizing, and the pH value of the fluid increases, which is accompanied by the formation of amorphous silica. Finally, neither pyrite nor marcasite has been observed in association with pyrrhotite in the NHF sulfides, likely indicating that the amount of sulfur was limited in this hydrothermal system, and most of the residual Fe was incorporated into the sphalerite. This suggests that the later pyrite and/or marcasite precipitation in the seafloor hydrothermal sulfide deposit is controlled by the sulfur content of the fluid. Furthermore, it is possible to use hydrothermal sulfides and their inclusions to trace subseafloor fluid circulation processes.

Keywords: element enrichment; hydrothermal vent; in-situ chemical and isotopic compositions; seafloor hydrothermal sulfides

Citation: Zeng, Z.; Chen, Z.; Qi, H.; Zhu, B. Chemical and Isotopic Composition of Sulfide Minerals from the Noho Hydrothermal Field in the Okinawa Trough. *J. Mar. Sci. Eng.* **2022**, *10*, 678. https://doi.org/10.3390/jmse10050678

Academic Editors: Dimitris Sakellariou, Anabela Oliveira and János Kovács

Received: 1 April 2022
Accepted: 9 May 2022
Published: 16 May 2022

Publisher's Note: MDPI stays neutral with regard to jurisdictional claims in published maps and institutional affiliations.

1. Introduction

The majority of seafloor hydrothermal sulfide deposits contain a mixture of sulfide minerals (e.g., pyrrhotite, pyrite, marcasite, chalcopyrite, isocubanite, sphalerite, wurtzite, and galena), secondary Cu sulfides (bornite, covellite, digenite, and chalcocite), sulfosalts (tennantite-tetrahedrite among many others) [1], sulfates (anhydrite and barite), and amorphous silica in different abundances. For example, pyrrhotite, which contains Cu, Co, Mn, Ni, Sn, Pb, Cr, Ti, Ag, and Se [2–4], is commonly found as a major phase in many seafloor hydrothermal sulfide deposits (e.g., [5–10]). Its in-situ chemical composition and genesis have been investigated in several hydrothermal fields, including the Bent Hill field on the northern Juan de Fuca Ridge (JdFR) [11], the Broken Spur vent field on the Mid-Atlantic Ridge (MAR) [12], the Lucky Strike vent field on the MAR [13], and the Ashadze hydrothermal field on the MAR [14]. However, the trace element contents of pyrrhotite are variable [1]. It is known that the pyrrhotite from the southern JdFR has higher silver (Ag) contents (0.04–0.07 wt.%; [15]) than that from the Endeavour Segment of the JdFR (<0.04 wt.%; [16]) and the Rainbow hydrothermal field on the MAR (0.02 wt.%; [17]). In addition, the Cu content of pyrrhotite from the Endeavour Segment of the JdFR (0.065 wt.%; [16]) is higher than that from the southern JdFR (0.01–0.06 wt.%; [15]).

Sphalerite, also one of the most common minerals in seafloor hydrothermal sulfides and chimneys (e.g., [18–21]), contains Mn, Fe, Co, Ni, Cu, Ag, Sn, Pb, Hg, Cd, Ge, Ga, In, and Sb [22–24]. Its Cd content can be very high (200–11,900 ppm) [1]. Sphalerite associated with arc-related geologic settings generally contains higher levels of As (1000–10,000 s ppm), Pb (1000–10,000 s ppm), and Sb (100–1000 s ppm) than that in mid-ocean ridge (MOR) hydrothermal systems (10–100 s ppm) [1]. The FeS content of sphalerite can be useful in determining its formation temperature and the sulfur fugacity of the fluid in the chimneys in seafloor hydrothermal fields [20].

The presence of significant amounts of galena is characteristic of arc and back-arc basin (BAB) hydrothermal sulfide deposits [10,20,25–27]. Galena contains Cu, Ag, Sb, Bi, Cd, As, In, Mn, Tl, and Zn [28]. Particularly high Cd (800–1100 ppm) and Zn (2000–3400 ppm) contents have been measured in the galena from the Palinuro Volcanic Complex, Hook Ridge, and the Endeavour Segment of the JdFR [16,29]. Furthermore, the highest Sb (2160 ppm) and Ag (9400 ppm) contents in galena have been observed in arc-related seafloor hydrothermal fields [1].

However, the in-situ element contents and their distributions in sulfide minerals from the Noho hydrothermal field (NHF) in the Okinawa Trough (OT) are poorly documented. To reveal the formation processes and metal sources of the sulfide minerals in the seafloor hydrothermal systems, we studied and characterized the mineralogical, chemical, and isotopic compositions of the sulfide minerals from the seafloor hydrothermal sulfide deposits in the NHF.

2. Geological Setting

The OT is an active BAB located behind the Ryukyu trench–arc system, extending approximately 1200 km from Taiwan island to Kyushu Island in the western Pacific (Figure 1). It was formed by the active subduction of the Philippine Sea Plate under the Eurasian Plate in the Early Pleistocene (2–1.5 Ma; [30–32]). However, the OT is divided into the southern (SOT), middle (MOT), and northern (NOT) segments by the Kerama and Tokara faults [33]. The latest phase of rifting started <100 ka ago according to the ages of the zircon in the OT volcanic rocks, but the exact ages of the earlier rifting phases may be considerably older (108 Ma to 2.7 Ga) [34]. Furthermore, the SOT is characterized by a sedimentary cover that is much thinner (1–3 km thick) than the very thick cover in the NOT (~8 km) [31,35,36]. The OT volcanic rocks vary from tholeiitic basalt to calc-alkaline andesite to rhyolite [37], including basaltic andesite and dacite in the MOT [38].

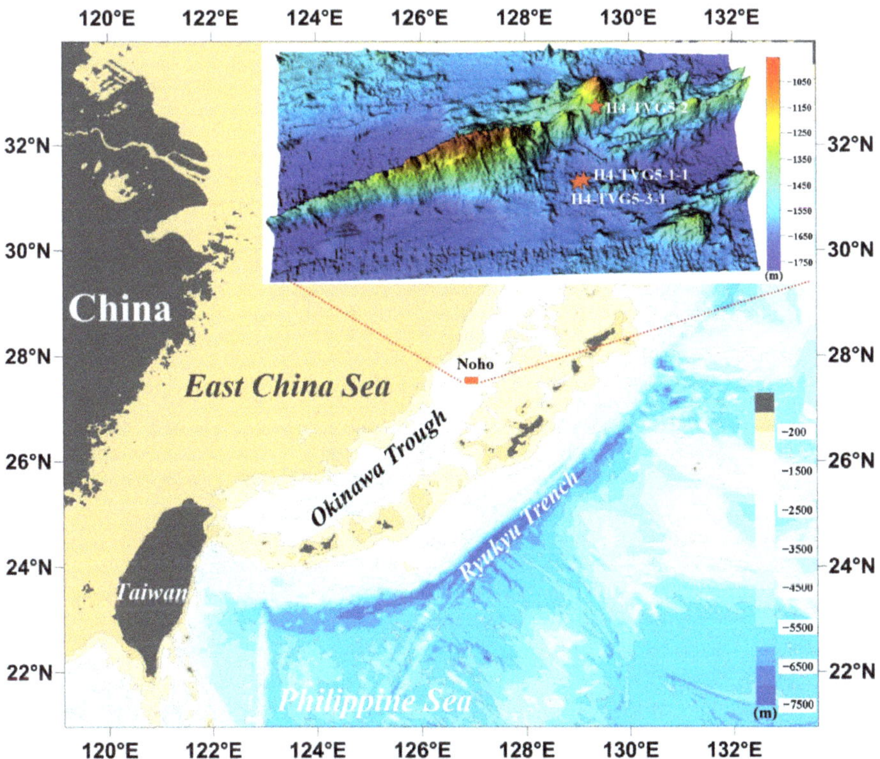

Figure 1. Locations of seafloor hydrothermal sulfide samples H4-TVG5-1-1 and H4-TVG5-3-1 in the NHF and trachyandesite samples H4-TVG5-2 (Li et al. [39]) near the Iheya Ridge in the OT.

However, the presence of high ^{3}He contents in the seawater [40] and a high heat flow [41] suggest that there are numerous active seafloor hydrothermal fields in the MOT [42]. Indeed, several seafloor hydrothermal fields have been discovered in the MOT [43–47]. For example, the NHF (27°31.1′ N, 126°59.0′ E, 1581 m) was surveyed in 2016 during seafloor hydrothermal activity off the Iheya Ridge in the MOT using a television (TV)-log grabber (Figure 1) during the HOBAB4 cruise. The survey was conducted near the Noho site (27°31.3′ N, 126°59.1′ E, 1595 m), approximately 3 km SE of the Clam hydrothermal field in the Sakai field discovered by the Japan Oil, Gas, and Metals National Corporation (JOGMEC) ([48]; JOGMEC news release on 4 December 2014; NT15-13 cruise report available at http://www.godac.jamstec.go.jp/darwin/cruise/Natsushima/NT15-13/) The exact position of the Noho (JOGMEC) site had not yet been published at that time [42]. The NHF likely corresponded to the acoustic water column anomaly site [42]. It is mainly associated with basaltic and trachyandesitic rocks (H4-TVG5-2; Figure 1) which are located on the outside slope of the Iheya Ridge and are covered by several meters of pumice-rich sediments. The site includes active hydrothermal vents, chimneys, hydrothermal sulfide deposits, and chemosynthetic communities. However, there is a coupling relationship between the tectonic setting, magmatism, mineralization, fluid-rock interactions, and sedimentary processes in the NHF hydrothermal system and the response, adaptation, record, and activity of organisms. It is interesting to investigate the synergetic metallogenic mechanism of the magma, fluid, rocks, sediments, seawater, and organisms in the NHF hydrothermal field. In the future, it will be possible to use hydrothermal sulfide and its inclusions to trace the deep circulation processes of the fluids in the NHF. Furthermore, we can combine geophysical prospecting, geochemical exploration, and biological exploration techniques with drilling techniques to explore the seafloor polymetallic sulfides and hydrothermal vents, and we can use the four systems of exploration techniques to understand the resource potential of the seafloor polymetallic sulfide deposits in the NHF.

3. Sampling and Methods

Seafloor hydrothermal sulfide samples H4-TVG5-1-1 and H4-TVG5-3-1 were collected from the NHF using a TV grabber (Figure 1; Table S1). To determine the morphology, structure, and chemical compositions of the minerals, reflected and transmitted light microscopy, scanning electron microscopy (SEM), and energy dispersive spectrometry (EDS) were used to analyze thin sections of selected samples. Furthermore, the in-situ major and trace element contents of the samples were analyzed, as well as their S and Pb isotopic compositions.

3.1. SEM Analysis

Pyrrhotite, isocubanite, sphalerite, galena, and amorphous silica were studied using the TESCAN VEGA 3 LMH SEM with an Oxford INCA X-Max EDS at the Institute of Oceanology, Chinese Academy of Sciences, Qingdao, China. The standard analytical operating conditions included an accelerating voltage of 20 kV, a beam intensity of 15 nA, and a working distance of ~15 mm. The following standards were used for the EDS measurements and calibrations: (1) sulfides–pyrrhotite (Fe, S), galena (Pb, S), sphalerite (Zn, S), chalcopyrite (Cu, Fe, S), greenote (Cd), stibnite (Sb), and proustite (Ag, As); (2) silicates–olivine (Ni) and basalt glass (Ti, Si); (3) carbonates–calcite (O); (4) metals–cobalt (Co); (5) oxides–Cr-spinel (Mn); and (6) system standard library (Au, Se). The EDS detection limits for Au, S, Pb, Ag, Cd, Sb, O, Se, As, Zn, Cu, Ni, Co, Fe, Mn, and Ti were 0.1 wt.% (Table S2).

3.2. Major Element Analysis

The major element compositions of the pyrrhotite, sphalerite, and galena were analyzed using the JXA-8230 electron microprobe analyzer (EMPA) at the State Key Laboratory of Continental Dynamics, Northwest University, Xi'an, China. The instrument was operated with a 15 kV acceleration voltage, 10 nA beam current, and 2 μm beam diameter. The following reference materials were used for the wavelength-dispersive spectrometry measurements and calibrations: (1) metals–Co, Au, Se, Ni, Mn, and Ti and (2) natural sulfides–FeS_2, $CuFeS_2$, ZnS, PbS, Sb_2S_3, Ag_3AsS_3, and CdS. The counting time for the peak and background of Au, S, Pb, Ag, Cd, Sb, Se, As, Zn, Cu, Ni, Co, Fe, Mn, and Ti are 20–60 s and 15 s, respectively. The detection limits of these elements were 0.02–0.11 wt.% (Table S3).

3.3. Trace Element Analysis

The trace element contents of the pyrrhotite, sphalerite, and galena were determined using the laser ablation inductively coupled plasma mass spectrometer (LA-ICP-MS) at the State Key Laboratory of Biogeology and Environmental Geology, China University of Geosciences, Wuhan. A GeoLas Pro 193 nm ArF excimer laser was used to analyze the samples. The laser energy was 80 mJ, the frequency was 6 Hz, and the diameter ablation spot was 44 μm. The ion-signal intensities were acquired using an Agilent 7500 x ICP-MS instrument. Helium was used as the carrier gas which was mixed with argon via a T-connector before entering the ICP-MS. Each measurement incorporated approximately 30 s of background acquisition (gas blank) followed by 50 s of data acquisition from the sample. The ICPMSDataCal software was used to quantitatively calibrate the trace element contents [49]. Fe, Cu, Zn, and Pb were selected as the internal standards for the data reduction. Reference glass NIST 610 was used as the external standard, which was analyzed every 10 spots to monitor the instrument drift. The accuracy was determined with respect to reference glass NIST 610 and was assessed to be better than 10% (1σ).

3.4. In-Situ S Isotope Analysis

The in-situ S isotope compositions of the sphalerite, pyrrhotite, and galena were determined using double side-polished slices (DSPS) and the 193 nm femtosecond laser-ablation multi-collector ICP-MS (fs-LA-MC-ICP-MS) at the State Key Laboratory of Continental Dynamics (SKLCD), Northwest University, Xi'an, China. The equipment consisted of a 193 nm NWRfemto (RESOlution M-50, ASI) laser ablation system coupled with a Nu Plasma II MC-ICP-MS (NU Instruments, Ltd., Wrexham, UK). The ablated materials were transported into the plasma using He as the carrier gas. In a cyclone coaxial mixer, Ar gas was mixed with the carrier gas before being transported into the ICP's torch. The energy fluence of the laser was approximately 3.5–4 J/cm^2. The beam diameter was 35 μm with a laser repletion rate of 3–4 Hz for a single spot analysis. Standard PSPT-3 (sphalerite; $\delta^{34}S_{V\text{-}CDT}$ = 26.5 ± 0.2‰) was used as the certified reference material. The obtained $\delta^{34}S_{V\text{-}CDT}$ of Standard PSPT-3 in this study is 26.36 ± 0.29‰ (2SD; n = 21; Table S4), which is consistent with the recommended values ($\delta^{34}S_{V\text{-}CDT}$ = 26.5 ± 0.2‰; [50]). The detailed analysis parameters are described in Bao et al. [50], Chen et al. [51], and Yuan et al. [52].

3.5. In-Situ Pb Isotope Analysis

The in-situ Pb isotope compositions of the galena were determined using DSPSs and the fs-LA-MC-ICP-MS at the SKLCD. To remove potential contamination, anhydrous ethanol was used to carefully clean the surfaces of the DSPSs prior to the laser ablation analysis. The analytical spots were carefully selected to prevent the possible influence of inclusions and impurities. High temperature-activated carbon was used to filter the Hg contained in the carrier gas, which lowered both the Hg background and the detection limit. Internal Tl isotope reference NIST SRM997 was used in conjunction with external reference NIST SRM 610 to correct for fractionation and mass discrimination effects. The Tl solution was introduced through the CETAC Aridus II desolvation nebulizer system. The exponential law correction method for Tl normalization with an optimally adjusted Tl ratio was used to obtain Pb isotope data with high precision and accuracy [53]. The measured isotope ratios were in agreement with the reference and the published values within 2σ measurement uncertainties [50,52–54]. Standard CBI-3 (Natural galena) was used as certified reference material. The measured Pb isotope ratios of standard CBI-3 in this study ($^{208}Pb/^{206}Pb$ = 2.1246 ± 0.0010, $^{207}Pb/^{206}Pb$ = 0.8696 ± 0.00006, $^{206}Pb/^{204}Pb$ = 17.963 ± 0.0017, $^{207}Pb/^{204}Pb$ = 15.621 ± 0.0025, and $^{208}Pb/^{204}Pb$ = 38.165 ± 0.0048; SD; N = 5; Table S4) were consistent with the recommended values ($^{208}Pb/^{206}Pb$ = 2.1247 ± 0.0000, $^{207}Pb/^{206}Pb$ = 0.8694 ± 0.00001, $^{206}Pb/^{204}Pb$ = 17.964 ± 0.0014, $^{207}Pb/^{204}Pb$ = 15.622 ± 0.0013, and $^{208}Pb/^{204}Pb$ = 38.168 ± 0.0028; [52]).

4. Results

4.1. Mineralogy of Hydrothermal Sulfide Samples

Seafloor hydrothermal sulfide ore samples H4 TVG5-1-1 (~8 cm in size) and H4-TVG5-3-1 (~24 cm in size) are black to dark grey in color and contained pores up to 0.2 mm (H4-TVG5-1-1) and 1 mm (H4-TVG5-3-1) in diameter. They are composed of sulfide minerals (Figure 2), i.e., pyrrhotite, galena, sphalerite, and isocubanite, as well as amorphous silica. The coarser-grained aggregates of pyrrhotite (up to 100 μm), sphalerite (up to 50 μm), galena (up to 30 μm), and isocubanite (up to 20 μm) comprise >70% of the sulfide samples by volume in both samples (Figures 2 and 3).

4.1.1. Pyrrhotite

Pyrrhotite is the most abundant mineral. It occurs as randomly oriented laths that range in size from a few to hundreds of micrometers. The laths form boxwork textures (Figure 2) shaped similar to blades or hexagonal plates (Figures 2 and 4a). Some of the pyrrhotite boxwork laths contain interstitial sphalerite, minor isocubanite, and late galena (Figure 2). The whitish rim around pyrrhotite is likely a thin oxidized pyrrhotite-rim, similar to the rims commonly observed in pyrrhotite from other hydrothermal fields (Figure 2f).

4.1.2. Sphalerite

The sphalerite from the NHF is red and Fe-rich (Table S3). It is intimately connected with the pyrrhotite and galena (Figures 2 and 4), intergrown with the pyrrhotite and isocubanite (Figure 2h), overgrown by amorphous silica, and contains local inclusions of galena. The sphalerite commonly occurs as hydrothermal aggregates that are intimately intergrown within the interstices between the pyrrhotite laths where it forms rounded or elongated anhedral grains up to several tens of micrometers across (Figure 2f).

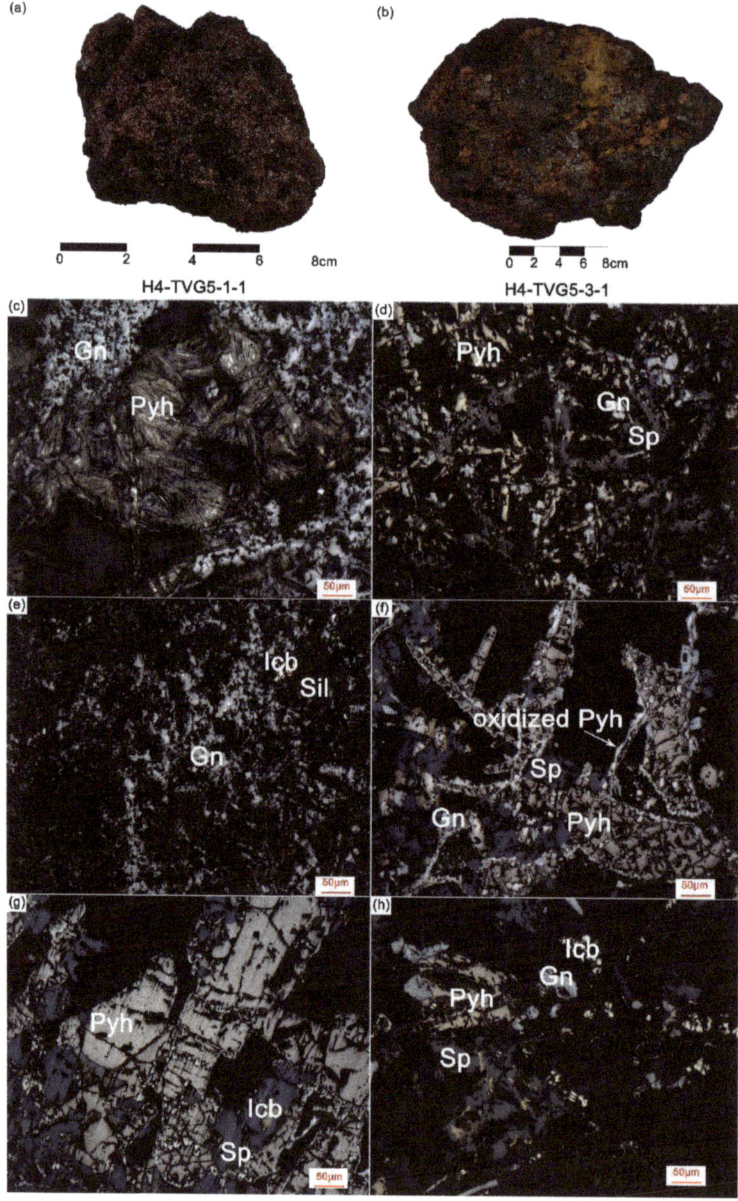

Figure 2. (**a**) Brown H4-TVG5-1-1 sample with a fluid channel and pores; (**b**) grey-green H4-TVG5-3-1 sample with reddish-brown Fe-Mn oxides; (**c**) bladed pyrrhotite (Pyh) surrounded by galena (Gn);

(**d**) pyrrhotite and black-grey sphalerite with a lattice texture and later coarse and fine galena filling the cavities; (**e**) fine isocubanite (Icb) inclusion in the galena and amorphous silica (Sil); (**f**) pyrrhotite (Pyh) aggregates surrounded by sphalerite (Sp) and galena (Gn). Oxidized pyrrhotite appears as a thin whitish rim around the pyrrhotite (Pyh) lath; (**g**) coarse bladed pyrrhotite and fine isocubanite (Icb) surrounded by black-grey sphalerite; and (**h**) pyrrhotite partially surrounded by galena and black-grey sphalerite with fine isocubanite (Icb).

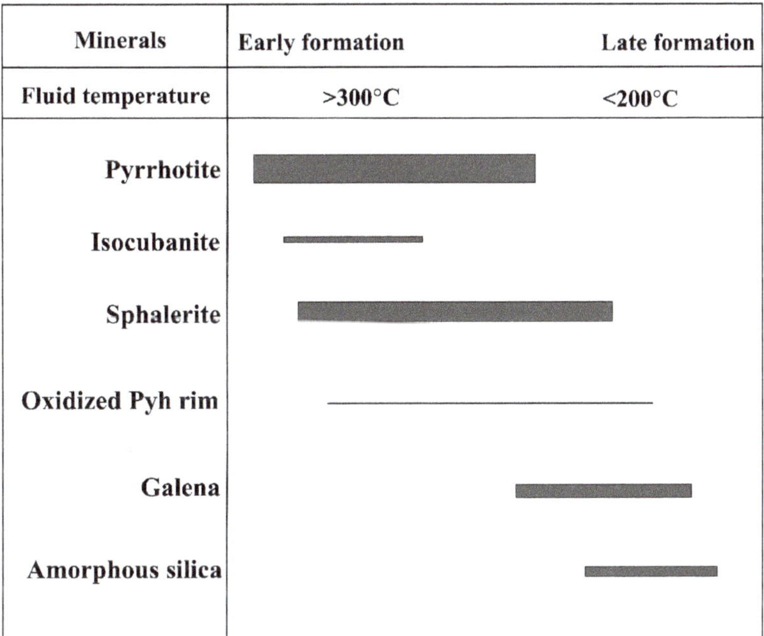

Minerals	Early formation	Late formation
Fluid temperature	**>300°C**	**<200°C**
Pyrrhotite		
Isocubanite		
Sphalerite		
Oxidized Pyh rim		
Galena		
Amorphous silica		

Figure 3. Sequence of sulfide mineralization in the seafloor hydrothermal sulfide deposits of the NHF in the MOT. The thicknesses of the horizontal bars indicate the relative abundances of the minerals. Pyh means pyrrhotite.

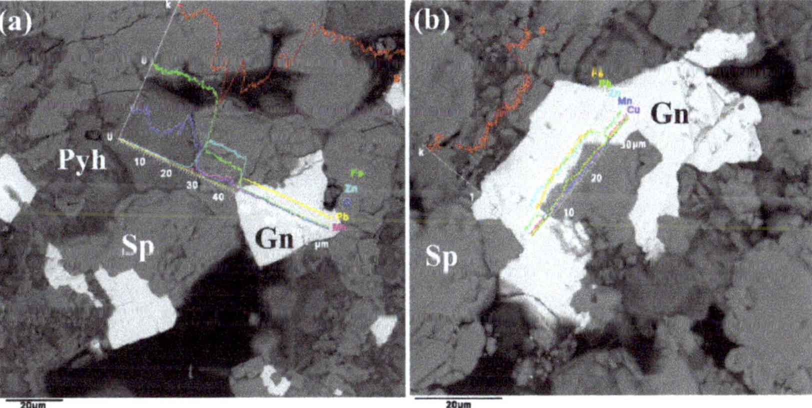

Figure 4. SEM BSE images of selected pyrrhotite (Pyh), sphalerite (Sp), and galena (Gn) grains from the seafloor hydrothermal sulfide deposits in the NHF: (**a**) SEM and EDS line analysis of pyrrhotite,

sphalerite, and galena in sample H4-TVG5-3-1. The pyrrhotite was oxidized with a high O content. The Mn and S contents of the sphalerite vary; (**b**) SEM and EDS line analysis of the sphalerite and galena in sample H4-TVG5-3-1.

4.1.3. Galena

Galena is the major sulfide mineral in both samples (Figures 2 and 4). It is disseminated throughout the samples, appearing as subhedral or anhedral crystals (Figure 2), and ranges in size from a few to tens of micrometers (Figures 2 and 4). Most of the drop-like, oval, or elongated inclusions of galena crystals up to 25 μm in size are found in the sphalerite (Figure 2d). This indicates that there is a positive correlation between the Zn and Pb in the sulfide minerals [10]. The galena is often surrounded by amorphous silica (Figure 2e). It is also found as intergrowths of minute (5–40 μm) euhedral to subhedral crystals in the pores within the pyrrhotite–sphalerite aggregates (Figure 2h).

4.1.4. Isocubanite and Amorphous Silica

The isocubanite coexists with the sphalerite and galena in hydrothermal sulfide samples H4-TVG5-1-1 and H4-TVG5-3-1 (Figure 2). When present, amorphous silica occurs as a thin late-stage rim around the sulfide minerals and may overgrow the sphalerite, pyrrhotite, and galena (Figure 2).

4.2. Major and Trace Element Contents of Sulfide Minerals

4.2.1. Major Element Contents of Sulfide Minerals

The major element concentrations of the sulfide minerals are presented in Tables S2 and S3. According to Belzile et al. [55] and Fallon et al. [1], pyrrhotite can be monoclinic (less Fe-deficient: 45.9–46.8 mol% Fe) or hexagonal (more Fe-deficient: 47.2–48.0 mol% Fe) (Table S2). Most sphalerite has low Cu contents (0.10–0.46 wt.%) and very high Fe (12.46–19.53 wt.%) and Mn (0.10–6.27 wt.%) contents (Figure 4; Table S2). The FeS contents of the sphalerite slabs range from 21.95 to 34.78 mol% (Table S2). In contrast, the Fe contents (1.09 to 2.61 wt.%) of the FeS-bearing sphalerite from the sulfide chimneys in the North Konll, Iheya Ridge, are reportedly very low (~4 mol%; Ueno et al. [20]). The major elements in the galena are Pb (83.68–87.17 wt.%) and S (13.31–13.69 wt.%) (Tables S2 and S3), and those in the isocubanite are Zn (0.80–2.12 wt.%) and Pb (0.38–2.35 wt.%) (Table S2), while most of the amorphous silica contains significant amounts of Fe (0.46–9.64 wt.%) and Zn (0.13–4.86 wt.%) (Table S2).

4.2.2. Trace Element Contents of Sulfide Minerals

The trace element concentrations of the sulfide minerals are presented in Table S5. The pyrrhotite is relatively enriched in Ge (28.0–97.3 ppm) and Mo (1.21–3.40 ppm). The Ge content (28.0 to 97.3 ppm) of the pyrrhotite samples (H4-TVG5-1-1) from the NHF are noticeably higher than those of the galena (0.51–1.09 ppm) and sphalerite (3.63–4.84 ppm) crystals (Table S5; Figure 5).

The sphalerite contains Cd (318–388 ppm), Sn (145–182 ppm), Au (0.67–0.82 ppm), Th (4.85–5.09 ppm), Mo (0.08–0.34 ppm), Tl (0.11–0.67 ppm), and Bi (0.03–0.15 ppm). The Th contents of the sphalerite samples from the NHF are usually higher than those of the pyrrhotite and galena from sample H4-TVG5-1-1 (Table S5; Figure 5).

The Mo (0.07–2.44 ppm), Cd (3.11–70.1 ppm), V (0.41–22.3 ppm), Cr (2.82–369 ppm), Ni (0.45–6.48 ppm), and Cu (12.9–4420 ppm) contents of the galena from the NHF are much more variable than those of the pyrrhotite and sphalerite, covering the largest range (Tables S2, S3 and S5; Figure 5). The As (1.74–2233 ppm), Se (17.0–109 ppm), Fe (3128–101,010 ppm), and Zn (829–127,104 ppm) contents of the galena samples from the NHF also span large ranges (Tables S2, S3 and S5). Se is concentrated in the galena (17.0–109 ppm) and pyrrhotite (13.2–52.7 ppm) (Table S5), whereas Tl was detected in the galena (0.49–4.09 ppm), pyrrhotite (0.12–0.64 ppm), and sphalerite (0.11–0.67 ppm) (Table S5). The galena is more enriched in Ag (693–1940 ppm) and Sb (1608–5129 ppm) than the pyrrhotite and sphalerite (Table S5; Figure 5). In particular, the Ag contents (693–1940 ppm) of most of the galena samples are significantly higher than those of the pyrrhotite (0.49–9.28 ppm) and sphalerite (42.4–53.9 ppm) samples (Figure 5). There is a positive correlation between the Ag and Bi contents of the galena. Most of the pyrrhotite (Th 0.08–2.38 ppm) and sphalerite (Th 4.85–5.09 ppm) contain more Th than the galena (Th 0.04–2.37 ppm), while the galena (7.52–16.8 ppm) has significantly higher Rh contents than the pyrrhotite (0.08–0.14 ppm) and sphalerite (0.04–0.05 ppm) (Table S5; Figure 5). Furthermore, a significant positive correlation was observed between the Au and Mo contents of the galena from the NHF.

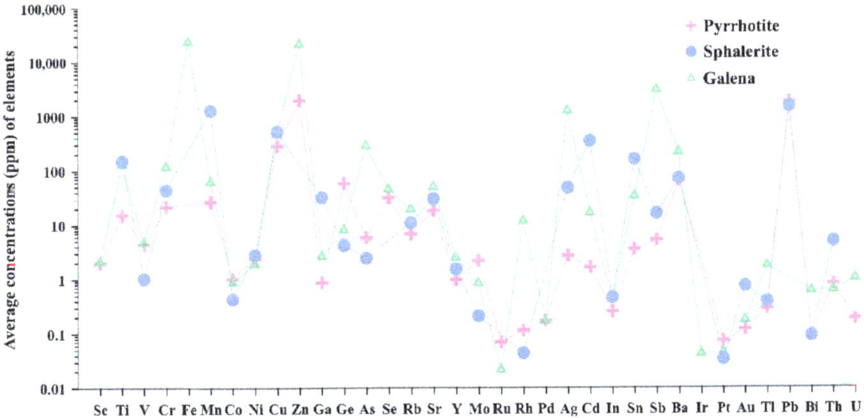

Figure 5. Average element contents of sulfide minerals from the NHF hydrothermal sulfide samples. The element contents of the pyrrhotite, sphalerite, and galena were determined via LA-ICP-MS.

4.3. Rare Earth Element Compositions of Sulfide Minerals

The rare earth element (REE) contents were normalized to Chondrite values (subscript CN) [56]. The Eu and Ce anomalies were assessed using the following equations: $(Eu/Eu^*)_{CN} = (2 \times Eu_{CN})/(Sm_{CN} + Gd_{CN})$ and $(Ce/Ce^*)_{CN} = (2 \times Ce_{CN})/(La_{CN} + Pr_{CN})$. The REE data for the Yonaguni Knoll IV vent fluids, Iheya Ridge volcanic rocks, and MOT sediments used for comparison are from Hongo et al. [57], Li et al. [39], and Xu et al. [58], respectively

The total REE contents (ΣREEs) of the pyrrhotite, sphalerite, and galena from the NHF are highly variable (2.01–36.7 ppm) (Table S5). Most of the ΣREE contents of the pyrrhotite (2.32–36.5 ppm) and galena (2.01–36.7 ppm) are higher than those of the sphalerite (10.2–13.2 ppm) (Table S5). Among the studied samples, the pyrrhotite from the NHF had the highest ΣREE contents (36.5 ppm, H4-TVG5-1-1, point 2–3) (Table S5).

The Chondrite-normalized REE distribution patterns of the pyrrhotite, sphalerite, and galena from the NHF are presented in Figure 6. The REE patterns of most of the sulfide minerals exhibit light rare earth element (LREE) enrichment (LREE/HREE = 8.09–32.0), variable La_{CN}/Lu_{CN} ratios (0.98 and 50.0), and Eu ($(Eu/Eu^*)_{CN} = 0.47$–4.13) and Ce ($(Ce/Ce^*)_{CN} = 1.70$–3.85) anomalies (Figure 6a–c). Compared with the Eu anomalies of the pyrrhotite ($(Eu/Eu^*)_{CN} = 1.44$–1.68) and sphalerite ($(Eu/Eu^*)_{CN} = 3.61$–4.13) from the NHF, the galena has smaller Eu anomalies ($(Eu/Eu^*)_{CN} = 0.47$–0.83). Moreover, most HREE contents are around the detection limit, which has been marked in dash lines (Figure 6) and were only used to qualitatively understand their low contents.

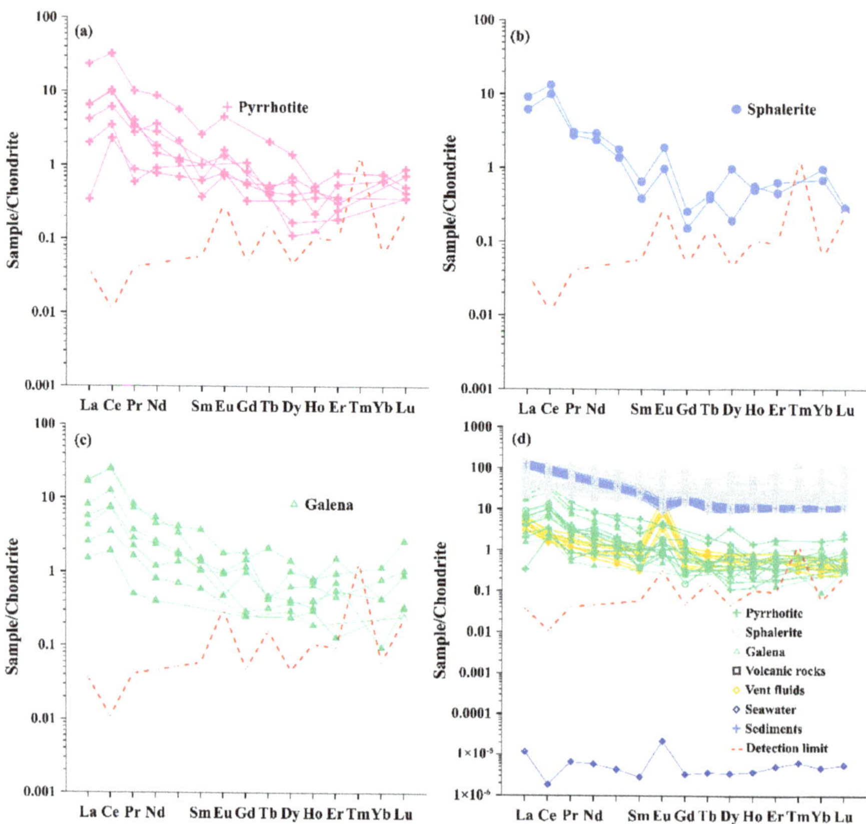

Figure 6. REE patterns of the sulfide minerals in the hydrothermal sulfide samples including pyrrhotite (**a**), sphalerite (**b**), and galena (**c**) from the NHF in the MOT compared to those of volcanic rocks, vent fluids, sediments, and seawater (**d**). The normalization data are from Evensen et al. [56]. The data for the Iheya Ridge volcanic rocks, Yonaguni Knoll IV vent fluids, MOT sediments, and seawater data are from Li et al. [39], Hongo et al. [57], Xu et al. [58], and Steele et al. [59]. The red dash lines are the detection limit of REEs.

4.4. In-Situ S and Pb Isotopic Compositions of the Sulfide Minerals

Five sulfur isotope analyses were conducted on the sphalerite, pyrrhotite, and galena grains from sample H4-TVG5-1-1. The δ^{34}S values obtained for the sulfide minerals are consistent and range from 3.58 to 5.69‰. One sphalerite sample from hydrothermal sulfide sample H4-TVG5-1-1 yielded the highest δ^{34}S values (5.69‰). The galena and pyrrhotite are characterized by lower sulfur isotope values (<4.16‰). The lowest δ^{34}S value (3.58‰) was for a galena sample (Table S6).

The galena from sample H4-TVG5-1-1 from the NHF has uniform Pb isotopic compositions, with a narrow range of Pb isotopic values (Table S6; Figure 7). The ^{207}Pb/^{204}Pb and ^{208}Pb/^{204}Pb ratios of the galena from the NHF are within or close to the range of values for the Iheya Ridge volcanic rocks (basalt and trachyandesite) and the JADE sediments in the MOT (Figure 7).

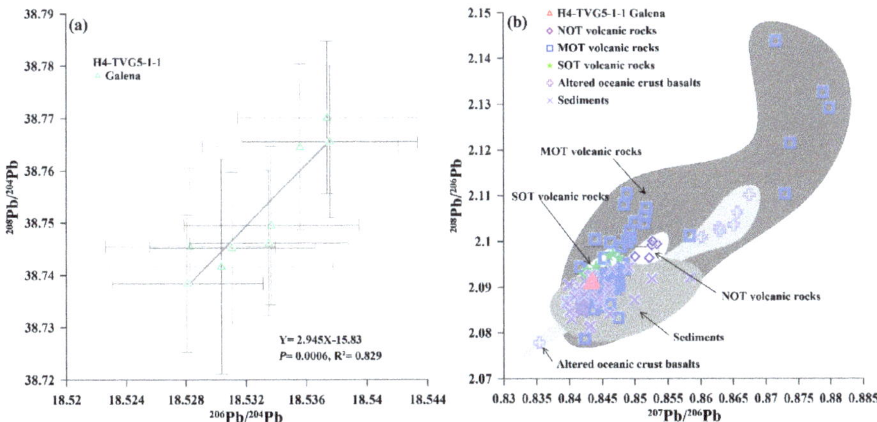

Figure 7. Plot of (**a**) ^{208}Pb/^{204}Pb vs. ^{206}Pb/^{204}Pb, and (**b**) ^{208}Pb/^{206}Pb vs. ^{207}Pb/^{206}Pb for the galena from the NHF. The data for the NOT, MOT, and SOT volcanic rocks are from Halbach et al. [38], Shu et al. [60], and Li et al. [39]. The data for the sediments from DSDP Sites 294, 295, 296, 442B, 443, and 444 in the western Philippine sea, the JADE sediments in the MOT, and the altered oceanic crust basalt from DSDP Sites 294 and 442B are from Halbach et al. [38] and Shu et al. [60]. As we have compared our in situ Pb isotope data to the end members, to avoid being hard to understand, we do not add the previous whole-rock Pb isotopes of sulfides to the same figure.

5. Discussion

5.1. Seawater–Fluid Mixing and Element Enrichments

The NHF sulfide minerals often contain well-developed pyrrhotite blades, xenomorphic isocubanite, interstitial sphalerite, and sharp-edged galena crystals, which are the result of rapid cooling during seawater–fluid mixing (Figure 2) [10]. Mo-enriched sulfide is characteristic of high-temperature paragenetic mineral associations [61]. The Mo contents (1.21–3.40 ppm) of the pyrrhotite samples from the NHF are mainly higher than those of the sphalerite (0.08–0.34 ppm) and galena (0.07–2.44 ppm) in sample H4-TVG5-1-1 (Table S5; Figure 5). This suggests that the Mo-enrichment of the pyrrhotite may have occurred at high temperatures.

The sphalerite grains from the NHF are Fe-rich (Tables S2 and S3); their FeS mol% contents are significantly higher than those from the PACMANUS hydrothermal field (0.00–2.22) in the eastern Manus Basin, SW Pacific [62]. These values are identical to previously reported values [63] for Vienna Woods (3.35–10.65) and sediment-poor JADE (0.00–1.93) samples and are consistent with those (23.48–50.81) reported for the sediment-rich Hakurei samples from the OT [62]. Such FeS contents indicate low-sulfidation conditions [64]. The sphalerite from the NHF is enriched in Th compared to the pyrrhotite and galena (Table S5; Figure 5). This suggests that the abundance of sphalerite controls the Th enrichment in seafloor hydrothermal sulfides. The Th/U ratios (1.86–4.76) of the pyrrhotite from the NHF are consistent with those of the Iheya Ridge basalt (2.64; Li et al. [39]), trachyandesite (1.13; Li et al. [39]), and the MOT metalliferous sediments (0.18–6.11; Yang et al. [65]), so pyrrhotite serves as the source of the Th and U in the subseafloor hydrothermal fluids. Furthermore, the Ge contents (0.51–97.3 ppm) of the sulfide minerals from the NHF in the OT are higher than those (Ge 0.0055 ppm) of seawater [59]. The positive correlation between the Ge and Pd contents indicates an affinity between these elements and the pyrrhotite (Figure 8a).

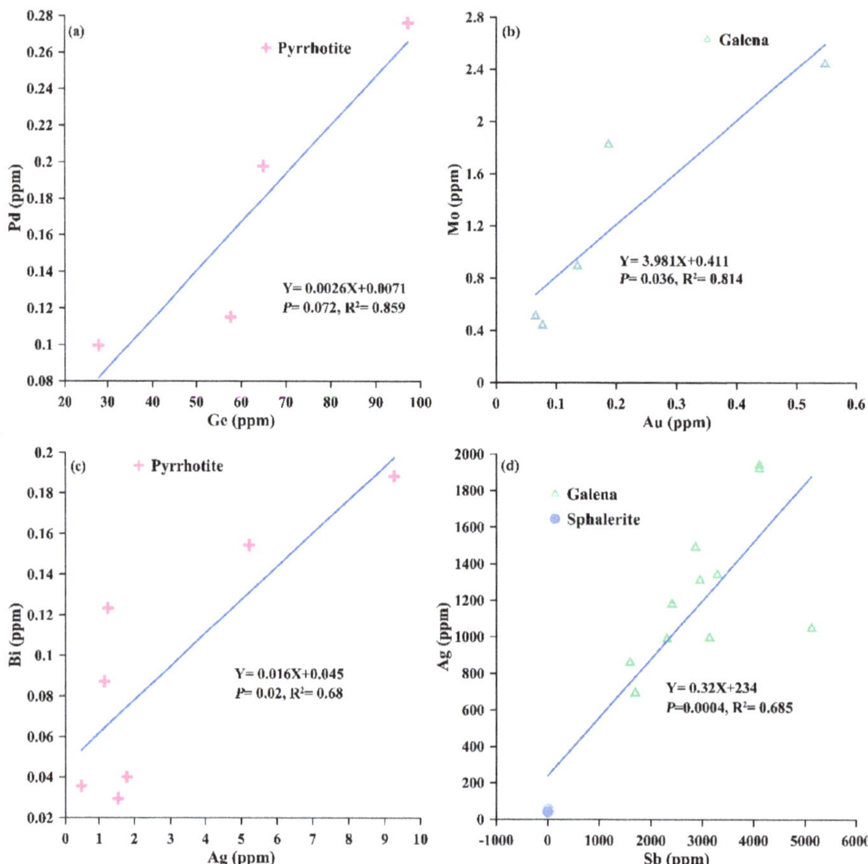

Figure 8. (**a**) Plot of Pd vs. Ge for pyrrhotite; (**b**) Plot of Mo vs. Au for galena; (**c**) Plot of Bi vs. Ag for pyrrhotite; and (**d**) Plot of Ag vs. Sb for galena and sphalerite from the NHF in the MOT.

The galena from the NHF is commonly associated with sphalerite and was observed to form partial rims around the sphalerite crystals (Figure 2d). The galena inclusions in the sphalerite indicate that the galena post-dates the sphalerite (Keith et al., 2016), and it may have formed through the following reaction: ZnS (s) + Pb^{2+} (aq) → PbS (s) + Zn^{2+} (aq). This would result in relatively Zn-rich galena (up to 1.83 wt.%) (Tables S2 and S3) and needs to be confirmed in a future study. However, this indicates a varying fluid chemistry and a change in the solubility as the temperature decreases or increased dilution during seawater–fluid mixing [66]. In the latter case, pyrrhotite would form later than sphalerite/galena, since the fluid is in the interior of both samples H4-TVG5-1-1 and H4-TVG5-3-1.

The Ag contents (0.49–9.28 ppm) of the pyrrhotite from sample H4-TVG5-1-1 (Table S5) are lower than those from the Rainbow hydrothermal vent field on the MAR (200 ppm), the southern JdFR (400–700 ppm), and the Endeavour Segment of the JdFR (<400 ppm) [15–17]. In addition, the Ag contents exhibit a positive correlation with the higher Bi contents (Figure 8c).

The Rh (7.52–16.8 ppm), Bi (0.14–1.38 ppm), Ag (693–1940 ppm), Tl (0.49–4.09 ppm), and Sb (1608–5129 ppm) contents of the galena from the NHF are variable and are noticeably higher than the Rh contents of the sulfides from the PACMANUS hydrothermal field (0.0004–0.3556 ppm) in the eastern Manus Basin, southwestern (SW) Pacific [67], the Turtle Pits, and the Logatchev hydrothermal field on the MAR (0.0012–0.149 ppm) [68]. The values of the galena from the NHF suggest that the chemical compositions of the fluids changed with time during the galena deposition. These values are also significantly higher than those of the sphalerite and pyrrhotite from the NHF (Table S5; Figure 5), the vent fluid (Bi = 0.000647 ppm) in the OT [69], and seawater (Bi = 0.00003 ppm, Ag = 0.002 ppm, Tl = 0.013 ppm, Sb = 0.2 ppm) [59]. Furthermore, the enrichment of Rh (7.52–16.8 ppm), Bi (0.14–1.38 ppm), Ag (693–1940 ppm), Tl (0.49–4.09 ppm), and Sb (1608–5129 ppm) in the galena may be the result of the galena precipitating later from late-stage fluids and rapid cooling due to seawater–fluid mixing (Table S5; Figure 5). However, Rh is more compatible in a monosulfide-solid solution (MSS) than in an intermediate solid solution (ISS) and has been found to be more soluble than Pt and Pd [70]. Due to these properties, Rh may crystallize as chalcogen (Se, Te, As, Sb, Bi)-rich discrete precipitate (e.g., RhSeS or RhSbS) before or synchronously with other chalcogen-bearing phases that are associated with late-stage low-sulfur precious metal haloes around hydrothermal sulfides. Similar associations have been documented at various locations around the Sudbury metallogenic deposit, Ontario [71,72]. Consequently, the Ag content is positively correlated with the Sb content for the galena from the NHF and in neighboring sphalerite (Figure 8d), which may also indicate the presence of Sb with the Ag incorporated into the galena from the sphalerite during mineralization. The Rh (7.52–16.8 ppm), Bi (0.14–1.38 ppm), Ag (693–1490 ppm), Tl (0.49–4.09 ppm), and Sb (1608–5129 ppm) contents of the galena are higher than those of the original hydrothermal sphalerite and the earlier pyrrhotite in the NHF (Table S5; Figure 5), which indicates that the incorporation of these trace elements into the galena, pyrrhotite, and sphalerite is controlled by their partition coefficients. This implies that the back-arc galena-rich hydrothermal sulfide deposit may be an Rh-Ag-Sb enriched deposit.

Compared with the PACMANUS hydrothermal mineralization (average Se content of pyrite is 5.97–7.39 ppm; [1]), most of the Noho sulfide minerals are enriched in Se (13.2–109 ppm) (Table S5), which was concentrated in the galena and pyrrhotite via seawater–fluid mixing (Table S5). This is similar to the seafloor hydrothermal mineralization in the Kulo Lasi volcano, Futuna, and SW Pacific. This implies that the Noho basalt-, trachyandesite-, and sediment-hosted hydrothermal mineralization may be enriched in Se [10] based on the fact that it was found in large quantities in the pyrrhotite and based on the reducing conditions during the fluid-rock and/or sediment interaction.

In addition, the Pb contents of the pyrrhotite (up to 1.33 wt.%) and sphalerite (up to 2.93 wt.%) are comparable to other BAB mineralization in the PACMANUS field (0.176 wt.% to 2.35 wt.% Pb in sphalerite) in the SW Pacific [73,74], but they are higher than those in the Trans-Atlantic Geotraverse (TAG) hydrothermal field (<0.1 to 0.16 wt.% Pb in sphalerite) on the MAR [1,75]. However, some of the sphalerite grains exhibit noticeably high Pb contents (up to 2.93 wt.%) (Table S2) which may indicate spectral contamination from nearby galena grains, or true chemical contamination, caused by the diffusion of the chemical components across grain boundaries. The seafloor pyrrhotite (Fe = 58.5–62.0 wt.%; Pb = 0.02–1.33 wt.%; S = 37.8–40.9 wt.%; Zn = 632–5300 ppm; Cu = 130–529 ppm) and galena (Fe = 0.10–1.32 wt.%; Pb = 81.1–88.8 wt.%; S = 12.7–14.5 wt.%; Zn = 829–127,104 ppm; Cu = 12.9–4420 ppm) from the NHF contain different proportions of major (Fe, Pb, and S) and minor (Cu and Zn) elements which may also indicate probable spectral contamination from the adjacent sphalerite and subsurface Cu-sulfide inclusions (isocubanite, up to 30.83 wt.% Cu) (Table S2).

5.2. Variable REE Contents and Sources of REEs

The ∑REEs contents of the pyrrhotite analyzed in this study vary significantly (2.01–36.7 ppm), and they do not exhibit systematic variations with the Fe (58.5–59.8 wt.%) and S (39.5–40.3 wt.%) contents. The degree of LREE and heavy rare earth element (HREE) fractionation in the NHF pyrrhotite (LREE/HREE) is highly variable, up to 32.0 (Table S5). The ∑REE contents and ranges of the pyrrhotite (Table S5) exceed those of the sphalerite. The substitutions of REEs into the pyrrhotite are similar to those involving galena and sphalerite [76,77], indicating the significant influence of the REEs' larger ionic radii.

The REE patterns of the pyrrhotite and sphalerite from the NHF contain positive Eu anomalies (1.44–4.13) (Table S5; Figure 6), similar to the values for the vent fluids in the OT [57,78] and almost all hydrothermal fields globally. These similarities indicate that the pyrrhotite and sphalerite inherited the positive Eu anomalies of the vent fluids [77,79–81]. The stability of the soluble Eu^{2+} species increases in contact with Cl^- complexation, low- to high-temperature acidic fluids, and under reducing conditions [77,82–86]. Therefore, the positive Eu anomalies of the pyrrhotite and sphalerite with high Eu content (0.04 to 0.27 ppm) may have been induced by Cl-complexation under high-temperature, low-pH, and strongly reducing fluid conditions [77,81].

The REE patterns of the galena from the NHF are characterized by negative Eu anomalies (0.47–0.83) (Table S5; Figure 6c), which are indicative of relatively low-temperature seawater [87]. A large proportion of the Eu in the low-temperature fluid is trivalent because divalent Eu is stable at temperatures of greater than ~250 °C [83]. The decreasing formation temperature of galena strongly correlates with the decreasing Eu^{2+}/Eu^{3+} ratios of the vent fluids [77]. The accumulation of Eu^{2+} in galena formed at medium to low temperatures is also reduced. The stability of the soluble Eu^{2+} species has been reported to decrease in association with SO_4^{2-} complexation, high- to low-temperature acidic fluids, and under reducing conditions [83]. Therefore, the negative Eu anomaly of the galena were most likely induced by lower Eu contents, medium (300–200 °C) to low (<200 °C) temperatures (Figure 3), and/or mixing between fluids and seawater and the galena chemistry. Consequently, the Eu content of the fluids may have influenced the Eu anomaly in the precipitated galena and its depositional conditions and processes. The negative Eu anomalies are related to the lower Eu contents of the galena (0.03–0.14 ppm) (Table S5), which have been interpreted to have formed at medium to low temperatures and from less concentrated Eu-bearing- fluids [80,81,88].

The REEs in the sulfide minerals from the NHF may indicate the sources and evolution of the seafloor hydrothermal fluids. Previous studies of seafloor hydrothermal sulfides from the Rainbow, Broken Spur, and TAG hydrothermal fields on the MAR have revealed that the REEs were mainly derived from the fluids [79,89] and were incorporated into the sulfide minerals during seawater–fluid mixing (e.g., Bau and Dulski [90] and Zeng et al. [81]). The REE patterns of the sulfide minerals from the NHF are comparable to those of the Yonaguni Knoll IV vent fluids (Figure 6d) [57]. Thus, the REEs in the sulfide minerals from the NHF were most likely derived from fluids which leached the REEs from the local sub-seafloor volcanic rocks and/or sediments and deposited them in the Noho hydrothermal sulfide deposits (Figure 6d) [58,81,91]. Admittedly, more detailed works focusing on how REE enters sulfides need to be conducted in the future.

5.3. Sources of Sulfur in Sulfide Minerals

The sulfur isotope compositions of the sulfides and their possible geological reservoirs are important indicators that constrain the sources of the reduced sulfur [92,93]. The NHF hydrothermal sulfide deposit is hosted in volcanic rocks and sediments. Igneous rock-derived sulfur, including that in the MORBs, mantle peridotites, and calc-alkaline volcanic rocks (e.g., andesites or rhyolites; Zeng et al. [21]) show $\delta^{34}S$ values very close to 0‰ (+0.1 ± 0.5‰) [94–97]. However, the $\delta^{34}S$ values of the Quaternary volcanic rocks from the Japanese Island Arc (+4.4 ± 2.1‰; Ueda and Sakai [98]) and the volcanic rocks from the northern Mariana Trough (+2.0 to +20.7‰; Woodhead et al. [99]) are significantly higher. Although volcanic sulfur is rapidly leached by high-temperature (300–500 °C) fluids [100], no significant isotopic fractionation occurs during the leaching, transport, and reprecipitation of sulfide minerals [101]. The seawater sulfur in the NHF was most likely derived from seawater and/or fluids that leached sulfur from the sediments. Corresponding to volcanic sulfur, seawater sulfate has a $\delta^{34}S$ value of +21‰ [102] and H_2S-containing seafloor hydrothermal fluids related to sulfide and sulfate mineralization usually have $\delta^{34}S$ values of 1.5‰ to 7‰ (e.g., Michard et al. [103]; Butterfield et al. [104]; Shanks et al. [96]; and Shanks [105]). These values are between the $\delta^{34}S$ values of volcanic rocks

(+0.1 ± 0.5‰; [94–97], the Japanese Island Arc volcanic rocks (+4.4 ± 2.1‰; Ueda and Sakai [98]), the northern Mariana Trough volcanic rocks (+2.0 to +20.7‰; Woodhead et al. [99]), and sediments (δ^{34}S values similar to those of seawater sulfate, which can be affected by microorganisms >+21‰) and seawater (+21‰). Therefore, the sulfur isotopic compositions of sulfide minerals can serve as evidence of the sources of the sulfur [21,106].

The δ^{34}S values of the SO_4 and H_2S from the OT vent fluids range from 20.6‰ to 25.7‰ (average of 22.4‰, $n = 9$) and –0.2‰ to 12‰ (average of 5.9‰, $n = 26$), respectively [69,78,107–111]. The δ^{34}S values of the sulfide mineral samples from the NHF are all within a narrow range (3.58–5.69‰; Table S6) which overlaps with the overall range for sulfide minerals from global seafloor hydrothermal fields (0.0–9.6‰; $n = 1901$; Zeng et al. [21]). This indicates a heavy sulfur contribution from reduced seawater sulfate, falling within the range between the seawater δ^{34}S values for SO_4 (20.99‰; [102]) and vent fluid H_2S (−0.2 to 12‰; [69,78,107–111]).

The sulfur isotope compositions of the sulfide minerals from the NHF may have been influenced by the mixing of seawater and/or sediment SO_4 with volcanic rock-derived sulfur during the sulfide formation according to the following reaction:

$$M^{2+} \text{ (aq)} + SO_4^{2-} \text{ (aq)} + 4H_2S \text{ (aq)} \rightarrow MSx \text{ (s)} + yS \text{ (s)} + 4H_2O \text{ (aq)}$$

where M^{2+} is Fe^{2+}, Pb^{2+}, or Zn^{2+}; and MS is pyrite ($x = 2$, $y = 3$), galena ($x = 1$, $y = 4$), or sphalerite ($x = 1$, $y = 4$).

According to a simple two end-member mixing model,

$$(\delta^{34}S\text{mix} = X \times \delta^{34}S_{\text{seawater and/or sediment}} SO_4 + (1 - X) \times \delta^{34}S_{\text{volcanic rock}}$$

where X is the amounts of seawater and/or the sediment SO_4 component, and $\delta^{34}S_{\text{mix}}$, $\delta^{34}S_{\text{seawater and/or sediment}} SO_4$ (21‰), and $\delta^{34}S_{\text{volcanic rock}}$ (0.1‰, volcanic rock-derived sulfur) are the sulfur isotope compositions of sulfide minerals or vent fluids, seawater and/or sediment SO_4 and volcanic rock-derived sulfur, respectively, the sulfur isotope composition of sulfide minerals can be obtained via mixing sulfur from seawater and/or sediment SO_4 (17–19%, $n = 5$) with sulfur from volcanic rocks (81–83%, $n = 5$). This mixing likely occurred in the downwelling limb of the hydrothermal convection cell in the NHF. This further suggests that the proportions of the sulfur in NHF sulfide minerals contributed by seawater and/or sediment SO_4 were less than that contributed by the volcanic rock-derived sulfur. The sulfide minerals in sample H4-TVG5-1-1 precipitated in the NHF primarily where the hydrothermal fluids are dominant along the tube interiors. In addition, disulfides and sulfates tend to be more abundant where mixing with seawater promotes oxidation of S^{2-} in the distal portions of the sample. The δ^{34}S values of the sphalerite from the NHF are higher than those of the galena and pyrrhotite from the NHF (Table S6), and they are significantly lower than those of the sphalerite from the JADE hydrothermal field which has been reported to be +5 to +6‰ [20] and +6.6 to +7.1‰ [112]. These values have been significantly influenced by the reduction in seawater and/or sediment sulfate under highly oxidizing fluid conditions during sphalerite formation [113,114].

In addition, the decrease in the δ^{34}S values from the sphalerite to the galena [115] suggests that the sphalerite-galena pair (point 8-9-3 for sphalerite and point 8-9-5 for galena) in the NHF represent sulfur isotopic equilibrium. Based on the known temperature-dependent fractionation factors [93,116], the formation temperature of the coexisting sphalerite-galena pair was estimated to be 314 °C and 318 °C using the equation for the sulfur isotope geothermometer (Table S6).

5.4. Fluid-Rock Interactions and Source of Lead in Galena

The majority of the Pb isotope ratios of the galena plot were in or near the fields of the basalts and andesites from the NHF and the JADE sediments from the MOT (Figure 7). This suggests that the MOT volcanic rocks and/or sediments are the main source of the Pb in the seafloor vent fluids (e.g., Vidal and Clauer [117]; Chen [118]; Hegner and Tatsumoto [119];

Hinkley and Tatsumoto [120]; Fouquet and Marcoux [121]; Halbach et al. [38]; Charlou et al. [122]; Yao et al. [123]; and Zeng et al. [21]). The range of the Pb isotope compositions of the galena (Table S6) is small compared to those of the volcanic rocks and sediments in the OT. The Pb isotope ratios of the galena from the NHF are the same as those of the volcanic rocks and sediments in the OT, which indicates that the heterogeneous Pb isotopes (e.g., Allègre et al. [124] and Hamelin et al. [125]) were extracted from the volcanic rocks and/or sediments via fluid-rock and/or sediment interactions and were homogenized during fluid circulation, supporting the conclusions of previous studies [21,121,126]. This indicates that the Pb isotope compositions of the galena can be used to infer the Pb isotope compositions of the local volcanic rocks, sediments, and fluids. The Pb isotope ratios of the galena from the NHF are close to the average Pb isotope ratios of the volcanic rocks and sediments in the MOT (Figure 7b), suggesting that the volcanic rock and/or sediments were likely the sources of the Pb in the galena from the NHF. In addition, these ratios plot in or near the field of the west Philippine Sea sediments and altered oceanic crust basalts (Figure 7b) [60], implying that sedimentary and subducted oceanic crust components were a possible source of the Pb in the galena from the MOT.

5.5. Hydrothermal Temperature and Redox Conditions during Sulfide Formation

The equilibrium sulfur isotope temperature (314 °C and 318 °C) of the sphalerite-galena pair (Table S6) suggests that the dominance of sphalerite is a characteristic of the Noho hydrothermal sulfides with formation temperatures of >300 °C. Based on the observed microtextural relationships of the minerals and the interpreted paragenetic sequences (Figure 3), the sulfide mineral crystallization sequence was as follows. First, the pyrrhotite formed, followed by isocubanite and sphalerite. In the final stage, galena, and then amorphous silica is precipitated. The amorphous silica is concluded to have formed during the late stage since it precipitates at low temperatures (i.e., from ~50 °C to <200 °C; Figure 3) [10]. The overall mineral crystallization sequence from pyrrhotite to isocubanite to sphalerite to galena and finally to amorphous silica is indicative of hot, reduced hydrothermal fluids mixing with cooler, more oxygenated seawater [18]. This also emphasizes the importance of the interactions between the mineralizing fluid and the coexisting seawater during the formation of this hydrothermal sulfide deposit in the NHF. However, replacement of fine-grained sulfides with coarse-crystalline sulfides indicates a reaction with hotter and, at least in part, more reduced ascending fluids from which pyrrhotite and sphalerite may also have precipitate [127,128]. In addition, the decrease in the Fe/S ratio of the pyrrhotite from 0.96 to 0.92 (Table S2) indicates an increase in the level of pyrrhotite oxidation (fO_2 increase or fS_2 decrease) [1].

Interestingly, no pyrite or marcasite was observed in samples H4-TVG5-1-1 or H4-TVG5-3-1. The lack of these minerals sheds important light on the formation processes of the seafloor hydrothermal sulfides in the NHF. For instance, it is known that pyrite is much less soluble than pyrrhotite (approximately 6–10 orders of magnitude less soluble at 200–250 °C; [129]). Just after the pyrrhotite precipitated, the residual Fe precipitated to form Fe-rich sphalerite. This did not leave enough sulfur to form abundant pyrite and/or marcasite. Thus, the lack of pyrite and/or marcasite when pyrrhotite is present merely indicates low sulfur activity. However, according to the mineral paragenesis, the pyrrhotite, isocubanite, and sphalerite were formed during the early ore-forming stage, while the sphalerite, galena, and amorphous silica were formed during the late stage. The pyrrhotite, which represents the earliest mineralization stage, is occasionally overgrown by sphalerite and late-stage galena (Figures 2 and 3).

Even so, the accurate sulfide mineralizations model requires drilling samples and data. Therefore, our TV-grabber collected samples are focused on exploring the element sources (e.g., REE, S, and Pb) and physicochemical conditions (e.g., temperature, fO_2, and fS_2) of the sulfide mineral formation.

6. Conclusions

The mineral assemblage of the seafloor hydrothermal sulfide ores from the NHF is dominated by pyrrhotite, sphalerite, galena, minor isocubanite, and amorphous silica. The early sulfides (pyrrhotite, isocubanite, and sphalerite) are partly or completely surrounded by the later minerals (sphalerite, galena, and amorphous silica). The late overgrowth of the pyrrhotite by sphalerite and the sphalerite by galena indicates an unstable hydrothermal system.

Most of the measured Rh, Bi, Ag, Tl, Se, and Sb contents of the pyrrhotite and sphalerite are significantly lower than those of the galena, indicating enrichment of these trace elements in the galena during the seawater–fluid mixing. This suggests that the galena-rich hydrothermal sulfide deposit in this BAB may be Rh-Ag-Sb-enriched.

The REEs in the sulfide minerals from the NHF were likely sourced from the seafloor hydrothermal fluids. The REE contents and patterns of the sulfide minerals are related to the mineral chemistry, but they were also influenced by the physiochemical compositions, REE contents and patterns of the fluids, the degree of mixing between the fluids and seawater, and interactions with sub-seafloor rocks and/or sediments. The pyrrhotite and the sphalerite precipitated at higher temperatures in association with acidic or reducing fluids have positive Eu anomalies. Thus, the Eu anomaly values may indicate the Eu content and temperature of the source fluids.

The sphalerite has higher δ^{34}S values than the pyrrhotite or galena. The higher δ^{34}S values of the sulfide minerals from the NHF indicate a heavy sulfur contribution from reduced seawater and/or sediment sulfate, which may indicate that the sulfur in the sulfides was mainly derived from the volcanic rocks.

The Pb isotope compositions of the galena are similar to those of the associated volcanic rocks and/or sediments in the NHF and the JADE sediments in the MOT, suggesting that the galena inherited the Pb isotope compositions of the host rocks in the Noho sub-seafloor hydrothermal systems during the galena formation via seawater-fluid mixing and fluid-rock and/or sediment interactions. The Pb isotope compositions of the galena from the NHF are very homogenous and have a narrow range, plotting in or near the field of the large Pb isotope dataset for the MOT volcanic rocks, western Philippine Sea sediments, and altered oceanic crust basalts. This implies that the isotopic composition of the Pb in the galena is important for understanding the influence of plate subduction on back-arc hydrothermal and magmatic systems.

The formation temperature of the coexisting sphalerite and galena in the NHF was estimated to be 314 °C and 318 °C. The higher sulfur fugacity of the fluid resulted in lower formation temperatures and greater Fe contents in the sphalerite. Altogether, these facts suggest that the sphalerite was deposited under reducing fluid conditions. After the pyrrhotite precipitated, most of the remaining Fe was used for the formation of Fe-rich sphalerite precipitates. These processes consumed all of the available sulfur and prevented the formation of pyrite and/or marcasite.

Supplementary Materials: The following supporting information can be downloaded at: https://www.mdpi.com/article/10.3390/jmse10050678/s1, Table S1: Sampling data for this study, Table S2: Oxford INCA X-Max energy dispersive spectrometry (EDS) analyses of the sulfide minerals and amorphous silica in samples H4-TVG5-1-1 and H4-TVG5 3-1 from the NHF (wt.%), Table S3: Electron microprobe analyses of the sulfide minerals in sample H4-TVG5-1-1 from the NHF (wt %) and their atoms per formula unit for elements, Table S4: Detective limit of the trace elements analyzed via LA-ICPMS, and standards of the S-Pb isotopes analyzed via LA-ICPMS. Table S5: Trace element contents of the sulfide minerals in sample H4-TVG5-1-1 from the NHF determined via LA-ICP-MS (in ppm), Table S6: Sulfur and lead isotopic compositions of the sulfide minerals in sample H4-TVG5-1-1 from the NHF determined via LA-MC-ICP-MS and the formation temperature of the coexisting sphalerite and galena.

Author Contributions: Z.Z.—Conceptualization, Data curation, formal analysis, funding acquisition, writing, and editing. Z.C.—data curation, formal analysis, validation. H.Q.—Data curation, methodology, validation. B.Z.—Data curation. All authors have read and agreed to the published version of the manuscript.

Funding: This research was supported by the NSFC Major Research Plan for West-Pacific Earth System Multispheric Interactions [project number 91958213], the Strategic Priority Research Program of the Chinese Academy of Sciences [grant number XDB42020402], the National Programme on Global Change and Air-Sea Interaction [grant number GASI-GEOGE-02], the International Partnership Program of the Chinese Academy of Sciences [grant number 133137KYSB20170003], the Special Fund for the Taishan Scholar Program of Shandong Province [grant number ts201511061], and the National Key Basic Research Program of China [grant number 2013CB429700].

Institutional Review Board Statement: Not applicable.

Informed Consent Statement: Not applicable.

Data Availability Statement: Data can be obtained from the corresponding author and are available online at Supplementary Materials.

Acknowledgments: We would like to thank the crews of the R/V Kexue during the HOBAB 4 cruise for their help with the sample collection. We are most grateful for the detailed and constructive comments and suggestions provided by two anonymous reviewers, which greatly improved an earlier version of the manuscript.

Conflicts of Interest: The authors declare no conflict of interest.

References

1. Fallon, E.K.; Petersen, S.; Brooker, R.A.; Scott, T.B. Oxidative dissolution of hydrothermal mixed-sulphide ore: An assessment of current knowledge in relation to seafloor massive sulphide mining. *Ore Geol. Rev.* **2017**, *86*, 309–337. [CrossRef]
2. Hawley, J.E.; Nichol, I. Trace elements in pyrite, pyrrhotite and chalcopyrite of different ores. *Econ. Geol.* **1961**, *56*, 467–487. [CrossRef]
3. Arnold, R. Range in composition and structure of 82 natural terrestrial pyrrhotites. *Can. Mineral.* **1967**, *9*, 31–50.
4. Cabri, L.J.; Campbell, J.L.; Laflamme, J.G.; Leigh, R.G.; Maxwell, J.A.; Scott, J.D. Proton-microprobe analysis of trace elements in sulfides from some massive-sulfide deposits. *Can. Mineral.* **1985**, *23*, 133–148.
5. Törmänen, T.O.; Koski, R.A. Gold enrichment and the Bi-Au association in pyrrhotite-rich massive sulfide deposits, Escanaba Trough, Southern Gorda Ridge. *Econ. Geol.* **2005**, *100*, 1135–1150. [CrossRef]
6. Berkenbosch, H.A.; De Ronde, C.E.J.; Gemmell, J.B.; McNeill, A.W.; Goemann, K. Mineralogy and formation of black smoker chimneys from Brothers submarine volcano, Kermadec arc. *Econ. Geol.* **2012**, *107*, 1613–1633. [CrossRef]
7. Tivey, M.; Becker, E.; Beinart, R.; Fisher, C.; Girguis, P.; Langmuir, C.; Michael, P.; Reysenbach, A.-L. Reysenbach. Links from mantle to microbe at the Lau Integrated Study Site: Insights from a back-arc spreading center. *Oceanography* **2012**, *25*, 62–77. [CrossRef]
8. Melekestseva, I.Y.; Tret'Yakov, G.A.; Nimis, P.; Yuminov, A.M.; Maslennikov, V.V.; Maslennikova, S.P.; Kotlyarov, V.A.; Beltenev, V.E.; Danyushevsky, L.; Large, R. Barite-rich massive sulfides from the Semenov-1 hydrothermal field (Mid-Atlantic Ridge, 13 30.87′ N): Evidence for phase separation and magmatic input. *Mar. Geol.* **2014**, *349*, 37–54. [CrossRef]
9. Evans, G.N.; Tivey, M.K.; Seewald, J.S.; Wheat, C. Wheat. Influences of the Tonga subduction zone on seafloor massive sulfide deposits along the eastern Lau spreading center and Valu Fa Ridge. *Geochim. Cosmochim. Acta* **2017**, *215*, 214–246. [CrossRef]
10. Fouquet, Y.; Pelleter, E.; Konn, C.; Chazot, G.; Dupré, S.; Alix, A.-S.; Chéron, S.; Donval, J.-P.; Guyader, V.; Etoubleau, J. Volcanic and hydrothermal processes in submarine calderas: The Kulo Lasi example (SW Pacific). *Ore Geol. Rev.* **2018**, *99*, 314–343. [CrossRef]
11. Goodfellow, W.D.; Franklin, J.M. Geology, mineralogy, and chemistry of sediment-hosted clastic massive sulfides in shallow cores, Middle Valley, Northern Juan de Fuca Ridge. *Econ. Geol.* **1993**, *88*, 2037–2068. [CrossRef]
12. Bogdanov, Y.A.; Lein, A.Y.; Sagalevich, A.M.; Ul'yanov, A.A.; Dorofeev, S.; Ul'yanova, N.V. Hydrothermal sulfide deposits of the Lucky Strike vent field, Mid-Atlantic Ridge. *Geochem. Int.* **2006**, *44*, 403–418. [CrossRef]
13. Bogdanov, Y.A.; Lein, A.Y.; Maslennikov, V.V.; Li, S.; Ul'Yanov, A.A. Mineralogical-geochemical features of sulfide ores from the Broken Spur hydrothermal vent field. *Oceanology* **2008**, *48*, 679–700. [CrossRef]
14. Mozgova, N.N.; Trubkin, N.V.; Borodaev, Y.S.; Cherkashev, G.A.; Stepanova, T.V.; Semkova, T.A.; Uspenskaya, T.Y. Mineralogy of massive sulfides from the Ashadze hydrothermal field, 13 N, Mid-Atlantic Ridge. *Can. Mineral.* **2008**, *46*, 545–567. [CrossRef]
15. Paradis, S.; Jonasson, I.R.; le Cheminant, G.M.; Watkinsons, D.H. Two zinc-rich chimneys from the Plume Site, southern Juan de Fuca Ridge. *Can. Mineral.* **1988**, *26*, 637–654.

16. Tivey, M.K.; Stakes, D.S.; Cook, T.L.; Hannington, M.D.; Petersen, S. A model for growth of steep-sided vent structures on the Endeavour Segment of the Juan de Fuca Ridge: Results of a petrologic and geochemical study. *J. Geophys. Res. Solid Earth* **1999**, *104*, 22859–22883. [CrossRef]

17. Marques, A.F.A.; Barriga, F.; Chavagnac, V.; Fouquet, Y. Mineralogy, geochemistry, and Nd isotope composition of the Rainbow hydrothermal field, Mid-Atlantic Ridge. *Miner. Depos.* **2006**, *41*, 52–67. [CrossRef]

18. Hannington, M.D.; Jonasson, I.R.; Herzig, P.M.; Petersen, S. Physical and chemical processes of seafloor mineralization at mid-ocean ridges. In *Seafloor Hydrothermal Systems: Physical, Chemical, Biological, and Geological Interactions*; Washington DC American Geophysical Union Geophysical Monograph Series; American Geophysical Union: Washington, DC, USA, 1995; Volume 91, pp. 115–157.

19. Monecke, T.; Petersen, S.; Hannington, M.D.; Grant, H.; Samson, I.M. The minor element endowment of modern sea-floor massive sulfides and comparison with deposits hosted in ancient volcanic successions. *Rev. Econ. Geol.* **2016**, *18*, 245–306.

20. Ueno, H. Ore and gangue minerals of sulfide chimneys from the North Knoll, Iheya Ridge, Okinawa Trough, Japan. *JAMSTEC Deep-Sea Res.* **2003**, *22*, 49–62.

21. Zeng, Z.; Ma, Y.; Chen, S.; Selby, D.; Wang, X.; Yin, X. Sulfur and lead isotopic compositions of massive sulfides from deep-sea hydrothermal systems: Implications for ore genesis and fluid circulation. *Ore Geol. Rev.* **2017**, *87*, 155–171. [CrossRef]

22. Tong, X.; Song, S.X.; He, J. Research on Mineral Processing of Marmatite Ore. *Metal Mine* **2006**, *6*, 8–12.

23. Cook, N.J.; Ciobanu, C.L.; Pring, A.; Skinner, W.; Shimizu, M.; Danyushevsky, L.; Saini-Eidukat, B.; Melcher, F. Trace and minor elements in sphalerite: A LA-ICPMS study. *Geochim. Cosmochim. Acta* **2009**, *73*, 4761–4791. [CrossRef]

24. Chen, Y.; Chen, J.; Guo, J. A DFT study on the effect of lattice impurities on the electronic structures and floatability of sphalerite. *Miner. Eng.* **2010**, *23*, 1120–1130. [CrossRef]

25. Waida, C.; Ueno, H. Ore and gangue minerals of seafloor hydrothermal deposits in the Mariana trough. *JAMSTEC Rep. Res. Dev.* **2005**, *1*, 1–12.

26. Suzuki, R.; Ishibashi, J.-I.; Nakaseama, M.; Konno, U.; Tsunogai, U.; Gena, K.; Chiba, H. Diverse range of mineralization induced by phase separation of hydrothermal fluid: Case study of the Yonaguni Knoll IV Hydrothermal Field in the Okinawa Trough Back-Arc Basin. *Resour. Geol.* **2008**, *58*, 267–288. [CrossRef]

27. Petersen, S.; Monecke, T.; Westhues, A.; Hannington, M.D.; Gemmell, J.B.; Sharpe, R.; Peters, M.; Strauss, H.; Lackschewitz, K.; Augustin, N.; et al. Drilling shallow-water massive sulfides at the Palinuro volcanic complex, Aeolian island arc, Italy. *Econ. Geol.* **2014**, *109*, 2129–2158. [CrossRef]

28. Chen, J.; Wang, L.; Chen, Y.; Guo, J. A DFT study of the effect of natural impurities on the electronic structure of galena. *Int. J. Miner. Process.* **2011**, *98*, 132–136. [CrossRef]

29. Petersen, S.; Herzig, P.M.; Schwarz-Schampera, U.; Hannington, M.D.; Jonasson, I.R. Hydrothermal precipitates associated with bimodal volcanism in the Central Bransfield Strait, Antarctica. *Miner. Depos.* **2004**, *39*, 358–379. [CrossRef]

30. Lee, C.-S.; Shor, G.G.; Bibee, L.; Lu, R.S.; Hilde, T.W. Okinawa Trough: Origin of a back-arc basin. *Mar. Geol.* **1980**, *35*, 219–241. [CrossRef]

31. Sibuet, J.-C.; Deffontaines, B.; Hsu, S.-K.; Thareau, N.; Le Formal, J.-P.; Liu, C.-S. Okinawa trough backarc basin: Early tectonic and magmatic evolution. *J. Geophys. Res. Solid Earth* **1998**, *103*, 30245–30267. [CrossRef]

32. Kimura, M. Genesis and formation of the Okinawa Trough, Japan. *Mem. Geol. Soc.* **1990**, *34*, 77–88.

33. Shinjo, R.; Kato, Y. Geochemical constraints on the origin of bimodal magmatism at the Okinawa Trough, an incipient back-arc basin. *Lithos* **2000**, *54*, 117–137. [CrossRef]

34. Zeng, Z.-G.; Chen, Z.-X.; Zhang, Y.-X. Zhang. Zircon record of an Archaean crustal fragment and supercontinent amalgamation in quaternary back-arc volcanic rocks. *Sci. Rep.* **2021**, *11*, 12367. [CrossRef] [PubMed]

35. Sibuet, J.-C.; Letouzey, J.; Barbier, F.; Charvet, J.; Foucher, J.-P.; Hilde, T.W.C.; Kimura, M.; Chiao, L.-Y.; Marsset, B.; Muller, C.; et al. Back arc extension in the Okinawa Trough. *J. Geophys. Res. Solid Earth* **1987**, *92*, 14041–14063. [CrossRef]

36. Gena, K.; Chiba, H.; Kase, K.; Nakashima, K.; Ishiyama, D. The Tiger Sulfide Chimney, Yonaguni Knoll IV Hydrothermal Field, Southern Okinawa Trough, Japan: The First Reported Occurrence of Pt-Cu-Fe-Bearing Bismuthinite and Sn-Bearing Chalcopyrite in an Active Seafloor Hydrothermal System. *Resour. Geol.* **2013**, *63*, 360–370. [CrossRef]

37. Shinjo, R.; Chung, S.-L.; Kato, Y.; Kimura, M. Geochemical and Sr-Nd isotopic characteristics of volcanic rocks from the Okinawa Trough and Ryukyu Arc: Implications for the evolution of a young, intracontinental back arc basin. *J. Geophys. Res. Solid Earth* **1999**, *104*, 10591–10608. [CrossRef]

38. Halbach, P.; Hansmann, W.; Köppel, V.; Pracejus, B. Whole-rock and sulfide lead-isotope data from the hydrothermal JADE field in the Okinawa back-arc trough. *Miner. Depos.* **1997**, *32*, 70–78. [CrossRef]

39. Li, X.; Zeng, Z.; Chen, S.; Ma, Y.; Yang, H.; Zhang, Y.; Chen, Z. Geochemical and Sr-Nd-Pb isotopic compositions of volcanic rocks from the Iheya Ridge, the middle Okinawa Trough: Implications for petrogenesis and a mantle source. *Acta Oceanolog. Sin.* **2018**, *37*, 73–88. [CrossRef]

40. Ishibashi, J.-I.; Gamo, T.; Sakai, H.; Nojiri, Y.; Igarashi, G.; Shitashima, K.; Tsubota, H. Geochemical evidence for hydrothermal activity in the Okinawa Trough. *Geochem. J.* **1988**, *22*, 107–114. [CrossRef]

41. Yamano, M.; Uyeda, S.; Foucher, J.-P.; Sibuet, J.-C. Heat flow anomaly in the middle Okinawa Trough. *Tectonophysics* **1989**, *159*, 307–318. [CrossRef]

42. Nakamura, K.; Kawagucci, S.; Kitada, K.; Kumagai, H.; Takai, K.; Okino, K. Water column imaging with multibeam echo-sounding in the mid-Okinawa Trough: Implications for distribution of deep-sea hydrothermal vent sites and the cause of acoustic water column anomaly. *Geochem. J.* **2015**, *49*, 579–596. [CrossRef]
43. Halbach, P.; Nakamura, K.-I.; Wahsner, M.; Lange, J.; Sakai, H.; Käselitz, L.; Hansen, R.-D.; Yamano, M.; Post, J.; Prause, B.; et al. Prause. Probable modern analogue of Kuroko-type massive sulphide deposits in the Okinawa Trough back-arc basin. *Nature* **1989**, *338*, 496–499. [CrossRef]
44. Sakai, H.; Gamo, T.; Kim, E.-S.; Tsutsumi, M.; Tanaka, T.; Ishibashi, J.; Wakita, H.; Yamano, M.; Oomori, T. Venting of carbon dioxide-rich fluid and hydrate formation in mid-Okinawa trough backarc basin. *Science* **1990**, *248*, 1093–1096. [CrossRef] [PubMed]
45. Gamo, T. Growth mechanism of the hydrothermal mounds at the CLAM Site, Mid Okinawa Trough, inferred from their morphological, mineralogical and chemical characteristics. *JAMSTEC Deep Sea Res.* **1991**, *7*, 163–184.
46. Kawagucci, S.; Toki, T.; Ishibashi, J.; Takai, K.; Ito, M.; Oomori, T.; Gamo, T. Isotopic variation of molecular hydrogen in 20–375 C hydrothermal fluids as detected by a new analytical method. *J. Geophys. Res. Biogeosci.* **2010**, *115*, G03021. [CrossRef]
47. Ishibashi, J.-I.; Ikegami, F.; Tsuji, T.; Urabe, T. Hydrothermal activity in the Okinawa Trough back-arc basin: Geological background and hydrothermal mineralization. In *Subseafloor Biosphere Linked to Hydrothermal Systems: TAIGA Concept*; Ishibashi, J.-I., Okino, K., Sunamura, M., Eds.; Springer: Tokyo, Japan, 2015; pp. 337–359.
48. Chiba, M.; Koizumi, A.; Oshika, J.; Tanahashi, M.; Ueda, S.; Ishikawa, N.; Okamoto, N.; Ishizuka, O.; Shimoda, G. *Discovery of Sulfide Mounds in the Okinawa Trough*; Abstracts with Programs; The Society of Resource: Tokyo, Japan, 2014; Volume 64, p. 24.
49. Liu, Y.; Hu, Z.; Gao, S.; Günther, D.; Xu, J.; Gao, C.; Chen, H. In situ analysis of major and trace elements of anhydrous minerals by LA-ICP-MS without applying an internal standard. *Chem. Geol.* **2008**, *257*, 34–43. [CrossRef]
50. Bao, Z.; Chen, L.; Zong, C.; Yuan, H.; Chen, K.; Dai, M. Development of pressed sulfide powder tablets for in situ sulfur and lead isotope measurement using LA-MC-ICP-MS. *Int. J. Mass Spectrom.* **2017**, *421*, 255–262. [CrossRef]
51. Chen, L.; Chen, K.; Bao, Z.; Liang, P.; Sun, T.; Yuan, H. Preparation of standards for in situ sulfur isotope measurement in sulfides using femtosecond laser ablation MC-ICP-MS. *J. Anal. At. Spectrom.* **2017**, *32*, 107–116. [CrossRef]
52. Yuan, H.; Liu, X.; Chen, L.; Bao, Z.; Chen, K.; Zong, C.; Li, X.-C.; Qiu, J.W. Simultaneous measurement of sulfur and lead isotopes in sulfides using nanosecond laser ablation coupled with two multi-collector inductively coupled plasma mass spectrometers. *J. Asian Earth Sci.* **2018**, *154*, 386–396. [CrossRef]
53. Chen, K.; Yuan, H.; Bao, Z.; Zong, C.; Dai, M. Precise and accurate in situ determination of lead isotope ratios in NIST, USGS, MPI-DING and CGSG glass reference materials using femtosecond laser ablation MC-ICP-MS. *Geostand. Geoanal. Res.* **2014**, *38*, 5–21.
54. Yuan, H.; Yin, C.; Liu, X.; Chen, K.; Bao, Z.; Zong, C.; Dai, M.; Lai, S.; Wang, R.; Jiang, S. High precision in-situ Pb isotopic analysis of sulfide minerals by femtosecond laser ablation multi-collector inductively coupled plasma mass spectrometry. *Sci. China Earth Sci.* **2015**, *58*, 1713–1721. [CrossRef]
55. Belzile, N.; Chen, Y.-W.; Cai, M.; Li, Y. A review on pyrrhotite oxidation. *J. Geochem. Explor.* **2004**, *84*, 65–76. [CrossRef]
56. Evensen, N.; Hamilton, P.; O'Nions, R. Rare-earth abundances in chondritic meteorites. *Geochim. Cosmochim. Acta* **1978**, *42*, 1199–1212. [CrossRef]
57. Hongo, Y.; Obata, H.; Gamo, T.; Nakaseama, M.; Ishibashi, J.; Konno, U.; Saegusa, S.; Ohkubo, S.; Tsunogai, U. Rare Earth Elements in the hydrothermal system at Okinawa Trough back-arc basin. *Geochem. J.* **2007**, *41*, 1–15. [CrossRef]
58. Xu, Z.; Lim, D.; Li, T.; Kim, S.; Jung, H.; Wan, S.; Chang, F.; Cai, M. REEs and Sr-Nd isotope variations in a 20 ky-sediment core from the middle Okinawa Trough, East China Sea: An in-depth provenance analysis of siliciclastic components. *Mar. Geol.* **2019**, *415*, 105970. [CrossRef]
59. Steele, J.H.; Thorpe, S.A.; Turekian, K.K. *Marine Chemistry and Geochemistry: A Derivative of Encyclopedia of Ocean Sciences*, 2nd ed.; Elsevier Science: London, UK, 2010; pp. 601–602.
60. Shu, Y.; Nielsen, S.G.; Zeng, Z.; Shinjo, R.; Blusztajn, J.; Wang, X.; Chen, S. Tracing subducted sediment inputs to the Ryukyu arc-Okinawa Trough system: Evidence from thallium isotopes. *Geochim. Cosmochim. Acta* **2017**, *217*, 462–491. [CrossRef]
61. Hannington, M.; Herzig, P.; Scott, S.; Thompson, G.; Rona, P. Comparative mineralogy and geochemistry of gold-bearing sulfide deposits on the mid-ocean ridges. *Mar. Geol.* **1991**, *101*, 217–248. [CrossRef]
62. Kawasumi, S.; Chiba, H. Redox state of seafloor hydrothermal fluids and its effect on sulfide mineralization. *Chem. Geol.* **2017**, *451*, 25–37. [CrossRef]
63. Moss, R.; Scott, S.D. Geochemistry and mineralogy of gold-rich hydrothermal precipitates from the eastern Manus Basin, Papua New Guinea. *Can. Mineral.* **2001**, *39*, 957–978. [CrossRef]
64. Scott, S.D.; Kissin, S.A. Sphalerite composition in the Zn-Fe-S system below 300 degrees C. *Econ. Geol.* **1973**, *68*, 475–479. [CrossRef]
65. Yang, B.; Liu, J.; Shi, X.; Zhang, H.; Wang, X.; Wu, Y.; Fang, X. Mineralogy and sulfur isotope characteristics of metalliferous sediments from the Tangyin hydrothermal field in the southern Okinawa Trough. *Ore Geol. Rev.* **2020**, *120*, 103464. [CrossRef]
66. Keith, M.; Häckel, F.; Haase, K.M.; Schwarz-Schampera, U.; Klemd, R. Trace element systematics of pyrite from submarine hydrothermal vents. *Ore Geol. Rev.* **2016**, *72*, 728–745. [CrossRef]

67. Pašava, J.; Vymazalová, A.; Petersen, S.; Herzig, P. PGE distribution in massive sulfides from the PACMANUS hydrothermal field, eastern Manus basin, Papua New Guinea: Implications for PGE enrichment in some ancient volcanogenic massive sulfide deposits. *Miner. Depos.* **2004**, *39*, 784–792. [CrossRef]
68. Pašava, J.; Vymazalová, A.; Petersen, S. PGE fractionation in seafloor hydrothermal systems: Examples from mafic-and ultramafic-hosted hydrothermal fields at the slow-spreading Mid-Atlantic Ridge. *Miner. Depos.* **2007**, *42*, 423–431. [CrossRef]
69. Gamo, T. Wide variation of chemical characteristics of submarine hydrothermal fluids due to secondary modification processes after high temperature water-rock interaction: A review. In *Biogeochemical Processes and Ocean Flux in the Western Pacific*; Sakai, H., Nozaki, Y., Eds.; Terra Scientific Publishing Company (TERRAPUB): Tokyo, Japan, 1995; pp. 425–451.
70. Chen, L.-M.; Song, X.-Y.; Danyushevsky, L.; Wang, Y.-S.; Tian, Y.-L.; Xiao, J.-F. A laser ablation ICP-MS study of platinum-group and chalcophile elements in base metal sulfide minerals of the Jinchuan Ni–Cu sulfide deposit, NW China. *Ore Geol. Rev.* **2015**, *65*, 955–967. [CrossRef]
71. Farrow, C.E.; Watkinson, D.H. Diversity of precious-metal mineralization in footwall Cu-Ni-PGE deposits, Sudbury, Ontario; implications for hydrothermal models of formation. *Can. Mineral.* **1997**, *35*, 817–839.
72. Liu, Y.; Brenan, J. Partitioning of platinum-group elements (PGE) and chalcogens (Se, Te, As, Sb, Bi) between monosulfide-solid solution (MSS), intermediate solid solution (ISS) and sulfide liquid at controlled fO_2–fS_2 conditions. *Geochim. Cosmochim. Acta* **2015**, *159*, 139–161. [CrossRef]
73. Binns, R.A.; Scott, S.D. Actively forming polymetallic sulfide deposits associated with felsic volcanic rocks in the eastern Manus back-arc basin, Papua New Guinea. *Econ. Geol.* **1993**, *88*, 2226–2236. [CrossRef]
74. Wohlgemuth-Ueberwasser, C.C.; Viljoen, F.; Petersen, S.; Vorster, C. Distribution and solubility limits of trace elements in hydrothermal black smoker sulfides: An in-situ LA-ICP-MS study. *Geochim. Cosmochim. Acta* **2015**, *159*, 16–41. [CrossRef]
75. Tivey, M.; Humphris, S.E.; Thompson, G.; Hannington, M.D., Rona, P.A. Deducing patterns of fluid flow and mixing within the TAG active hydrothermal mound using mineralogical and geochemical data. *J. Geophys. Res. Solid Earth* **1995**, *100*, 12527–12555. [CrossRef]
76. Alt, J.C. The chemistry and sulfur isotope composition of massive sulfide and associated deposits on Green Seamount, eastern Pacific. *Econ. Geol.* **1988**, *83*, 1026–1033. [CrossRef]
77. Mills, R.; Elderfield, H. Rare earth element geochemistry of hydrothermal deposits from the active TAG Mound, 26 N Mid-Atlantic Ridge. *Geochim. Cosmochim. Acta* **1995**, *59*, 3511–3524. [CrossRef]
78. Ishibashi, J.-I.; Urabe, T. Hydrothermal activity related to arc-backarc magmatism in the western Pacific. In *Backarc Basins*; Springer: New York, NY, USA, 1995; pp. 451–495.
79. Barrett, T.J.; Jarvis, I.; Jarvis, K.E. Rare earth element geochemistry of massive sulfides-sulfates and gossans on the Southern Explorer Ridge. *Geology* **1990**, *18*, 583–586. [CrossRef]
80. Gillis, K.M.; Smith, A.D.; Ludden, J.N. Trace element and Sr isotopic contents of hydrothermal clays and sulfides from the Snakepit hydrothermal field: ODP site 649. *Proc. ODP Sci. Res.* **1990**, *106*, 315–319.
81. Zeng, Z.; Ma, Y.; Yin, X.; Selby, D.; Kong, F.; Chen, S. Factors affecting the rare earth element compositions in massive sulfides from deep-sea hydrothermal systems. *Geochem. Geophys. Geosyst.* **2015**, *16*, 2679–2693. [CrossRef]
82. Schade, J.; Cornell, D.H.; Theart, H.F.J. Rare earth element and isotopic evidence for the genesis of the Prieska massive sulfide deposit, South Africa. *Econ. Geol.* **1989**, *84*, 49–63. [CrossRef]
83. Sverjensky, D.A. Europium redox equilibria in aqueous solution. *Earth Planet. Sci. Lett.* **1984**, *67*, 70–78. [CrossRef]
84. Wood, S.A.; Williams-Jones, A.E. The aqueous geochemistry of the rare-earth elements and yttrium 4. Monazite solubility and REE mobility in exhalative massive sulfide-depositing environments. *Chem. Geol.* **1994**, *115*, 47–60. [CrossRef]
85. Haas, J.R.; Shock, E.L.; Sassani, D.C. Rare earth elements in hydrothermal systems: Estimates of standard partial molal thermodynamic properties of aqueous complexes of the rare earth elements at high pressures and temperatures. *Geochim. Cosmochim. Acta* **1995**, *59*, 4329–4350. [CrossRef]
86. Allen, D.E.; Seyfried, W., Jr. REE controls in ultramafic hosted MOR hydrothermal systems: An experimental study at elevated temperature and pressure. *Geochim. Cosmochim. Acta* **2005**, *69*, 675–683. [CrossRef]
87. De Baar, H.J.; Brewer, P.G.; Bacon, M.P. Anomalies in rare earth distributions in seawater: Gd and Tb. *Geochim. Cosmochim. Acta* **1985**, *49*, 1961–1969. [CrossRef]
88. Michard, A.; Albarede, F. The REE content of some hydrothermal fluids. *Chem. Geol.* **1986**, *55*, 51–60. [CrossRef]
89. Rimskaya-Korsakova, M.N.; Dubinin, A.V. Rare earth elements in sulfides of submarine hydrothermal vents of the Atlantic Ocean. *Dokl. Earth Sci.* **2003**, *389*, 432–436.
90. Bau, M.; Dulski, P. Comparing yttrium and rare earths in hydrothermal fluids from the Mid-Atlantic Ridge: Implications for Y and REE behaviour during near-vent mixing and for the Y/Ho ratio of Proterozoic seawater. *Chem. Geol.* **1999**, *155*, 77–90. [CrossRef]
91. Langmuir, C.; Humphris, S.; Fornari, D.; Van Dover, C.; Von Damm, K.; Tivey, M.; Colodner, D.; Charlou, J.-L.; Desonie, D.; Wilson, C.; et al. Hydrothermal vents near a mantle hot spot: The Lucky Strike vent field at 37 N on the Mid-Atlantic Ridge. *Earth Planet. Sci. Lett.* **1997**, *148*, 69–91. [CrossRef]
92. Rye, R.O.; Ohmoto, H. Sulfur and carbon isotopes and ore genesis: A review. *Econ. Geol.* **1974**, *69*, 826–842. [CrossRef]
93. Ohomoto, H.; Rye, R.O. Isotopes of sulfur and carbon. In *Geochemistry of Hydrothermal Ore Deposits*, 2nd ed.; Barnes, H.L., Ed.; J Wiley and Sons: New York, NY, USA, 1979; pp. 509–567.

94. Sakai, H.; Marais, D.; Ueda, A.; Moore, J. Concentrations and isotope ratios of carbon, nitrogen and sulfur in ocean-floor basalts. *Geochim. Cosmochim. Acta* **1984**, *48*, 2433–2441. [CrossRef]

95. Alt, J.C.; Anderson, T.F.; Bonnell, L. The geochemistry of sulfur in a 1.3 km section of hydrothermally altered oceanic crust, DSDP Hole 504B. *Geochim. Cosmochim. Acta* **1989**, *53*, 1011–1023. [CrossRef]

96. Shanks, W.C.; Böhlke, J.K.; Seal, R.R. Stable isotopes in mid-ocean ridge hydrothermal systems: Interactions between fluids, minerals, and organisms. *Geophys. Monogr. Am. Geophys. Union* **1995**, *91*, 194.

97. Alt, J.C.; Shanks, W.C., III. Serpentinization of abyssal peridotites from the MARK area, Mid-Atlantic Ridge: Sulfur geochemistry and reaction modeling. *Geochim. Cosmochim. Acta* **2003**, *67*, 641–653. [CrossRef]

98. Ueda, A.; Sakai, H. Sulfur isotope study of Quaternary volcanic rocks from the Japanese Islands Arc. *Geochim. Cosmochim. Acta* **1984**, *48*, 1837–1848. [CrossRef]

99. Woodhead, J.D.; Harmon, R.S.; Fraser, D. O, S, Sr, and Pb isotope variations in volcanic rocks from the Northern Mariana Islands: Implications for crustal recycling in intra-oceanic arcs. *Earth Planet. Sci. Lett.* **1987**, *83*, 39–52. [CrossRef]

100. Mottl, M.J.; Holland, H.D.; Corr, R.F. Chemical exchange during hydrothermal alteration of basalt by seawater—II. Experimental results for Fe, Mn, and sulfur species. *Geochim. Cosmochim. Acta* **1979**, *43*, 869–884. [CrossRef]

101. Shanks, W.; Jeffrey, N. Sulfur isotope studies of hydrothermal anhydrite and pyrite, Deep Sea Drilling Project leg 64, Guaymas Basin, Gulf of California. *Init. Rep. Deep Sea Drill. Proj.* **1982**, *64*, 1137–1142.

102. Rees, C.; Jenkins, W.; Monster, J. The sulphur isotopic composition of ocean water sulphate. *Geochim. Cosmochim. Acta* **1978**, *42*, 377–381. [CrossRef]

103. Michard, G.; Albarede, F.; Michard, A.; Minster, J.-F.; Charlou, J.-L.; Tan, N. Chemistry of solutions from the 13 N East Pacific Rise hydrothermal site. *Earth Planet. Sci. Lett.* **1984**, *67*, 297–307. [CrossRef]

104. Butterfield, D.; Massoth, G.J.; McDuff, R.E.; Lupton, J.E.; Lilley, M.D. Geochemistry of hydrothermal fluids from Axial Seamount hydrothermal emissions study vent field, Juan de Fuca Ridge: Subseafloor boiling and subsequent fluid-rock interaction. *J. Geophys. Res. Solid Earth* **1990**, *95*, 12895–12921. [CrossRef]

105. Shanks, W.C., III. Stable isotopes in seafloor hydrothermal systems: Vent fluids, hydrothermal deposits, hydrothermal alteration, and microbial processes. *Rev. Mineral. Geochem.* **2001**, *43*, 469–525. [CrossRef]

106. Fouquet, Y.; Knott, R.; Cambon, P.; Fallick, A.; Rickard, D.; Desbruyères, D. Formation of large sulfide mineral deposits along fast spreading ridges. Example from off-axial deposits at 12 43′ N on the East Pacific Rise. *Earth Planet. Sci. Lett.* **1996**, *144*, 147–162. [CrossRef]

107. Gamo, T.; Sakai, H.; Kim, E.-S.; Shitashima, K.; Ishibashi, J.-I. High alkalinity due to sulfate reduction in the CLAM hydrothermal field, Okinawa Trough. *Earth Planet. Sci. Lett.* **1991**, *107*, 328–338. [CrossRef]

108. Nakagawa, S.; Takai, K.; Inagaki, F.; Chiba, H.; Ishibashi, J.-I.; Kataoka, S.; Hirayama, H.; Nunoura, T.; Horikoshi, K.; Sako, Y. Variability in microbial community and venting chemistry in a sediment-hosted backarc hydrothermal system: Impacts of subseafloor phase-separation. *FEMS Microbiol. Ecol.* **2005**, *54*, 141–155. [CrossRef] [PubMed]

109. Kawagucci, S.; Chiba, H.; Ishibashi, J.-I.; Yamanaka, T.; Toki, T.; Muramatsu, Y.; Ueno, Y.; Makabe, A.; Inoue, K.; Yoshida, N.; et al. Hydrothermal fluid geochemistry at the Iheya North field in the mid-Okinawa Trough: Implication for origin of methane in subseafloor fluid circulation systems. *Geochem. J.* **2011**, *45*, 109–124. [CrossRef]

110. Kawagucci, S. Fluid geochemistry of high-temperature hydrothermal fields in the Okinawa Trough. In *Subseafloor Biosphere Linked to Hydrothermal Systems: TAIGA Concept*; Ishibashi, J.-I., Okino, K., Sunamura, M., Eds.; Springer: Tokyo, Japan, 2015; pp. 387–403.

111. Takai, K.; Nakagawa, S.; Nunoura, T. Comparative investigation of microbial communities associated with hydrothermal activities in the Okinawa trough. In *Subseafloor Biosphere Linked to Hydrothermal Systems: TAIGA Concept*; Ishibashi, J.-I., Okino, K., Sunamura, M., Eds.; Springer: Tokyo, Japan, 2015; pp. 421–435.

112. Lüders, V.; Pracejus, B.; Halbach, P. Fluid inclusion and sulfur isotope studies in probable modern analogue Kuroko-type ores from the JADE hydrothermal field (Central Okinawa Trough, Japan). *Chem. Geol.* **2001**, *173*, 45–58. [CrossRef]

113. Alt, J.C.; Shanks, W.C.; Bach, W.; Paulick, H.; Garrido, C.J.; Beaudoin, G. Hydrothermal alteration and microbial sulfate reduction in peridotite and gabbro exposed by detachment faulting at the Mid-Atlantic Ridge, 15 20′ N (ODP Leg 209): A sulfur and oxygen isotope study. *Geochem. Geophys. Geosyst.* **2007**, *8*, Q08002. [CrossRef]

114. Delacour, A.; Früh-Green, G.L.; Bernasconi, S.M.; Kelley, D.S. Sulfur in peridotites and gabbros at Lost City (30 N, MAR): Implications for hydrothermal alteration and microbial activity during serpentinization. *Geochim. Cosmochim. Acta* **2008**, *72*, 5090–5110. [CrossRef]

115. Sakai, H. Isotopic properties of sulfur compounds in hydrothermal processes. *Geochem. J.* **1968**, *2*, 29–49. [CrossRef]

116. Li, Y.; Liu, J. Calculation of sulfur isotope fractionation in sulfides. *Geochim. Cosmochim. Acta* **2006**, *70*, 1789–1795. [CrossRef]

117. Vidal, P.; Clauer, N. Pb and Sr isotopic systematics of some basalts and sulfides from the East Pacific Rise at 21 N (project RITA). *Earth Planet. Sci. Lett.* **1981**, *55*, 237–246. [CrossRef]

118. Chen, J. U, Th, and Pb isotopes in hot springs on the Juan de Fuca Ridge. *J. Geophys. Res. Solid Earth* **1987**, *92*, 11411–11415. [CrossRef]

119. Hegner, E.; Tatsumoto, M. Pb, Sr, and Nd isotopes in basalts and sulfides from the Juan de Fuca Ridge. *J. Geophys. Res. Solid Earth* **1987**, *92*, 11380–11386. [CrossRef]

120. Hinkley, T.K.; Tatsumoto, M. Metals and isotopes in Juan de Fuca Ridge hydrothermal fluids and their associated solid materials. *J. Geophys. Res. Solid Earth* **1987**, *92*, 11400–11410. [CrossRef]

121. Fouquet, Y.; Marcoux, E. Lead isotope systematics in Pacific hydrothermal sulfide deposits. *J. Geophys. Res. Solid Earth* **1995**, *100*, 6025–6040. [CrossRef]
122. Charlou, J.; Donval, J.; Fouquet, Y.; Jean-Baptiste, P.; Holm, N. Geochemistry of high H_2 and CH_4 vent fluids issuing from ultramafic rocks at the Rainbow hydrothermal field (36 14′ N, MAR). *Chem. Geol.* **2002**, *191*, 345–359. [CrossRef]
123. Yao, H.-Q.; Zhou, H.-Y.; Peng, X.-T.; Bao, S.-X.; Wu, Z.-J.; Li, J.-T.; Sun, Z.-L.; Chen, Z.-Q.; Li, J.-W.; Chen, G.-Q. Metal sources of black smoker chimneys, Endeavour Segment, Juan de Fuca Ridge: Pb isotope constraints. *Appl. Geochem.* **2009**, *24*, 1971–1977. [CrossRef]
124. Allègre, C.J.; Hamelin, B.; Dupré, B. Statistical analysis of isotopic ratios in MORB: The mantle blob cluster model and the convective regime of the mantle. *Earth Planet. Sci. Lett.* **1984**, *71*, 71–84. [CrossRef]
125. Hamelin, B.; Dupré, B.; Allègre, C.J. Lead-strontium isotopic variations along the East Pacific Rise and the Mid-Atlantic Ridge: A comparative study. *Earth Planet. Sci. Lett.* **1984**, *67*, 340–350. [CrossRef]
126. Andrieu, A.; Honnorez, J.; Lancelot, J. Lead isotope compositions of the TAG mineralization, Mid-Atlantic Ridge, 26 08′ N. In *Ocean Drilling Program*; Scientific Results; Elsevier Science: Amsterdam, The Netherlands, 1998; pp. 101–109.
127. Thompson, G.; Humphris, S.E.; Schroeder, B.; Sulanowska, M.; Rona, P.A. Active vents and massive sulfides at 26 degrees N (TAG) and 23 degrees N (Snakepit) on the Mid-Atlantic Ridge. *Can. Mineral.* **1988**, *26*, 697–711.
128. Davis, A.S.; Clague, D.A.; Zierenberg, R.; Wheat, C.; Cousens, B.L. Sulfide formation related to changes in the hydrothermal system on Loihi Seamount, Hawai'i, following the seismic event in 1996. *Can. Mineral.* **2003**, *41*, 457–472. [CrossRef]
129. Murowchick, J.B. Marcasite inversion and the petrographic determination of pyrite ancestry. *Econ. Geol.* **1992**, *87*, 1141–1152. [CrossRef]

Journal of
*Marine Science
and Engineering*

Article

Two Processes of Anglesite Formation and a Model of Secondary Supergene Enrichment of Bi and Ag in Seafloor Hydrothermal Sulfide Deposits

Zhigang Zeng [1,2,3,*], **Zuxing Chen** [1,4] **and Haiyan Qi** [1,4]

1   Seafloor Hydrothermal Activity Laboratory, CAS Key Laboratory of Marine Geology and Environment, Institute of Oceanology, Chinese Academy of Sciences, Qingdao 266071, China; chenzuxing@qdio.ac.cn (Z.C.); qihaiyan@qdio.ac.cn (H.Q.)
2   Laboratory for Marine Mineral Resources, Qingdao National Laboratory for Marine Science and Technology, Qingdao 266071, China
3   College of Marine Sciences, University of Chinese Academy of Sciences, Beijing 100049, China
4   Center for Ocean Mega-Science, Chinese Academy of Sciences, 7 Nanhai Road, Qingdao 266071, China
*   Correspondence: zgzeng@qdio.ac.cn; Tel.: +86-532-8289-8525

Abstract: The in situ element concentrations and the sulfur (S), and lead (Pb) isotopic compositions in anglesite were investigated for samples from seafloor hydrothermal fields in the Okinawa Trough (OT), Western Pacific. The anglesite grains are of two kinds: (1) low Pb/high S primary hydrothermal anglesite (PHA), which is formed by mixing of fluid and seawater, and (2) high Pb/low S secondary supergene anglesite (SSA), which is the product of low-temperature (<100 °C) alteration of galena in the seawater environment. The Ag and Bi in the SSA go through a second enrichment process during the formation of high Pb/low S anglesite by galena alteration, indicating that the SSA and galena, which may be the major minerals host for considerable quantities of Ag and Bi, are potentially Ag-Bi-enriched in the back-arc hydrothermal field. Moreover, REEs, S and Pb in the OT anglesite are likely to have been leached by fluids from local sub-seafloor volcanic rocks and/or sediments. A knowledge of the anglesite is useful for understanding the influence of volcanic rocks, sediments and altered subducted oceanic plate in hydrothermal systems, showing how trace metals behave during the formation of secondary minerals.

Keywords: anglesite; element distribution and mobility; supergene Ag–Bi enrichment; sulfide; hydrothermal vent

Citation: Zeng, Z.; Chen, Z.; Qi, H. Two Processes of Anglesite Formation and a Model of Secondary Supergene Enrichment of Bi and Ag in Seafloor Hydrothermal Sulfide Deposits. *J. Mar. Sci. Eng.* **2022**, 10, 35. https://doi.org/10.3390/jmse10010035

Academic Editor: János Kovács

Received: 17 November 2021
Accepted: 19 December 2021
Published: 31 December 2021

Publisher's Note: MDPI stays neutral with regard to jurisdictional claims in published maps and institutional affiliations.

1. Introduction

Anglesite ($PbSO_4$) has been widely identified as a supergene weathering product of primary galena in oxidation zones in supergene weathering deposits and mine-waste sites [1–15]. Oxidation and acid-neutralization reactions partially dissociate the galena and promote the development of secondary supergene anglesite [16], which then surrounds and directly replaces galena contained within the anglesite [17], and the galena in situ appears to be directly oxidized and partly or totally replaced by anglesite along its cleavage planes and grain boundaries [11,18]. For example, in the Sidi Flah mine, Morocco, galena is locally surrounded by a halo of altered anglesite $PbSO_4$ [19]. Replacement also takes place at the rim of galena grains as well as along the typical cleavage planes [7], which is dissolved releasing Pb^{2+} and SO_4^{2-} ions at high pH (>9) oxidizing environment [20], and the solution interacting with the galena surface is strongly undersaturated with respect to both the primary galena and secondary supergene anglesite ($PbSO_4$) [21]. Supergene weathering of primary galena leads to the formation of thermodynamically more stable secondary anglesite. Whether or not the anglesite is stable depends on its pH, sulfur and lead activity, partial pressure of CO_2 (P_{CO2}) and temperature in the seawater [7,11]. Under oxidizing conditions, anglesite is stable between pH values of about 0.4 and 5.0 [22], and the reported value of the solubility constant for anglesite (K_{sp} = 10–7.79) indicates that it is relatively insoluble [23].

However, replacement of galena by anglesite does not produce acid [24]. Locally occurring iron sulfides undergoing oxidation acidify solutions and provide ferric ions, which are important oxidizing agents [10], and anglesite preferentially forms in such microenvironments [3]. In the first phase of this process, oxidation takes place along cleavage planes and grain boundaries, and galena weathers to anglesite [25]. Moreover, in the supergene Cu-Pb-Zn-V ores of the Oriental High Atlas, Morocco, anglesite is observed in fractures and cleavage planes in galena, and as a thin rim around it [26]. It is also found throughout the waste dumps at Callahan Cu-Zn-Pb mine in the Goose Pond tidal estuary, Maine, USA, as rims on masses of galena, and inside thin fractures that cut across grains of galena and surrounding minerals and in veinlets and fracture sets that redistribute Pb [6].

However, anglesite precipitation can effectively control the attenuation process of fluid Pb concentration [27]. Weathering reactions control the release and transport of Pb and other metals in acidic and near-neutral conditions in surface environments [6,28]. In the Butte mining district, Montana, the low Pb levels may be partly explained by the low solubility of anglesite, especially in more sulfate-rich waters, or by the low solubility of galena in the sulfidic waters of the flooded West Camp mining district in Butte [29].

Anglesite also acts an important solid-phase control on the aqueous mobility of Pb in the Santa Lucia mine, Cuba [16]. The main mechanism controlling the mobility of As, Ba and Pb is the precipitation of secondary supergene anglesite [16]. Geochemical modeling indicates that leaching solutions are supersaturated with anglesite over the entire pH range, showing that anglesite controls the As and Pb, and to a lesser degree the Zn and Cu, unless the pH drops below 3.0 [27]. Furthermore, electron microprobe analyses (EMPA) of anglesite in the waste rocks of the abandoned Seobo tungsten mine in Korea have shown notable Cu concentrations up to 4.6 wt.% [27].

Although the mineral characteristics and chemical compositions of anglesite provide important information about weathering reactions as well as fluid/solid interaction and supergene processes, little is known about the chemical composition of anglesites from seafloor hydrothermal systems, since it occurs only as a minor mineral in sulfide samples, and also due to its secondary supergene genesis in sulfide deposits. The study of element concentrations and isotopic compositions in anglesite from seafloor hydrothermal systems is an important means of understanding element enrichment during supergene oxidation of hydrothermal sulfide deposits, allowing both metal sources and reconstruction of the physicochemical conditions of their preservation to be inferred. In this study, in situ major, trace element abundance, sulfur and lead isotopic compositions in anglesite were measured for the first time in seafloor hydrothermal fields from the back-arc Okinawa Trough (OT) (Figure 1). The isotopic compositions of S and Pb in the anglesite and the abundance of in situ elements from seafloor sulfide deposits are described in combination with the chemical compositions of associated galena and other sulfide minerals. The hydrothermal and seawater contributions of sulfur and are described in an attempt to reveal the mechanisms of element enrichment and the isotope variations in the anglesite, as well as the anglesite formation processes in seafloor hydrothermal systems generally for understanding seafloor massive sulfide deposit preservation and supergene oxidation processes.

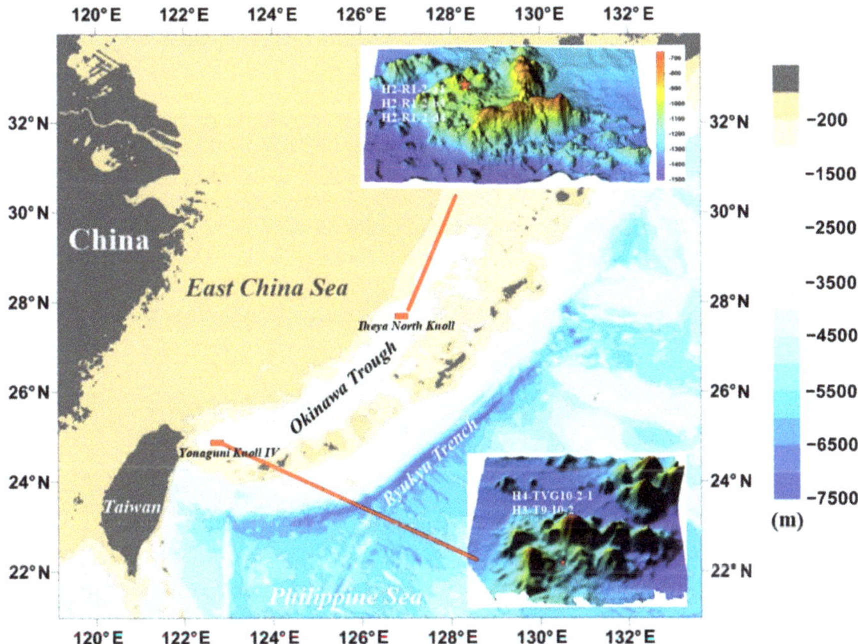

Figure 1. Locations of seafloor hydrothermal sulfide and sulfate samples from the Yonaguni Knoll IV, Irabu Knoll and Iheya North Knoll hydrothermal fields in the Okinawa Trough (red rectangle = location of seafloor hydrothermal field; red star = sulfide and sulfate sampling site).

2. Geological Setting

The OT is at the rifting-to-spreading stage of a back-arc basin, exhibiting the development of normal faulting of brittle crustal rocks and frequent magma intrusions. The OT topography changes markedly from the central to southern parts of the trough, with deepening of the trough floor and steepening of the continental slope, accompanied by en echelon intra-trough grabens [30,31]. The tectonic and geophysical properties of the OT also differ from north to south, with crustal thickness decreasing southwards. The southern part shows variable gravity and linear magnetic anomalies, with greatest heat flow in the central part. Most of the crust in the OT is transitional, although oceanic crust probably appears in some grabens in both the central and southern parts [32]. It provides a favorable geological environment for the development of seafloor hydrothermal systems [31–34].

As of 2015, the InterRidge Vents Database had recorded at least 15 deep-sea hydrothermal fields in the OT, including the Minami-Ensei [35]; Iheya North [36–38]; Jade [39,40]; Hakurei [41]; Irabu Knoll [42,43]; Hatoma [44,45]; Yonaguni Knoll IV [46–48] and Tangyin [34,49] hydrothermal fields. This study focuses on the Iheya North Knoll and Yonaguni Knoll IV hydrothermal fields.

The Iheya North Knoll hydrothermal field (27°47.2′ N, 126°53.9′ E) is located in about 1000 m water depth along the eastern slope of a small knoll, which is part of the Iheya North Knoll volcanic complex [31] (Figure 1). Seismic studies around the Iheya North Knoll have indicated relatively disordered seismic reflectors as deep as 400–500 mbsf (meters below seafloor) in the central valley, surrounded by small knolls [50], indicating the presence of pumiceous volcaniclastic flow deposits beneath hemipelagic surficial sediments, rather than massive igneous rocks [31]. About 10 active hydrothermal mounds, aligned north to south, are concentrated in a small region [31], many of which host active fluid vents and sulfide and sulfate mineralization [31]. A large mound (named North Big Chimney) more than 30 m high is associated with vigorous venting of clear fluid and with the highest temperature of 311 °C, and appears to mark the center of hydrothermal activity [36]. The vent fluids have lower alkalinity (0.5–3.6 mmol 1–1), NH^{4+} (1.9–2.6 mmol kg^{-1}) and Cl- (441–458 mmol kg^{-1}), indicating phase separation of the fluid beneath the seafloor [37]. Mineralization in the Iheya North Knoll included sulfide and sulfate mineralization [31]. For instance, the main sulfate minerals are barite, gypsum and anhydrite. Sulfide consists mainly of sphalerite, galena, pyrite, chalcopyrite, marcasite, wurtzite and tennantite–tetrahedrite. Luzonite, freieslebenite and covellite are also present in small amounts [31]. However, it is possible to use hydrothermal sulfide and its inclusions to trace the deep circulation processes of fluid in the future.

The Yonaguni Knoll IV hydrothermal field is situated in an elongated valley [48] (Figure 1). The valley is mostly covered with muddy sediment, except for the active vent field and volcanic rocks on its northern slope. Suzuki et al. [48] distinguished five types of mineralization: (1) anhydrite-rich chimneys; (2) massive Zn–Pb–Cu sulfides; (3) Ba–As chimneys; (4) Mn-rich chimneys and (5) a pavement of silicified sediment. In addition, diverse styles of fluid venting were found to occur within the hydrothermal field, including slightly Cl-enriched (614–635 mmol kg^{-1}) and depleted (376–491 mmol kg^{-1}) fluids associated with the discharge of liquid droplets of CO_2 [48]. However, there is a relationship between the tectonic setting, magmatism, mineralization, fluid–rock interaction and sedimentary processes in the seafloor hydrothermal system (e.g., Iheya North Knoll and Yonaguni Knoll IV) and the response, adaptation, record and action of organism, and it will be interesting to explore the synergetic metallogenic mechanism of magma, fluid, rock, sediment, seawater and organisms in the seafloor hydrothermal field in the future.

3. Sampling and Methods

3.1. Specimen and Analytical Techniques

Anglesite, galena and other sulfide mineral samples from seafloor hydrothermal sulfide deposits were collected in 2014 and 2016 during the HOBAB 2, 3 and 4 cruises of hydrothermal fields in the OT. Samples were taken from the Iheya North (samples H2-R1-2-a4, -b3, -d4), and the Yonaguni Knoll IV (samples H4-TVG10-2-1, and H3-T9-10-2) hydrothermal fields (Figures 1 and 2).

Figure 2. Photographs of hand specimens.

The anglesite, associated galena and other sulfide mineral samples from selected seafloor sulfide and sulfate deposits were analyzed using thin sections to determine the content of major and trace elements and S and Pb isotopic compositions. Reflected and transmitted light microscopy techniques were used to reveal the external morphology and internal structure of samples. Anglesite, associated galena and other sulfide minerals and microstructures were determined using back-scattered electron (BSE) images and an Oxford Instruments INCA X-Max energy dispersive spectrometer (EDS) on a TESCAN VEGA 3 LMH scanning electron microscope (SEM) at the Institute of Oceanology, Chinese Academy of Sciences.

3.2. In Situ Major Element Determination

The major element compositions of the anglesite, associated galena, pyrite, sphalerite and chalcopyrite were determined using a JXA-8230 electron microprobe analyzer (EMPA) at the State Key Laboratory of Continental Dynamics, Northwest University, Xian, China. The instrument was operated at a 15 kV acceleration voltage, 10 nA beam current and 2 μm beam diameter. The following reference materials were used for wavelength dispersive spectrometry measurements and calibrations: pyrite (for S and Fe); chalcopyrite (for Cu); sphalerite (for Zn); galena (for Pb) and skutterudite (for As). Pure metals were Au, Ag, Cd, Sb, Se, Ni, Co and Cr, and the silicates were diopside (for Ca, Mg and Si); albite (for Al); bustamite (for Mn); rutile (for Ti); apatite (for Sr) and barite (for Ba). The test accuracy for major elements in the sulfides and sulfates was better than 5%. The detection limits for selected trace elements were approximately within the range of 0.01 to 0.12 wt.% (Table S1).

3.3. In Situ Trace Element Determination

In situ measurements of trace elements in the anglesite, associated galena, pyrite, sphalerite and chalcopyrite were determined using a laser-ablation inductively coupled plasma mass spectrometer (LA-ICP-MS) at the State Key Laboratory of Biogeology and Environmental Geology, China University of Geosciences, Wuhan. A GeoLas Pro 193 nm ArF excimer laser was applied to analyze the samples. Laser energy was 80 mJ, and frequency was 6 Hz with a 44 μm diameter ablation spot. Acquisition of ion-signal intensity was performed using an Agilent 7500x ICP-MS instrument. Helium was used as the carrier gas, which was mixed with argon via a T-connector before entering the ICP-MS. Each measurement incorporated a background acquisition of approximately 30 s (gas blank) followed by 50 s of data acquisition from the sample, using ICPMSDataCal software to quantitatively calibrate trace element contents [51,52]. Fe, Cu, Zn and Pb were selected as the internal standards for data reduction. NIST 610 glass was used as the external standard, which was analyzed every 10 spots to monitor for any instrument drifts. The accuracy was determined with respect to the NIST 610 reference glass and was assessed to be better than 10% (1σ).

The rare earth element (REE) contents were normalized to Cl-chondrites (subscript $_{CN}$) [53]. The Eu and Ce anomalies were assessed as $(Eu/Eu^*)_{CN} = Eu_{CN}/(Sm_{CN} + Gd_{CN})^{0.5}$ and $(Ce/Ce^*)_{CN} = Ce_{CN}/(La_{CN} + Pr_{CN})^{0.5}$.

3.4. In Situ S Isotope Analysis

In situ measurements of S isotope ratios of the anglesite and associated galena, pyrite, sphalerite and chalcopyrite were conducted on double-sided-polished slices (DSPSs) of these minerals using a 193 nm femtosecond laser-ablation multi-collector inductively coupled plasma mass spectrometer (fs-LA-MC-ICP-MS) at the State Key Laboratory of Continental Dynamics (SKLCD), Northwest University, Xian, China. The equipment consisted of a 193 nm NWRfemto (RESOlution M-50, ASI) laser ablation system coupled with a Nu Plasma II MC-ICP-MS (NU Instruments Ltd., Wrexham, UK). The ablated materials were transported into the plasma using He as a carrier gas. In a cyclone coaxial mixer, Ar gas was mixed with the carrier gas before being transported into the ICP torch. The energy fluence of the laser was approximately 3.5–4 J/cm^2. The beam diameter was 35 μm with a laser repletion rate of 3–4 Hz for single spot analysis. Standard PSPT-3 (sphalerite; $\delta^{34}S_{V-CDT} = 26.5 \pm 0.2‰$) was used as certified reference material. Detailed analysis parameters are described in Bao et al. [54], Chen et al. [55] and Yuan et al. [56]. In this study, the S isotope ratios of the anglesite and associated galena, pyrite, sphalerite and chalcopyrite are given in standard notation (per mil) relative to Vienna Canyon Diablo Troilite.

3.5. In Situ Pb Isotope Analysis

In situ measurements of Pb isotope compositions of the anglesite and galena were analyzed on DSPSs using a 193 nm fs-LA-MC-ICP-MS at the SKLCD. To remove potential contamination, anhydrous ethanol was used to carefully clean the surfaces of the DSPSs prior to laser ablation analysis. Analytical spots were carefully selected to prevent the possible influence of inclusions and impurities. High-temperature-activated carbon was used to filter Hg contained in the carrier gas, which lowered both the Hg background and the detection limit. An internal Tl isotope reference NIST SRM997 in conjunction with an external reference NIST SRM 610 were used to correct for fractionation and mass discrimination effects, where the Tl solution is introduced through the CETAC Aridus II desolvation nebulizer system. The exponential law correction method for Tl normalization with optimally adjusted Tl ratio was used to obtain Pb isotopic data with high precision and accuracy [57]. The measured isotopic ratios were in agreement with the reference and published values within 2σ measurement uncertainty [54,56–58].

4. Results

4.1. Occurrence and Major and Trace Element Concentrations of Anglesite

Most seafloor hydrothermal sulfide and sulfate samples from the middle OT (MOT) Iheya North (samples H2-R1-2-a4, b3, d4) and from the southern OT (SOT) Yonaguni Knoll IV (samples H4-TVG10-2-1, and H3-T9-10-2) consist of major (>15 vol%) sphalerite, chalcopyrite, galena, pyrite and barite, with minor (<15 vol%) anglesite (Figure 3). The anglesite minerals occur as infills and veins with low Pb and high S contents in the host sulfide, as crusts with high Pb and low S contents around the margin of the relict galena, as in situ replacement or as euhedral crystals in cavities of the former, partially dissolved galena (Figures 4 and 5). This implies that low Pb and high S and high Pb and low S grains may be primary hydrothermal anglesite (PHA) and secondary supergene anglesite (SSA), respectively.

Figure 3. (**a**) Sphalerite forming large porous structures in sample H4-TVG10-2-1. Sphalerite with "dust" chalcopyrite disease condensing into larger particles. Pore centers with anglesite and galena infill indicates their later formation. (**b**) Intergrowth of subhedral to euhedral pyrite and sphalerite forming large porous structures in sample H4-TVG10-2-1; pore center filled with anglesite. Relict galena surrounded by anglesite indicates secondary origin of anglesite replacing galena. (Ang = anglesite; Sph = sphalerite; Py = pyrite; Ccp = chalcopyrite; Gn = galena).

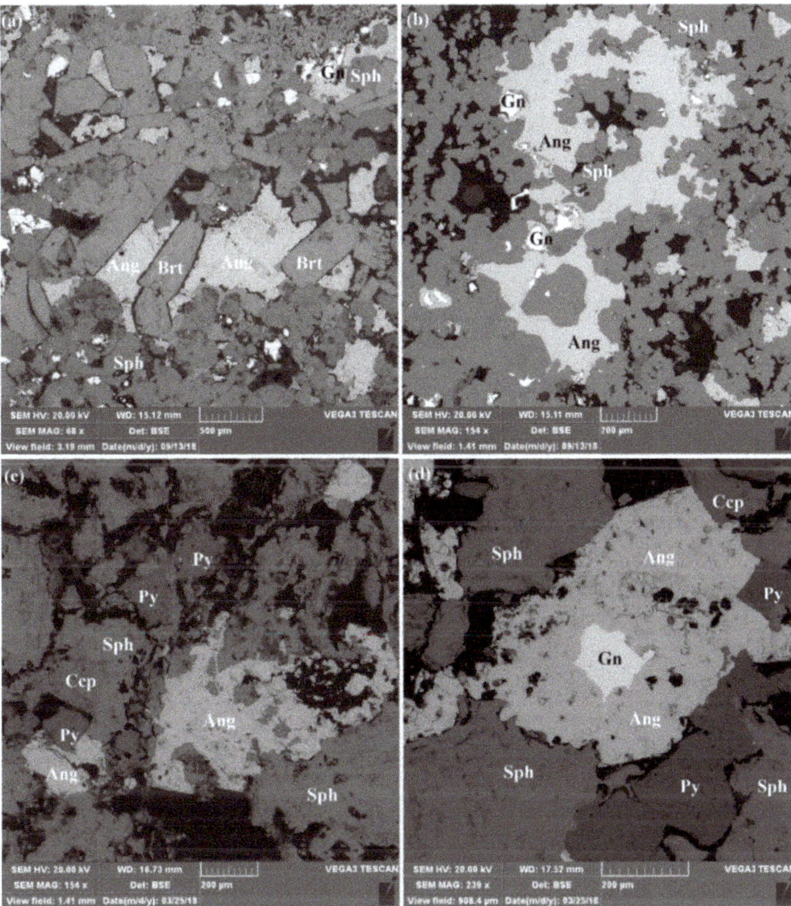

Figure 4. (**a**) SEM back-scattered electron (BSE) images of selected anglesite and sulfide minerals hosting Ag and Bi in seafloor hydrothermal sulfide deposits, Okinawa Trough. (**a**) Sample H3-T9-10-2: Intergrowth of large subhedral to euhedral barite, anglesite, sphalerite and galena forming porous structures. (**b**) Sample H4-TVG10-2-1: Anglesite replacing galena. Fluid channels with sphalerite infill indicate their later formation. (**c**) Sample H2-R1-2-a4: Intergrowth of sphalerite and pyrite forming large porous structures. Some pores filled with anglesite minerals indicate their later formation. (**d**) Sample H2-R1-2-b3: Intergrowth of sphalerite, pyrite and chalcopyrite forming porous structures. Relict galena surrounded by anglesite indicates a secondary origin of anglesite replacing galena (Ang = anglesite; Brt = barite; Sp = sphalerite; Py = pyrite; Ccp = chalcopyrite; Gn = galena).

Moreover, the PbO and SO_3 concentrations in the anglesite samples from the Iheya North hydrothermal field are more variable than those from the Yonaguni Knoll IV fields (Table S1; Figure 6a), suggesting that two types of anglesite occur in the OT back-arc basin: a high-Pb/low-S secondary supergene anglesite type (PbO > 72.50 wt.%, SO_3 < 27.00 wt.%) in the Iheya North hydrothermal field and a low-Pb/high-S type primary hydrothermal anglesite (PbO < 72.00 wt.%, SO_3 > 29.00 wt.%) in the Yonaguni Knoll IV hydrothermal fields (Figure 6a).

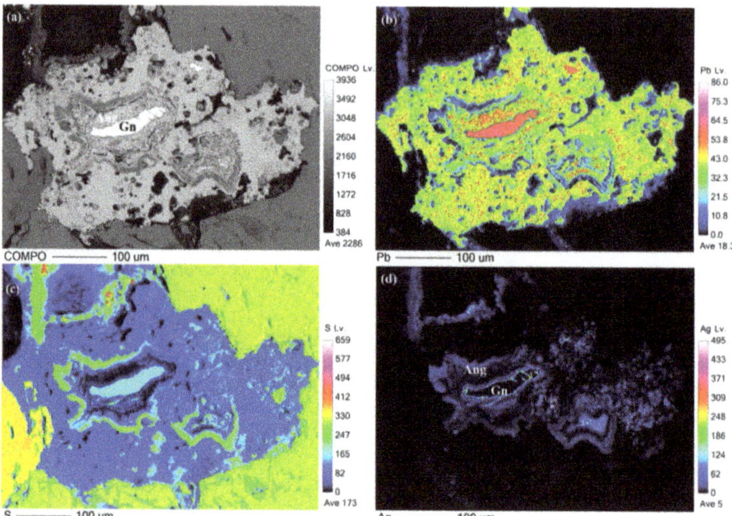

Figure 5. (**a**) BSE image of Ag–Bi-enriched anglesite and galena. X-ray elemental distribution maps after EPMA of BSE image, showing compositional heterogeneity of anglesite crystals in (**b**) Pb; (**c**) S; (**d**) galena with Ag-enriched anglesite rim and veins.

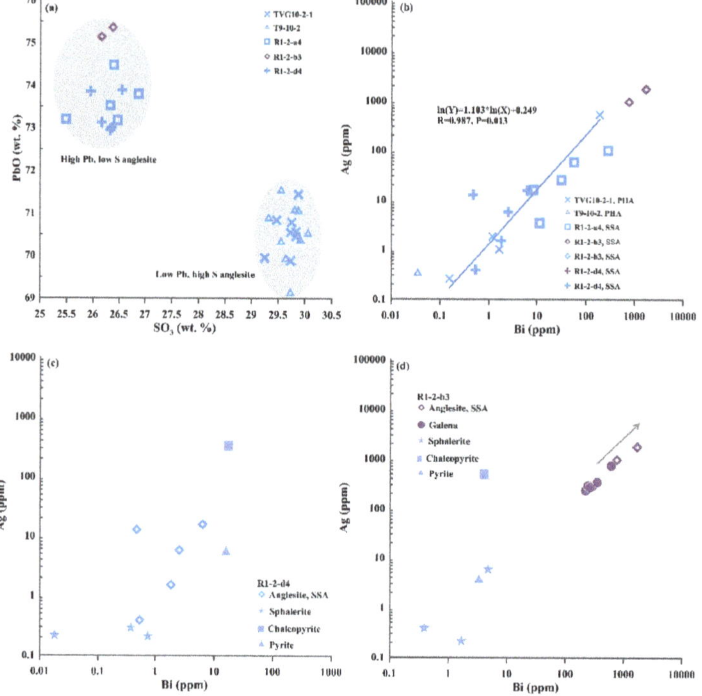

Figure 6. Plots of anglesite minerals in samples H2-R1-2-a4, b3, d4 (MOT hydrothermal field) and samples H3-T9-10-2 and H4-TVG10-2-1 (SOT hydrothermal field): (**a**) PbO vs. SO$_3$; (**b**) Ag vs. Bi; (**c**) Ag vs. Bi in sample H2-R1-2-d4; (**d**) Ag vs. Bi in sample H2-R1-2-b3. PHA—primary hydrothermal anglesite; SSA—secondary supergene anglesite.

Generally higher U, Cu, Th, Ni, Co and V with HREE concentrations are found in the high Pb and low S anglesites from the Iheya North than in the low Pb and high S anglesites from the Yonaguni Knoll IV hydrothermal fields (Tables S1 and S2). In addition, the Bi concentrations in the high Pb and low S anglesite samples (H2-R1-2-b3) from the Iheya North hydrothermal field are markedly more variable (0.04–3112 ppm) than for the low Pb and high S anglesite samples from the SOT hydrothermal fields (Table S2; Figure 6b). Most of the high Pb and low S anglesite from the MOT contain notably higher concentrations of both Ag and Bi than those in the high Pb and low S anglesite from the SOT (Table S2; Figure 6b). Notably, the Ag content showed a remarkable positive correlation with Bi in the anglesite in seafloor hydrothermal sulfide deposits (Figure 6b). The Mo concentrations in high Pb and low S anglesite from the Iheya North Knoll hydrothermal field are generally higher than those in the low Pb and high S anglesite from the Yonaguni Knoll IV field (Table S2; Figure 7b).

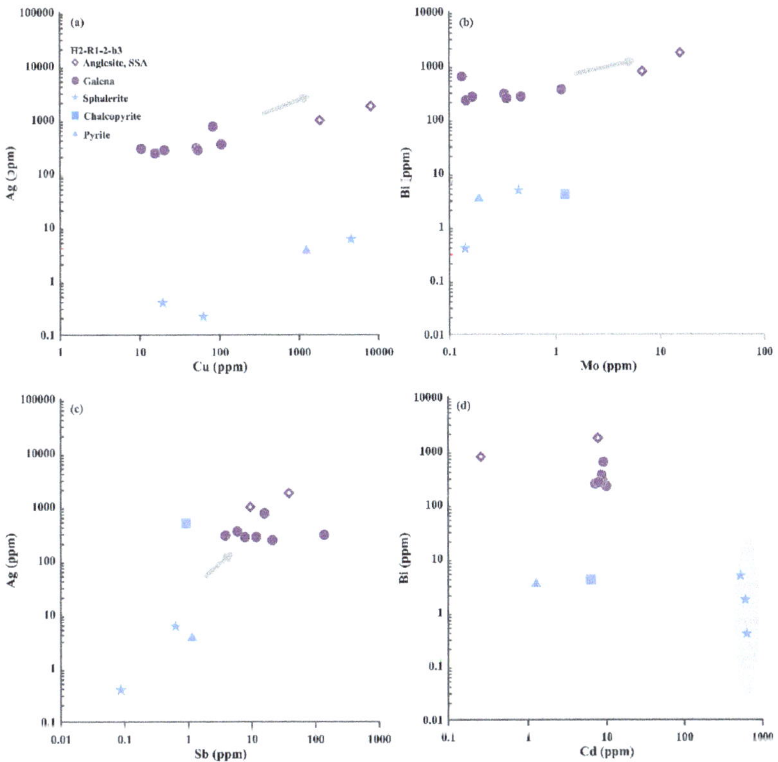

Figure 7. Plots of anglesite and sulfide minerals in sample H2-R1-2-b3: (**a**) Ag vs. Cu; (**b**) Bi vs. Mo; (**c**) Ag vs. Sb; (**d**) Bi vs. Cd. SSA—secondary supergene anglesite.

The Bi and Cu concentrations in galena associated anglesite in the Iheya North and Yonaguni Knoll IV hydrothermal fields are variable, covering the widest range (0.02–616 to 2.83–2895 ppm) (Table S2). The Sb and Ag concentrations in galena (sample H3 19 10-2) associated anglesite from the Yonaguni Knoll IV hydrothermal field is considerably more variable (53.3–5843 and 11.4–3515 ppm) than those in the galena associated anglesite from the Iheya North hydrothermal fields (Table S2). However, most of the anglesite samples have higher Ag and Bi concentrations than the associated sulfide minerals in the OT (Table S2; Figures 6 and 7), and most of the Ag and Bi concentrations in anglesite and associated galena are considerably higher than in pyrite, chalcopyrite and sphalerite (Figures 6 and 7). The Bi/Sb ratios of galena in samples from Yonaguni Knoll IV hydrothermal field are low than 1. In contrast, the Bi/Sb ratios of galena in samples from Iheya North hydrothermal field are much greater than 1 (Table S2).

4.2. Rare Earth Element Compositions in Anglesite

The total REE concentrations ($\sum$REEs) in the anglesite from the OT hydrothermal fields are highly variable (0.01–30.3 ppm) (Table S2). Most of the $\sum$REEs in the high Pb and low S anglesite from the MOT are higher than those in the low Pb and high S anglesite from the SOT (Figure 8a). Of the high Pb and low S anglesite in the present study, those from the MOT hydrothermal fields exhibited the highest $\sum$REEs (30.3 ppm, sample H2-R1-2-a4, point 12-6; Table S2). However, most of the $\sum$REEs in the OT anglesite were higher than those in the OT associated galena (0.01–0.33 ppm) and other sulfide minerals (Table S2; Figure 8a).

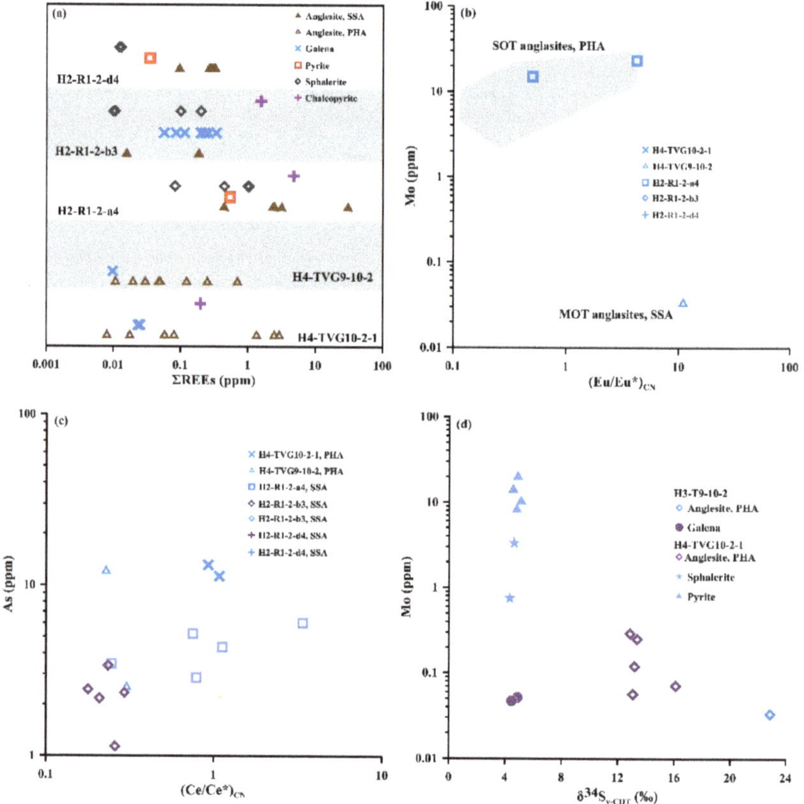

Figure 8. Plots of anglesite and sulfide minerals in OT hydrothermal sulfide deposits: (**a**) $\sum$REE distribution range; (**b**) Mo vs. (Eu/Eu*)$_{CN}$; (**c**) As vs. (Ce/Ce*)$_{CN}$; (**d**) Mo vs. $\delta^{34}S_{v\text{-}CDT}$. PHA—primary hydrothermal anglesite; SSA—secondary supergene anglesite.

The C1-chondrite-normalized REE distribution patterns of the anglesite from the OT hydrothermal fields are shown in Figure 9. The REE patterns of most of the anglesite show evidence of LREE enrichment (LREE/HREE ratios of 0.39–233), variable La$_{CN}$/Lu$_{CN}$ ratios between 0.32 and 10.7, Eu anomalies ((Eu/Eu*)$_{CN}$ ratios of 0.49–11.0) and minor Ce anomalies ((Ce/Ce*)$_{CN}$ ratios of 0.05–3.37) (Figure 8b,c). Most of the (Eu/Eu*)$_{CN}$ (0.49–4.21) and (Ce/Ce*)$_{CN}$ (0.05–3.37) ratios of the high Pb and low S anglesite from the MOT are lower and higher than those (3.45–11.0, 0.06–1.07) in the low Pb and high S anglesite from the SOT, respectively. In comparison, the Eu and Ce anomalies in the galena and other sulfides associated anglesite have a (Eu/Eu*)$_{CN}$ ratio of 0.83 and (Ce/Ce*)$_{CN}$ ratios of 0.06–1.38.

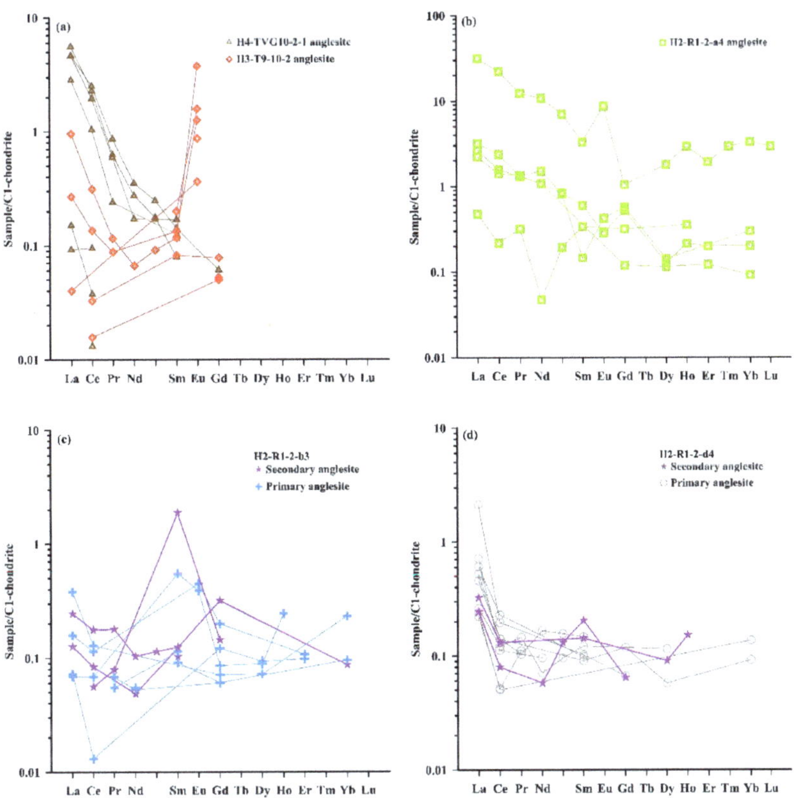

Figure 9. REE patterns of anglesite in samples (**a**) H4-TVG10-2-1 and H3-T9-10-2; (**b**) H2-R1-2-a4; (**c**) H2-R1-2-b3; and (**d**) H2-R1-2-d4 (Normalized data from Sun and McDonough, [53]).

4.3. In Situ S and Pb Isotopic Compositions of the Anglesite and Sulfides

Anglesite forming in the MOT (Iheya North, 11.04–17.42‰) and SOT (Yonaguni Knoll IV, 12.84–22.86‰) hydrothermal fields exhibits a range of δ^{34}S values of 11.04 to 22.86‰ (Table S3; Figure 10). Moreover, in the Okinawa Trough, the galena, pyrite, sphalerite and chalcopyrite have much lower sulfur isotope ratios than those in the associated anglesite (δ^{34}S values from −5.01 to 11.33‰, avg. 7.48‰, n = 60) (Table S3). Most of these are within the previously reported range 0.0–9.6‰ from previous analyses of seafloor hydrothermal sulfides [59,60]. In addition, the pyrite–sphalerite, sphalerite–chalcopyrite and sphalerite–galena pairs in samples H2-R1-2-a4, -b3 and -d4 display a sulfur isotopic equilibrium, according to known temperature-dependent fractionation factors, characteristic of most seafloor hydrothermal systems [61]. The formation temperatures of coexisting sulfide mineral pairs from the Iheya North Knoll hydrothermal field, as calculated by the sulfur isotope geothermometer formula [61,62], range from 248 to 333 °C (Table S4)

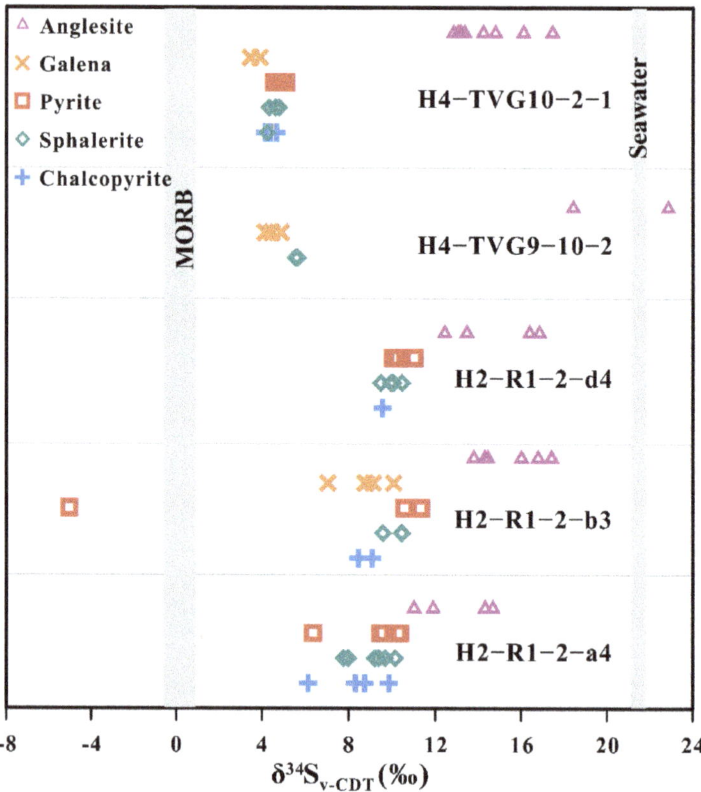

Figure 10. Sulfur isotope values for anglesite and sulfide minerals from the MOT and SOT hydrothermal sulfide deposits. MORB data from Sakai et al. [63], Alt et al. [64], Shanks et al. [65] and Alt and Shanks [66]. Seawater data from Rees et al. [67].

The anglesite in the OT hydrothermal fields have similar Pb isotopic compositions, and the Pb isotopic compositions lie within a narrow range (Table S3; Figure 11). In the Okinawa Trough, the Pb isotopic compositions of the low Pb and high S anglesite and associated galena samples from the Yonaguni Knoll IV hydrothermal fields are more radiogenic than those of the high Pb and low S anglesite and associated galena from the Iheya North hydrothermal field (Table S3; Figure 11). In addition, the ^{207}Pb/^{204}Pb and ^{208}Pb/^{204}Pb ratios of OT anglesite and associated galena are within or close to the range for the corresponding OT volcanic rocks and sediments (Figure 11).

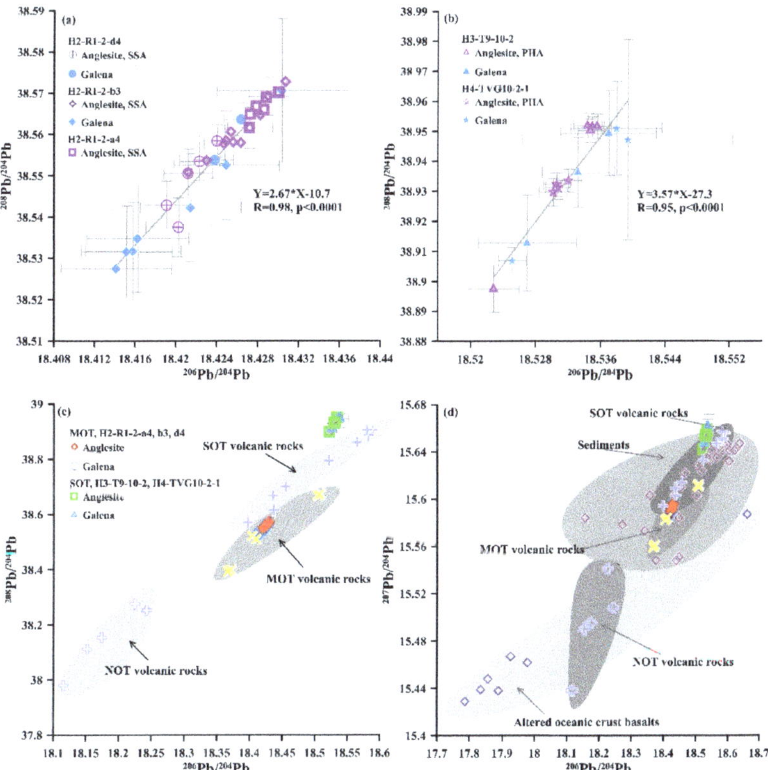

Figure 11. Plots of anglesite and galena in MOT (samples H2-R1-2-d4, b3, a4) and SOT (samples H3-T9-10-2, H4-TVG10-2-1): (**a–c**) $^{208}Pb/^{204}Pb$ vs. $^{206}Pb/^{204}Pb$; (**d**) $^{207}Pb/^{204}Pb$ vs. $^{206}Pb/^{204}Pb$. PHA—primary hydrothermal anglesite; SSA—secondary supergene anglesite. Data for NOT, MOT and SOT volcanic rock, sediment, and altered oceanic crust basalt from Shu et al. [68].

5. Discussion

5.1. Variable REE Compositions, Origin of Eu Anomalies and Sources of REEs

The $\sum$REEs in the anglesite in the present study vary considerably (0.01–30.3 ppm) (Table S2). They exhibit no systematic variation with Pb or S. The extent of fractionation of the LREEs and HREEs is highly variable in the anglesite from the OT hydrothermal fields (Table S2). The substitution of REEs into anglesite, analogous to the substitution of REEs into pyrite, chalcopyrite and sphalerite, appears to be strongly influenced by the larger ionic radii of the REEs [69,70]. Thus, greater LREE enrichment takes place in Pb-rich anglesite and associated galena than in pyrite, chalcopyrite and sphalerite minerals because of the similar ionic radii of Pb^{2+} (119 pm) and La^{3+} (103.2 pm) compared to the much smaller ionic radii of Fe^{2+} (55 pm), Cu^{2+} (73 pm) and Zn^{2+} (74 pm) [71,72].

Most of the REE patterns in low Pb and high S anglesite from the SOT hydrothermal fields (samples H3-T9-10-2, points 7-73, 2-1-18, 3-1-34, 8-1-82) exhibit positive Eu anomalies of 3.34 to 9.98 (Table S2; Figure 8b), similar to the values in the vent fluids at the Yonaguni Knoll IV, the East Pacific Rise near 13° N and the Logatchev hydrothermal fields [73–77]. These similarities suggest that this low Pb and high S anglesite is primary hydrothermal anglesite, which precipitated and inherited positive Eu anomalies from the vent fluids [70,78,79].

Most of the REE patterns in the high Pb and low S anglesite from the MOT hydrothermal fields are characterized by a negative or negligible Eu anomaly of 0.49 to 1.01 (Table S2; Figure 9), which is considered to signal low-temperature seawater [80]. This implies that the high Pb and low S anglesite is formed under low-temperature conditions. Furthermore, the accumulation of Eu^{2+} in anglesite that formed at low-temperatures is also consistent with the fact that negative or negligible Eu anomalies are related to lower Eu content in the anglesite (<0.51 ppm) (Table S2). Consequently, the negligible or negative Eu anomalies in the anglesite have been interpreted as the result from low temperatures and a less Eu-enriched environment [73,79]. Seafloor hydrothermal recrystallization (i.e., zone refining) and secondary enrichment of Bi and Ag may also have caused remobilization of REEs, particularly Eu^{2+}, producing negligible or negative Eu anomalies in the anglesite (Figure 8b; [70]). This suggests that high Pb and low S anglesite in the MOT can be formed by a secondary supergene process.

The REEs in the anglesite may reflect the sources and evolution of hydrothermal fluids [70]. Comparisons show that the REE patterns of the low Pb and high S anglesite from the SOT Yonaguni Knoll IV hydrothermal fields are similar to those of Yonaguni Knoll IV vent fluids, with the anglesite also exhibiting LREE enrichment and positive Eu anomalies (Figure 9; [76]). Thus, the REEs in the low Pb and high S anglesite are likely to have all been derived from hydrothermal fluids that leached REEs from local sub-seafloor volcanic rocks and/or sediments [76] and incorporated them into the low Pb and high S anglesite [81,82]. Furthermore, the anglesite from the MOT and SOT exhibit markedly low Nd/Pb and Ce/Pb ratios (Figure 12), and Ce/PbO ratios show notable negative and positive correlations with BaO/La and Nd/PbO ratios (Figure 12). This might indicate the effect of fluid–sediment interaction [83–85].

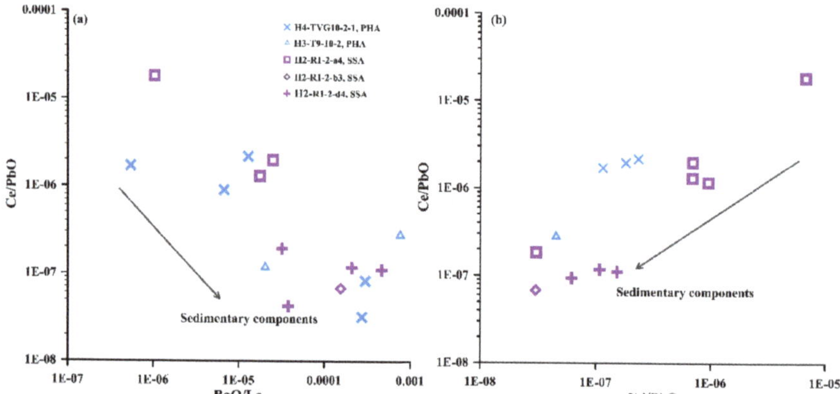

Figure 12. Plots of (**a**) Ce/PbO vs. BaO/La; and (**b**) Ce/PbO vs. Nd/PbO in anglesite minerals from the MOT (samples H2-R1-2-a4, b3, d4) and SOT (samples H3-T9-10-2, H4-TVG10-2-1) hydrothermal fields. PHA—primary hydrothermal anglesite; SSA—secondary supergene anglesite.

5.2. Sulfur Sources of Anglesite

It is known that the hydrothermal activity on the MOT and SOT hydrothermal fields is hosted by pumiceous volcaniclastic flow, sediment deposition [31], basaltic lava [42,43] and volcanic breccia [48]. The $\delta^{34}S$ values in anglesite from the OT hydrothermal fields are spread over a broad range (11.04–22.86‰; Table S3; Figure 10), which falls between the vent fluid $\delta^{34}S$ values for SO_4 (20.6–25.7‰), sediments ($\delta^{34}S$ values similar to seawater sulfate, +21‰; [67] and the vent fluid $\delta^{34}S$ values for H_2S (−0.2 to 12‰) [36,37,60,65,67,86–90], island-arc andesites and rhyolites (+4–+5‰; [91–94]), basalts (+0.1 ± 0.5‰; [63–66]) and Ryukyu Arc volcanic rocks (7.0 ± 1.9–12.9 ± 2.1‰; [91]). This suggests that the sulfur isotopic compositions in anglesite from the OT hydrothermal fields exhibit contributions of sulfur from seawater and/or sediment SO_4 and volcanic sulfur are likely to have been influenced by the mixing of seawater and/or sediment SO_4 and volcanic sulfur during anglesite formation.

However, the $\delta^{34}S$ values in the anglesite from the MOT hydrothermal fields are higher than in the associated sulfide minerals (Table S3; Figure 10), as the higher the oxidization state the more fractionation between different S species. Hence, the $\delta^{34}S$ values of anglesite are always heavier than those in the associated sulfide minerals (unless microbes are involved) and can also be thought to have been influenced considerably by the lowered seawater sulfate under more oxidizing conditions during anglesite formation [95,96].

5.3. Fluid–Rock and/or Sediment Interaction

Hydrothermal activity in the OT is hosted by volcanic rocks and sediments [31,42, 43,48,97–101]. Most of the Pb isotope ratios in anglesite in the OT hydrothermal fields lie within, or close to, the range of values for rhyolitic pumices, basaltic andesites, rhyolites and sediment from the MOT (Iheya North) and SOT (Yonaguni Knoll IV) hydrothermal fields (Figure 11c,d). This suggests that OT volcanic rocks and/or sediments are the principal source of Pb in the OT hydrothermal fluids (e.g., [102–108], and the Pb isotopic compositions in the hydrothermal fluid forming anglesite are similar to those of the fluid forming associated galena in the OT hydrothermal fields. However, the Pb isotopic composition in high Pb and low S anglesite is similar to that found in associated galena in the MOT hydrothermal field, indicating that the Pb in high Pb and low S anglesite may be from the associated galena. Furthermore, the Pb isotope data for anglesite and associated sulfide cover a smaller domain than in back-arc basin volcanic rocks and sediments overall. The mean Pb isotopic compositions of anglesite from the OT hydrothermal fields (Table S3) are plotted into the ranges of Pb isotopic compositions of volcanic rocks in the different hydrothermal fields. In turn, this suggests that the Pb isotopic composition in the anglesite signifies the Pb isotopic composition of the crust and hydrothermal fluid locally. The Pb isotopic ratios of anglesite from the MOT (Iheya North) hydrothermal field appear to be comparable with the average values for volcanic rocks in the MOT (Figure 11c,d), implying that seawater is not a possible source of the Pb in the MOT anglesite.

However, comparing the more radiogenic Pb isotope composition in anglesite and associated galena from the SOT hydrothermal fields with those from the MOT hydrothermal field (Table S3) suggests that the Pb in the SOT anglesite and associated galena is possibly related to volcanism and/or sediment containing a more radiogenic Pb isotope composition. Additionally, as the Bi/Sb ratios of galena are sensitive to the host rock with Bi/Sb ratio > 1 in basic and < 1 in felsic host rocks [15], the very low Bi/Sb ratios of galena in samples from the SOT (Yonaguni Knoll IV) hydrothermal field (Table S2) suggest it was hosted by felsic rocks. This is consistent with the geological background of the hydrothermal field, which was hosted by rhyolite [101] characterized by a more radiogenic Pb isotope composition [99]. This implies that the relatively high Pb isotope ratios of low Pb and high S anglesite in the SOT are due to less seawater–volcanic rock and/or sediment interaction, which resulted in variation in hydrothermal fluid Pb isotope ratios.

Furthermore, the Pb isotopic ratios in the OT anglesite and associated galena appear to lie within or close to the range in sediments or altered oceanic crust basalts in the Philippine Sea plate ([68]; Figure 11d). This implies that subducted sedimentary and oceanic crust components are also a possible source of the Pb in the OT anglesite and associated galena by means of fluid–sediment and/or oceanic crust interaction.

5.4. The Secondary Enrichment of Bi and Ag in Secondary Supergene Anglesite

The Ag and Bi concentrations of anglesite have large ranges (Table S2), covering the range 1709 to >2000 ppm of Ag enrichment in anglesite from the oxide zone of the Prairie Creek Deposit, NW Territories, Canada [109]. The Bi and Ag concentrations of associated galena are variable (Bi: 0.02–340 avg. 57.4 ppm, n = 11; Ag: 11.4–3515 avg. 1308 ppm, n = 13) and are considerably higher than those of the vent fluid (Bi 0.000647 ppm) in the OT [87] and seawater (Bi 0.00002 ppm, Ag 0.00028 ppm) [110], which is the initial enrichment of Bi and Ag in the associated galena (Table S2, Figures 5, 6d and 7). However, the anglesite is observed to partially replace and form rims around galena crystals

(Figures 4d and 5) resulting from the alteration of galena, the reaction being PbS (s) + 2O$_2$ (aq) $\rightarrow$ PbSO$_4$ (s), and the Bi and Ag concentrations in the secondary supergene anglesite (SSA) are generally higher than in neighboring galena (Table S2; Figures 3, 6d, 7a–b and 13), exhibiting secondary enrichment of Bi and Ag (Table S2; Figures 5, 6d, 7a–b and 13), resulting in most of the Ag and Bi concentrations in the high Pb and low S secondary supergene anglesite in the MOT being higher than in the low Pb and high S primary hydrothermal anglesite (Bi <200 ppm, Ag <600 ppm), which was formed by the reaction Pb^{2+} (aq) + SO$_4{}^{2-}$ (aq) $\rightarrow$ PbSO$_4$ (s) (Figures 6 and 7). The implication is that back-arc hydrothermal sulfide deposits may be a potential source of enriched Ag and Bi due to supergene processes (i.e., oxidization) of seafloor hydrothermal sulfides. However, in the prevailing acidic environment (pH $\approx$ 5.8), the fluid reacts with the galena with increasing lead activity and decreasing pH until the anglesite stability field is reached [7]. This explains the commonly observed in situ transformation of galena to anglesite in textural coexistence (Figures 4 and 5). The fluid precipitates anglesite that is influenced by the low P_{CO2} of seafloor hydrothermal vent fluid and the higher fluid/mineral ratios [7].

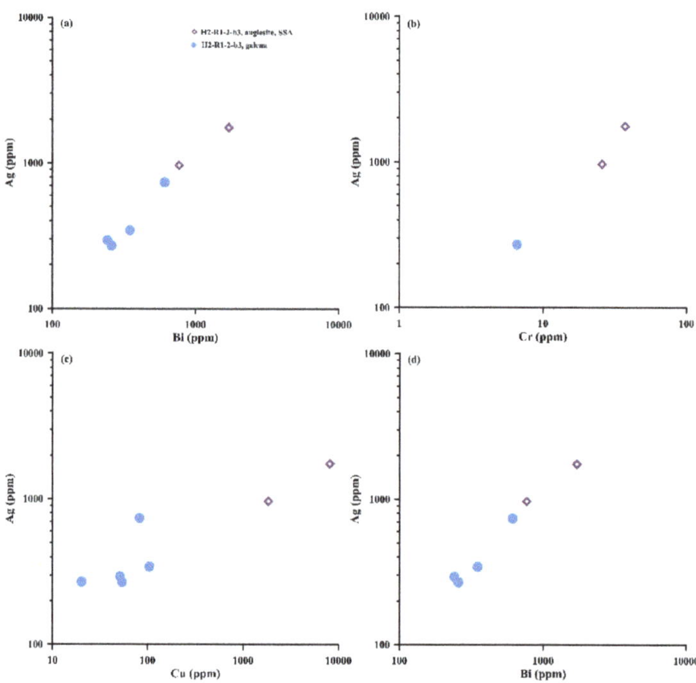

Figure 13. Plots of anglesite and neighboring galena from the MOT hydrothermal field (sample H2-R1-2-b3): (**a**) Ag vs. Sr; (**b**) Ag vs. Cr; (**c**) Ag vs. Cu; (**d**) Ag vs. Bi. SSA—secondary supergene anglesite.

Furthermore, the Pb isotopic compositions in the secondary supergene anglesite (SSA) are similar to those of associated galena in the MOT hydrothermal field (Figure 11a), suggesting that the SSA inherited the Pb isotopic compositions of galena during anglesite formation by galena alteration. However, a notable positive correlation is observed between Bi and Ag and between Mo and Cu concentrations (up to 8144 ppm) in the anglesite from the MOT hydrothermal fields (Figure 14). Cu-enriched sulfates (e.g., atacamite) are characteristic of oxidation of Cu-sulfide minerals by seawater [111], implying that Bi and Ag become enriched in anglesite in low-temperature (<100 °C) conditions. Most of the Ag, Cu, Ni, V and HREEs with Pb concentrations of SSA from the MOT (Iheya North) were generally higher than those of primary hydrothermal anglesite (PHA) from the SOT (Yonaguni Knoll IV) hydrothermal fields (Table S2), also indicating that the formation temperature of the SSA in the MOT is lower than those of the PHA in the SOT. Furthermore, the Bi and Ag in anglesite from the OT hydrothermal fields are positively correlated with Cu and Mo (Figure 14), which might indicate that Bi and Ag are incorporated into the PHA at higher temperatures, possibly 300 °C or more. The Ag concentrations also show positive correlations with Sr, Cr, Cu and Bi concentrations in SSA and neighboring galena from the MOT hydrothermal field (Figure 13). Most of Bi (<100 ppm), Ag (<600 ppm), Se (<60 ppm) and Sb (<180 ppm) concentrations in the PHA are significantly smaller than the concentrations in SSA, which perhaps suggests that Sr, Cr, Cu, Bi, Tl, Se, Sb and Ag are incorporated into the SSA from the neighboring galena during its alteration.

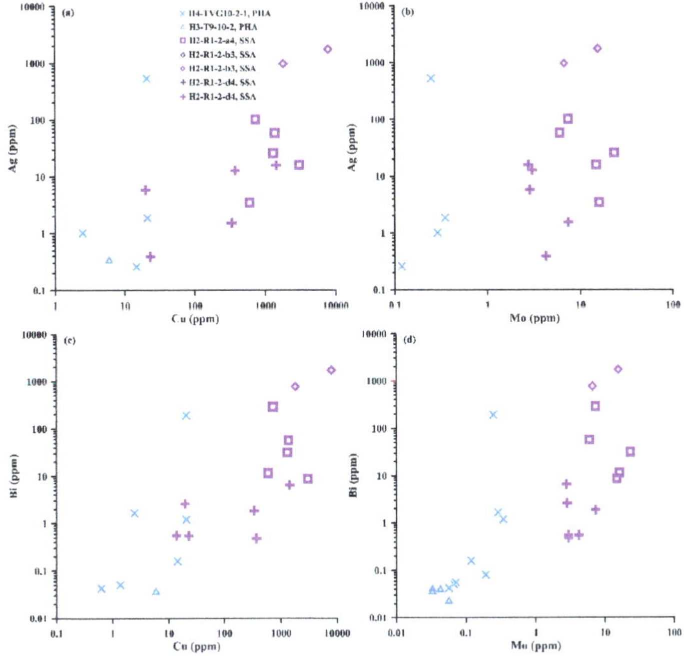

Figure 14. Plots of anglesite from MOT (samples H2-R1-2-a4, b3, d4) and SOT (samples H3-T9-10-2, H4-TVG10-2-1) hydrothermal fields: (**a**) Ag vs. Cu; (**b**) Ag vs. Mo; (**c**) Bi vs. Cu; (**d**) Bi vs. Mo. PHA—primary hydrothermal anglesite; SSA—secondary supergene anglesite.

6. Conclusions

There are two types of anglesite minerals in the OT seafloor hydrothermal fields, high Pb/low S secondary supergene anglesite and low Pb/high S primary hydrothermal anglesite. The primary hydrothermal anglesite occurs in the pores precipitated from hydrothermal fluid and intergrowth with pyrite and sphalerite in the SOT hydrothermal field. The secondary supergene anglesite is formed by the low-temperature alteration of galena, surrounding the relict galena in the MOT hydrothermal field. The galena alteration results in most of the Sr, Cr, Cu, Bi, Se, Sb and Ag being incorporated into the secondary supergene anglesite from the neighboring galena, and most of the Sr, Cr, Se, Sb, Pb with Bi and Ag concentrations in the primary hydrothermal anglesite being considerably lower than in secondary supergene anglesite. However, the secondary supergene anglesite exhibits notably higher Ag and Bi concentrations than in primary hydrothermal anglesite, galena, pyrite, sphalerite and chalcopyrite, resulting in a redistribution of Ag and Bi in seafloor sulfide deposits. Furthermore, the Ag and Bi concentrations in the galena (initially from Ag and Bi enrichment of hydrothermal fluid during galena deposition) are higher than in the primary hydrothermal anglesite. This suggests that Bi and Ag enrichment in the seafloor hydrothermal sulfide deposits is the result of Bi and Ag twice enrichment during ore-forming processes, which also offered new evidence for the element multi-enrichment hypothesis of global hydrothermal sulfide deposits. Thus, it is possible that a valuable Bi- and Ag-enriched anglesite and associated galena will be discovered in back-arc seafloor hydrothermal sulfide deposits, and it is necessary to explore Bi and Ag deposits in those locations.

However, anglesite from the OT seafloor hydrothermal fields contains a wide and varying range of $\sum$REEs. Most of the secondary supergene anglesite contains lower ΣREE concentrations than the primary hydrothermal anglesite. They are notably more enriched in LREEs than in HREEs. The REEs in the low Pb and high S anglesite, which exhibit positive Eu anomalies, similar to the values in the vent fluids, are likely to have all been derived from hydrothermal fluids that leached REEs from local volcanic rocks and/or sediments. The negative or negligible Eu anomalies of the REE patterns in the high Pb and low S anglesite from the MOT hydrothermal fields are considered to signal low-temperature seawater. However, secondary enrichment of Bi and Ag during secondary supergene anglesite formation may also have caused remobilization of Eu^{2+}, resulting in negligible or negative Eu anomalies in the secondary supergene anglesite.

The sulfur isotopic compositions in anglesite in the OT hydrothermal fields vary between 11.04 and 22.86‰ for $\delta^{34}S$. These are higher than the corresponding values in associated sulfide minerals in the same hydrothermal field and are similar to those for seawater and/or sediment sulfates. This may be an indication that the sulfur in the OT anglesite samples was derived mainly from a mixing of seawater and/or sediment sulfate and volcanic sulfur. In addition, the pyrite–sphalerite, sphalerite–galena and sphalerite–chalcopyrite pairs of sulfide minerals in the OT hydrothermal fields lie within the equilibrium temperature range, suggesting that the formation temperature of pyrite, sphalerite, galena and chalcopyrite is between 248 and 333 °C.

In contrast to the mixed-origin source of S in the anglesite, most of the Pb isotopic compositions in the anglesite are identical to OT volcanic rock values, falling within or close to the range of Pb isotopic compositions found in sediments and altered oceanic crust. This suggests that the Pb in the anglesite has mainly been leached from local volcanic rocks and/or sediments that host the sub-seafloor hydrothermal systems in the OT back-arc basin. However, the Pb isotope composition of the SSA was found to be consistent with that of the associated galena, suggesting that it inherited the Pb isotopic composition during the galena alteration process. Furthermore, the Pb isotopic compositions in the anglesite from the OT hydrothermal fields are very homogeneous and cover a narrow range within, or close to, the scope of the large dataset of Pb isotopic compositions in the OT volcanic rocks, subducted sediments and altered oceanic crustal basalts. This emphasizes the importance of determining Pb isotopic compositions in anglesite from the seafloor hydrothermal sulfide deposits and implies a new means of understanding the influence of plate subduction in back-arc hydrothermal and magmatic systems. Additionally, the crystallographic difference, if any, between primary and secondary anglesite is an interesting subject to be studied in the future by X-ray diffraction due to the non-stoichiometric balance between sulfur and lead.

Supplementary Materials: The following are available online at https://www.mdpi.com/article/10.3390/jmse10010035/s1, Table S1: Electron microprobe analyses of anglesite and sulfide minerals in seafloor hydrothermal sulfide samples from the Yonaguni Knoll IV and Iheya North hydrothermal fields (wt.%) and their atoms per formula unit for elements, Table S2: LA-ICP-MS trace element concentrations in anglesite and sulfide minerals in seafloor hydrothermal sulfide samples from the Yonaguni Knoll IV and Iheya North hydrothermal fields (ppm), Table S3: LA-MC-ICP-MS sulfur and lead isotopic compositions of anglesite and sulfide minerals in seafloor hydrothermal sulfide samples from the Yonaguni Knoll IV and Iheya North hydrothermal fields, Table S4: Sulfur isotopic compositions of coexisting sulfide mineral phases in seafloor hydrothermal sulfide samples from the Iheya North hydrothermal fields, and corresponding temperature (°C) calculated by geothermometers based on equilibrium sulfur isotope fractionation factors (A) of sulfides with respect to H_2S (after Ohmoto and Rye, [62]).

Author Contributions: Z.Z.—Conceptualization, Data curation, formal analysis, funding acquisition, writing and editing. Z.C.—Data curation, formal analysis, validation. H.Q.—Data curation, methodology, validation. All authors have read and agreed to the published version of the manuscript.

Funding: This research was funded by the NSFC Major Research Plan on West-Pacific Earth System Multispheric Interactions, grant number 91958213; the Strategic Priority Research Program of the Chinese Academy of Sciences, grant number XDB42020402; the Special Fund for the Taishan Scholar Program of Shandong Province, grant number ts201511061 and the National Key Basic Research Program of China, grant number 2013CB429700.

Institutional Review Board Statement: Not applicable.

Informed Consent Statement: Not applicable.

Data Availability Statement: Data can be obtained from the corresponding author and are available online at Supplementary Materials.

Acknowledgments: We would like to thank the crews of the R/V KEXUE during the HOBAB 2, 3 and 4 cruises for their help with sample collection. We are most grateful for the detailed and constructive comments and suggestions provided by three anonymous reviewers, which greatly improved an earlier version of the manuscript. We thank the editors of MDPI for their efficient editor works.

Conflicts of Interest: The authors declare no conflict of interest.

References

1. Barton, P.B.; Bethke, P.M.; Roedder, E. Environment of ore deposition in the Creede mining district, San Juan Mountains, Colorado; Part III, Progress toward interpretation of the chemistry of the ore-forming fluid for the OH Vein. *Econ. Geol.* **1977**, *72*, 1–24. [CrossRef]
2. Fanfani, L.; Zuddas, P.; Chessa, A. Heavy metals speciation analysis as a tool for studying mine tailings weathering. *J. Geochem. Explor.* **1997**, *58*, 241–248. [CrossRef]
3. Szczerba, M.; Sawłowicz, Z. Remarks on the origin of cerussite in the Upper Silesian Zn-Pb deposits, Poland. *Mineralogia* **2009**, *40*, 54–64. [CrossRef]
4. Lara, R.H.; Briones, R.; Monroy, M.G.; Mullet, M.; Humbert, B.; Dossot, M.; Naja, G.M.; Cruz, R. Galena weathering under simulated calcareous soil conditions. *Sci. Total Environ.* **2011**, *409*, 3971–3979. [CrossRef] [PubMed]
5. Nordstrom, D.K. Hydrogeochemical processes governing the origin, transport and fate of major and trace elements from mine wastes and mineralized rock to surface waters. *Appl. Geochem.* **2011**, *26*, 1777–1791. [CrossRef]
6. Ayuso, R.A.; Foley, N.K.; Seal II, R.R.; Bove, M.; Civitillo, D.; Cosenza, A.; Grezzi, G. Lead isotope evidence for metal dispersal at the Callahan Cu–Zn–Pb mine: Goose Pond tidal estuary, Maine, USA. *J. Geochem. Explor.* **2013**, *126*, 1–22. [CrossRef]
7. Keim, M.F.; Markl, G. Weathering of galena: Mineralogical processes, hydrogeochemical fluid path modeling, and estimation of the growth rate of pyromorphite. *Am. Miner.* **2015**, *100*, 1584–1594. [CrossRef]
8. Courtin-Nomade, A.; Waltzing, T.; Evrard, C.; Soubrand, M.; Lenain, J.-F.; Ducloux, E.; Ghorbel, S.; Grosbois, C.; Bril, H. Arsenic and lead mobility: From tailing materials to the aqueous compartment. *Appl. Geochem.* **2016**, *64*, 10–21. [CrossRef]
9. Paradis, S.; Simandl, G.J.; Keevil, H.; Raudsepp, M. Carbonate-Hosted Nonsulfide Pb-Zn Deposits of the Quesnel Lake District, British Columbia, Canada. *Econ. Geol.* **2016**, *111*, 179–198. [CrossRef]
10. Jazi, M.A.; Karimpour, M.H.; Shafaroudi, A.M. Nakhlak carbonate-hosted Pb(Ag) deposit, Isfahan province, Iran: A geological, mineralogical, geochemical, fluid inclusion, and sulfur isotope study. *Ore Geol. Rev.* **2017**, *80*, 27–47. [CrossRef]
11. Keim, M.F.; Gassmann, B.; Markl, G. Formation of basic lead phases during fire setting and other natural and man-made processes. *Am. Mineral. J. Earth Planet. Mater.* **2017**, *102*, 1482–1500. [CrossRef]
12. McClenaghan, M.; Parkhill, M.; Pronk, A.; Sinclair, W. Indicator mineral and till geochemical signatures of the Mount Pleasant W-Mo-Bi and Sn-Zn-In deposits, New Brunswick, Canada. *J. Geochem. Explor.* **2017**, *172*, 151–166. [CrossRef]
13. McClenaghan, M.B.; Paulen, R.C.; Oviatt, N.M. Geometry of indicator mineral and till geochemistry dispersal fans from the Pine Point Mississippi Valley-type Pb-Zn district, Northwest Territories, Canada. *J. Geochem. Explor.* **2018**, *190*, 69–86. [CrossRef]
14. Dill, H.G.; Weber, B.; Botz, R. Metalliferous duricrusts ("orecretes")-markers of weathering: A mineralogical and climatic-geomorphological approach to supergene Pb-Zn-Cu-Sb-P mineralization on different parent materials. *Neues Jahrb. Mineral. Abh.* **2013**, *190*, 123–195. [CrossRef]
15. Dill, H.G. Die Vererzung am Westrand der Bhmischen Masse-Metallogenese in einer ensialischen Orogenzone. *Geol. Jahrb.* **1985**, *73*, 3–461.
16. Romero, F.M.; Prol-Ledesma, R.M.; Canet, C.; Alvares, L.N.; Pérez-Vázquez, R.G. Acid drainage at the inactive Santa Lucia mine, western Cuba: Natural attenuation of arsenic, barium and lead, and geochemical behavior of rare earth elements. *Appl. Geochem.* **2010**, *25*, 716–727. [CrossRef]

17. Santoro, L.; Rollinson, G.K.; Boni, M.; Mondillo, N. Automated scanning electron microscopy (qemscan(r))-based mineral identification and quantification of the jabali zn-pb-ag nonsulfide deposit (yemen). *Econ. Geol.* **2015**, *110*, 1083–1099. [CrossRef]
18. Jébrak, M.; Marcoux, É.; Nasloubi, M.; Zaharaoui, M. From sandstone- to carbonate-hosted stratabound deposits: An isotope study of galena in the Upper-Moulouya District (Morocco). *Miner. Depos.* **1998**, *33*, 406–415. [CrossRef]
19. Bernardinetti, S.; Pieruccioni, D.; Mugnaioli, E.; Talarico, F.M.; Trotta, M.; Harroud, A.; Tufarolo, E. A pilot study to test the reliability of the ERT method in the identification of mixed sulphides bearing dykes: The example of Sidi Flah mine (Anti-Atlas, Morocco). *Ore Geol. Rev.* **2018**, *101*, 819–838. [CrossRef]
20. Batonneau, Y.; Brémard, C.; Laureyns, J.; Merlin, J.C. Microscopic and imaging Raman scattering study of PbS and its photo-oxidation products. *J. Raman Spectrosc.* **2000**, *31*, 1113–1119. [CrossRef]
21. De Giudici, G.; Zuddas, P. In situ investigation of galena dissolution in oxygen saturated solution: Evolution of surface features and kinetic rate. *Geochim. Cosmochim. Acta* **2001**, *65*, 1381–1389. [CrossRef]
22. Brookins, D.G. *Eh-pH Diagrams for Geochemistry*; Springer: Berlin/Heidelberg, Germany, 1988.
23. Eary, L. Geochemical and equilibrium trends in mine pit lakes. *Appl. Geochem.* **1999**, *14*, 963–987. [CrossRef]
24. Jeong, G.Y.; Lee, B.Y. Secondary mineralogy and microtextures of weathered sulfides and manganoan carbonates in mine waste-rock dumps, with implications for heavy-metal fixation. *Am. Miner.* **2003**, *88*, 1933–1942. [CrossRef]
25. Grondijs, H.F.; Schouten, C. A study of the Mount Isa ores [Queensland, Australia]. *Econ. Geol.* **1937**, *32*, 407–450. [CrossRef]
26. Verhaert, M.; Bernard, A.; Dekoninck, A.; Lafforgue, L.; Saddiqi, O.; Yans, J. Mineralogical and geochemical characterization of supergene Cu–Pb–Zn–V ores in the Oriental High Atlas, Morocco. *Miner. Depos.* **2017**, *52*, 1049–1068. [CrossRef]
27. Lee, P.-K.; Kang, M.-J.; Choi, S.-H.; Touray, J.-C. Sulfide oxidation and the natural attenuation of arsenic and trace metals in the waste rocks of the abandoned Seobo tungsten mine, Korea. *Appl. Geochem.* **2005**, *20*, 1687–1703. [CrossRef]
28. Vitkova, M.; Ettler, V.; Šebek, O.; Mihaljevič, M. Metal-contaminant leaching from lead smelter fly ash using pH-stat experiments. *Mineral. Mag.* **2008**, *72*, 521–524. [CrossRef]
29. Gammons, C.H.; Metesh, J.J.; Snyder, D.M. A Survey of the Geochemistry of Flooded Mine Shaft Water in Butte, Montana. *Mine Water Environ.* **2006**, *25*, 100–107. [CrossRef]
30. Kimura, M. Back-arc rifting in the Okinawa Trough. *Mar. Pet. Geol.* **1985**, *2*, 222–240. [CrossRef]
31. Ishibashi, J.-I.; Ikegami, F.; Tsuji, T.; Urabe, T. Hydrothermal Activity in the Okinawa Trough Back-Arc Basin: Geological Background and Hydrothermal Mineralization. In *Subseafloor Biosphere Linked to Hydrothermal Systems: TAIGA Concept*; Ishibashi, J.-I., Okino, K., Sunamura, M., Eds.; Springer: Tokyo, Japan, 2015; pp. 337–359.
32. Sibuet, J.-C.; Deffontaines, B.; Hsu, S.-K.; Thareau, N.; Le Formal, J.-P.; Liu, C.-S. Okinawa trough backarc basin: Early tectonic and magmatic evolution. *J. Geophys. Res. Space Phys.* **1998**, *103*, 30245–30267. [CrossRef]
33. Yan, Q.; Shi, X. Petrologic perspectives on tectonic evolution of a nascent basin (Okinawa Trough) behind Ryukyu Arc: A review. *Acta Oceanol. Sin.* **2014**, *33*, 1–12. [CrossRef]
34. Zeng, Z.; Wang, X.; Chen, C.-T.A.; Qi, H. Understanding the Compositional Variability of the Major Components of Hydrothermal Plumes in the Okinawa Trough. *Geofluids* **2018**, *2018*, 1536352. [CrossRef]
35. Kawagucci, S.; Ueno, Y.; Takai, K.; Toki, T.; Ito, M.; Inoue, K.; Makabe, A.; Yoshida, N.; Muramatsu, Y.; Takahata, N.; et al. Geochemical origin of hydrothermal fluid methane in sediment-associated fields and its relevance to the geographical distribution of whole hydrothermal circulation. *Chem. Geol.* **2013**, *339*, 213–225. [CrossRef]
36. Nakagawa, S.; Takai, K.; Inagaki, F.; Chiba, H.; Ishibashi, J.-I.; Kataoka, S.; Hirayama, H.; Nunoura, T.; Horikoshi, K.; Sako, Y. Variability in microbial community and venting chemistry in a sediment-hosted backarc hydrothermal system: Impacts of subseafloor phase-separation. *FEMS Microbiol. Ecol.* **2005**, *54*, 141–155. [CrossRef] [PubMed]
37. Kawagucci, S.; Chiba, H.; Ishibashi, J.-I.; Yamanaka, T.; Toki, T.; Muramatsu, Y.; Ueno, Y.; Makabe, A.; Inoue, K.; Yoshida, N.; et al. Hydrothermal fluid geochemistry at the Iheya North field in the mid-Okinawa Trough: Implication for origin of methane in subseafloor fluid circulation systems. *Geochem. J.* **2011**, *45*, 109–124. [CrossRef]
38. Kawagucci, S.; Miyazaki, J.; Nakajima, R.; Nozaki, T.; Takaya, Y.; Kato, Y.; Shibuya, T.; Konno, U.; Nakaguchi, Y.; Hatada, K. Post-drilling changes in fluid discharge pattern, mineral deposition, and fluid chemistry in the Iheya North hydrothermal field, Okinawa Trough. *Geochem. Geophys. Geosyst.* **2013**, *14*, 4774–4790. [CrossRef]
39. Sakai, H.; Gamo, T.; Kim, E.-S.; Shitashima, K.; Yanagisawa, F.; Tsutsumi, M.; Ishibashi, J.; Sano, Y.; Wakita, H.; Tanaka, T.; et al. Unique chemistry of the hydrothermal solution in the mid-Okinawa Trough Backarc Basin. *Geophys. Res. Lett.* **1990**, *17*, 2133–2136. [CrossRef]
40. Ishibashi, J.; Sano, Y.; Wakita, H.; Gamo, T.; Tsutsumi, M.; Sakai, H. Helium and carbon geochemistry of hydrothermal fluids from the Mid-Okinawa Trough Back Arc Basin, southwest of Japan. *Chem. Geol.* **1995**, *123*, 1–15. [CrossRef]
41. Ishibashi, J.-I.; Noguchi, T.; Toki, T.; Miyabe, S.; Yamagami, S.; Onishi, Y.; Yamanaka, T.; Yokoyama, Y.; Omori, E.; Takahashi, Y.; et al. Diversity of fluid geochemistry affected by processes during fluid upwelling in active hydrothermal fields in the Izena Hole, the middle Okinawa Trough back-arc basin. *Geochem. J.* **2014**, *48*, 357–369. [CrossRef]
42. Watanabe, M.; Hoshino, K.; Shiokawa, R.; Takaoka, Y.; Fukumoto, H.; Shibata, Y.; Shinjo, R.; Oomori, T. Metallic mineralization associated with pillow basalts in the Yacyama Central Graben, Southern Okinawa Trough, Japan. *JAMSTEC Rep. Res. Dev.* **2006**, *3*, 1–8. [CrossRef]

43. Fukuba, T.; Noguchi, T.; Fujii, T. The Irabu Knoll: Hydrothermal Site at the Eastern Edge of the Yaeyama Graben. In *Subseafloor biosphere linked to hydrothermal systems: TAIGA Concept*; Ishibashi, J.-I., Okino, K., Sunamura, M., Eds.; Springer: Tokyo, Japan, 2015; pp. 493–496.

44. Nakano, A. Geochemistry of hydrothermal fluids from the Hatoma Knoll in the South Okinawa Trough. *JAMSTEC J. Deep. Sea Res.* **2001**, *18*, 139–144.

45. Kawagucci, S.; Toki, T.; Ishibashi, J.; Takai, K.; Ito, M.; Oomori, T.; Gamo, T. Isotopic variation of molecular hydrogen in 20–375 °C hydrothermal fluids as detected by a new analytical method. *J. Geophys. Res. Space Phys.* **2010**, *115*, 03021. [CrossRef]

46. Kishida, K.; Sohrin, Y.; Okamura, K.; Ishibashi, J.-I. Tungsten enriched in submarine hydrothermal fluids. *Earth Planet. Sci. Lett.* **2004**, *222*, 819–827. [CrossRef]

47. Konno, U.; Tsunogai, U.; Nakagawa, F.; Nakaseama, M.; Ishibashi, J.I.; Nunoura, T.; Nakamura, K.I. Liquid CO_2 venting on the seafloor: Yonaguni Knoll IV hydrothermal system, Okinawa Trough. *Geophys. Res. Lett.* **2006**, *33*, L16607. [CrossRef]

48. Suzuki, R.; Ishibashi, J.-I.; Nakaseama, M.; Konno, U.; Tsunogai, U.; Gena, K.; Chiba, H. Diverse Range of Mineralization Induced by Phase Separation of Hydrothermal Fluid: Case Study of the Yonaguni Knoll IV Hydrothermal Field in the Okinawa Trough Back-Arc Basin. *Resour. Geol.* **2008**, *58*, 267–288. [CrossRef]

49. Zeng, Z. New hydrothermal field in the Okinawa Trough. In Proceedings of the 25th Goldschmidt Conference, Prague, Czech Republic, 16–21 August 2015; p. 3573.

50. Tsuji, T.; Takai, K.; Oiwane, H.; Nakamura, Y.; Masaki, Y.; Kumagai, H.; Kinoshita, M.; Yamamoto, F.; Okano, T.; Kuramoto, S. Hydrothermal fluid flow system around the Iheya North Knoll in the mid-Okinawa trough based on seismic reflection data. *J. Volcanol. Geotherm. Res.* **2012**, *213*, 41–50. [CrossRef]

51. Liu, Y.; Hu, Z.; Gao, S.; Günther, D.; Xu, J.; Gao, C.; Chen, H. In situ analysis of major and trace elements of anhydrous minerals by LA-ICP-MS without applying an internal standard. *Chem. Geol.* **2008**, *257*, 34–43. [CrossRef]

52. Liu, Y.; Zong, K.; Kelemen, P.; Gao, S. Geochemistry and magmatic history of eclogites and ultramafic rocks from the Chinese continental scientific drill hole: Subduction and ultrahigh-pressure metamorphism of lower crustal cumulates. *Chem. Geol.* **2008**, *247*, 133–153. [CrossRef]

53. Sun, S.-S.; McDonough, W.F. Chemical and isotopic systematics of oceanic basalts: Implications for mantle composition and processes. *Geol. Soc.* **1989**, *42*, 313–345. [CrossRef]

54. Bao, Z.; Chen, L.; Zong, C.; Yuan, H.; Chen, K.; Dai, M. Development of pressed sulfide powder tablets for in situ sulfur and lead isotope measurement using LA-MC-ICP-MS. *Int. J. Mass Spectrom.* **2017**, *421*, 255–262. [CrossRef]

55. Chen, L.; Chen, K.; Bao, Z.; Liang, P.; Sun, T.; Yuan, H. Preparation of standards for in situ sulfur isotope measurement in sulfides using femtosecond laser ablation MC-ICP-MS. *J. Anal. At. Spectrom.* **2017**, *32*, 107–116. [CrossRef]

56. Yuan, H.; Liu, X.; Chen, L.; Bao, Z.; Chen, K.; Zong, C.; Li, X.-C.; Qiu, J.W. Simultaneous measurement of sulfur and lead isotopes in sulfides using nanosecond laser ablation coupled with two multi-collector inductively coupled plasma mass spectrometers. *J. Asian Earth Sci.* **2018**, *154*, 386–396. [CrossRef]

57. Chen, K.; Yuan, H.; Bao, Z.; Zong, C.; Dai, M. Precise and accurate in situ determination of lead isotope ratios in NIST, USGS, MPI-DING and CGSG glass reference materials using femtosecond laser ablation MC-ICP-MS. *Geostand. Geoanal. Res.* **2014**, *38*, 5–21.

58. Yuan, H.; Yin, C.; Liu, X.; Chen, K.; Bao, Z.; Zong, C.; Dai, M.; Lai, S.; Wang, R.; Jiang, S. High precision in-situ Pb isotopic analysis of sulfide minerals by femtosecond laser ablation multi-collector inductively coupled plasma mass spectrometry. *Sci. China Earth Sci.* **2015**, *58*, 1713–1721. [CrossRef]

59. Hannington, M.D.; De Ronde, C.E.J.; Petersen, S.; Hedenquist, J.W.; Thompson, J.F.H.; Goldfarb, R.J.; Richards, J.P. Sea-Floor Tectonics and Submarine Hydrothermal Systems. In *One Hundredth Anniversary Volume*; Society of Economic Geologists: Littelton, Colorado, USA, 2005; pp. 111–141.

60. Zeng, Z.; Chen, S.; Ma, Y.; Yin, X.; Wang, X.; Zhang, S.; Zhang, J.; Wu, X.; Li, Y.; Dong, D.; et al. Chemical compositions of mussels and clams from the Tangyin and Yonaguni Knoll IV hydrothermal fields in the southwestern Okinawa Trough. *Ore Geol. Rev.* **2017**, *87*, 172–191. [CrossRef]

61. Ohmoto, H. Stable isotope geochemistry of ore deposits. *Rev. Mineral. Geochem.* **1986**, *16*, 491–560.

62. Ohmoto, H.; Rye, R.O. Isotopes of sulfur and carbon. In *Geochemistry of Hydrothermal Ore Deposits*, 2nd ed.; Barnes, H.L., Ed.; J Wiley and Sons: New York, NY, USA, 1979; pp. 509–567.

63. Sakai, H.; Marais, D.; Ueda, A.; Moore, J. Concentrations and isotope ratios of carbon, nitrogen and sulfur in ocean-floor basalts. *Geochim. Cosmochim. Acta* **1984**, *48*, 2433–2441. [CrossRef]

64. Alt, J.C.; Anderson, T.F.; Bonnell, L. The geochemistry of sulfur in a 1.3 km section of hydrothermally altered oceanic crust, DSDP Hole 504B. *Geochim. Cosmochim. Acta* **1989**, *53*, 1011–1023. [CrossRef]

65. Shanks, W.C.; Böhlke, J.K.; Seal, R.R. Stable isotopes in mid-ocean ridge hydrothermal systems: Interactions between fluids, minerals, and organisms. *Geophys. Monogr. Am. Geophys. Union* **1995**, *91*, 194.

66. Alt, J.C.; Shanks, W.C. Serpentinization of abyssal peridotites from the MARK area, Mid-Atlantic Ridge: Sulfur geochemistry and reaction modeling. *Geochim. Cosmochim. Acta* **2003**, *67*, 641–653. [CrossRef]

67. Rees, C.; Jenkins, W.; Monster, J. The sulphur isotopic composition of ocean water sulphate. *Geochim. Cosmochim. Acta* **1978**, *42*, 377–381. [CrossRef]

68. Shu, Y.; Nielsen, S.G.; Zeng, Z.; Shinjo, R.; Blusztajn, J.; Wang, X.; Chen, S. Tracing subducted sediment inputs to the Ryukyu arc-Okinawa Trough system: Evidence from thallium isotopes. *Geochim. Cosmochim. Acta* **2017**, *217*, 462–491. [CrossRef]

69. Alt, J.C. The chemistry and sulfur isotope composition of massive sulfide and associated deposits on Green Seamount, eastern Pacific. *Econ. Geol.* **1988**, *83*, 1026–1033. [CrossRef]

70. Mills, R.; Elderfield, H. Rare earth element geochemistry of hydrothermal deposits from the active TAG Mound, 26 N Mid-Atlantic Ridge. *Geochim. Cosmochim. Acta* **1995**, *59*, 3511–3524. [CrossRef]

71. Shannon, R.D. Revised effective ionic radii and systematic studies of interatomic distances in halides and chalcogenides. *Acta Crystallogr. Sect. A* **1976**, *32*, 751–767. [CrossRef]

72. Rimskaya-Korsakova, M.N.; Dubinin, A.V. Rare earth elements in Sulfides of submarine hydrothermal vents of the Atlantic Ocean. *Geochmica Cosmochim. Acta* **2003**, *389*, 432–436.

73. Michard, A.; Albarede, F. The REE content of some hydrothermal fluids. *Chem. Geol.* **1986**, *55*, 51–60. [CrossRef]

74. Klinkhammer, G.; Elderfield, H.; Edmond, J.; Mitra, A. Geochemical implications of rare earth element patterns in hydrothermal fluids from mid-ocean ridges. *Geochim. Cosmochim. Acta* **1994**, *58*, 5105–5113. [CrossRef]

75. Douville, E.; Bienvenu, P.; Charlou, J.L.; Donval, J.P.; Fouquet, Y.; Appriou, P.; Gamo, T. Yttrium and rare earth elements in fluids from various deep-sea hydrothermal systems. *Geochim. Cosmochim. Acta* **1999**, *63*, 627–643. [CrossRef]

76. Hongo, Y.; Obata, H.; Gamo, T.; Nakaseama, M.; Ishibashi, J.; Konno, U.; Saegusa, S.; Ohkubo, S.; Tsunogai, U. Rare Earth Elements in the hydrothermal system at Okinawa Trough back-arc basin. *Geochem. J.* **2007**, *41*, 1–15. [CrossRef]

77. Schmidt, K.; Koschinsky, A.; Garbe-Schönberg, D.; de Carvalho, L.M.; Seifert, R. Geochemistry of hydrothermal fluids from the ultramafic-hosted Logatchev hydrothermal field, 15°N on the Mid-Atlantic Ridge: Temporal and spatial investigation. *Chem. Geol.* **2007**, *242*, 1–21. [CrossRef]

78. Barrett, T.J.; Jarvis, I.; Jarvis, K.E. Rare earth element geochemistry of massive sulfides-sulfates and gossans on the Southern Explorer Ridge. *Geology* **1990**, *18*, 583–586. [CrossRef]

79. Gillis, K.; Smith, A.; Ludden, J. Trace element and Sr-isotopic contents of hydrothermal clays and sulfides from the Snake Pit hydrothermal field: ODP site 649. In *Proceedings of the Ocean Drilling Program, Mid-Atlantic Ridge; Covering Legs 106 and 109 of the Cruises of the Drilling Vessel JOIDES Resolution, St. John, Newfoundland to Malaga, Spain, Sites 648–649. Ocean Drilling Program*; Texas A&M University: College Station, TX, USA; 1990, pp. 315–319.

80. De Baar, H.J.; Brewer, P.G.; Bacon, M.P. Anomalies in rare earth distributions in seawater: Gd and Tb. *Geochim. Cosmochim. Acta* **1985**, *49*, 1961–1969. [CrossRef]

81. Piepgras, D.; Wasserburg, G. Strontium and neodymium isotopes in hot springs on the East Pacific Rise and Guaymas Basin. *Earth Planet. Sci. Lett.* **1985**, *72*, 341–356. [CrossRef]

82. Langmuir, C.; Humphris, S.; Fornari, D.; Van Dover, C.; Von Damm, K.; Tivey, M.; Colodner, D.; Charlou, J.-L.; Desonie, D.; Wilson, C.; et al. Hydrothermal vents near a mantle hot spot: The Lucky Strike vent field at 37°N on the Mid-Atlantic Ridge. *Earth Planet. Sci. Lett.* **1997**, *148*, 69–91. [CrossRef]

83. Pearce, J.A.; Peate, D.W. Tectonic implications of the composition of volcanic arc magmas. *Annu. Rev. Earth Planet. Sci.* **1995**, *23*, 251–285. [CrossRef]

84. Plank, T.; Langmuir, C.H. The chemical composition of subducting sediment and its consequences for the crust and mantle. *Chem. Geol.* **1998**, *145*, 325–394. [CrossRef]

85. Yang, Y.-Z.; Wang, Y.; Ye, R.-S.; Li, S.-Q.; He, J.-F.; Siebel, W.; Chen, F. Petrology and geochemistry of Early Cretaceous A-type granitoids and late Mesozoic mafic dikes and their relationship to adakitic intrusions in the lower Yangtze River belt, Southeast China. *Int. Geol. Rev.* **2017**, *59*, 62–79. [CrossRef]

86. Gamo, T.; Sakai, H.; Kim, E.-S.; Shitashima, K.; Ishibashi, J.-I. High alkalinity due to sulfate reduction in the CLAM hydrothermal field, Okinawa Trough. *Earth Planet. Sci. Lett.* **1991**, *107*, 328–338. [CrossRef]

87. Gamo, T. Wide variation of chemical characteristics of submarine hydrothermal fluids due to secondary modification processes after high temperature water-rock interaction: A review. In *Biogeochemical Processes and Ocean Flux in the Western Pacific*; Sakai, H., Nozaki, Y., Eds.; Terra Scientific Publishing Company (TERRAPUB): Tokyo, Japan, 1995; pp. 425–451.

88. Ishibashi, J.-I.; Urabe, T. Hydrothermal activity related to arc-backarc magmatism in the western Pacific. In *Backarc Basins*; Taylor, B., Ed.; Springer: Boston, MA, USA, 1995; pp. 451–495.

89. Kawagucci, S. Fluid geochemistry of high-temperature hydrothermal fields in the Okinawa Trough. In *Subseafloor Biosphere Linked to Hydrothermal Systems: TAIGA Concept*; Ishibashi, J.-I., Okino, K., Sunamura, M., Eds.; Springer: Tokyo, Japan, 2015; pp. 387–403.

90. Takai, K.; Nakagawa, S.; Nunoura, T. Comparative investigation of microbial communities associated with hydrothermal activities in the Okinawa trough. In *Subseafloor Biosphere Linked to Hydrothermal Systems: TAIGA Concept*; Ishibashi, J.-I., Okino, K., Sunamura, M., Eds.; Springer: Tokyo, Japan, 2015; pp. 421–435.

91. Ueda, A.; Sakai, H. Sulfur isotope study of Quaternary volcanic rocks from the Japanese Islands Arc. *Geochim. Cosmochim. Acta* **1984**, *48*, 1837–1848. [CrossRef]

92. Woodhead, J.D.; Harmon, R.S.; Fraser, D. O, S, Sr, and Pb isotope variations in volcanic rocks from the Northern Mariana Islands: Implications for crustal recycling in intra-oceanic arcs. *Earth Planet. Sci. Lett.* **1987**, *83*, 39–52. [CrossRef]

93. Herzig, P.M.; Hannington, M.D.; Arribas, A.A., Jr. Sulfur isotopic composition of hydrothermal precipitates from the Lau back-arc: Implications for magmatic contributions to seafloor hydrothermal systems. *Miner. Deposita* **1998**, *33*, 226–237. [CrossRef]

94. Chen, Z.; Zeng, Z.; Tamehe, L.S.; Wang, X.; Chen, K.; Yin, X.; Yang, W.; Qi, H. Magmatic sulfide saturation and dissolution in the basaltic andesitic magma from the Yaeyama Central Graben, southern Okinawa Trough. *Lithos* **2021**, *388*, 106082. [CrossRef]
95. Alt, J.C.; Shanks, W.C.; Bach, W.; Paulick, H.; Garrido, C.J.; Beaudoin, G. Hydrothermal alteration and microbial sulfate reduction in peridotite and gabbro exposed by detachment faulting at the Mid-Atlantic Ridge, 15°20′N (ODP Leg 209): A sulfur and oxygen isotope study. *Geochem. Geophys. Geosyst.* **2007**, *8*, 08002. [CrossRef]
96. Delacour, A.; Früh-Green, G.L.; Bernasconi, S.M.; Kelley, D.S. Sulfur in peridotites and gabbros at Lost City (30°N, MAR): Implications for hydrothermal alteration and microbial activity during serpentinization. *Geochim. Cosmochim. Acta* **2008**, *72*, 5090–5110. [CrossRef]
97. Zhang, Y.; Zeng, Z.; Chen, S.; Wang, X.; Yin, X. New insights into the origin of the bimodal volcanism in the middle Okinawa Trough: Not a basalt-rhyolite differentiation process. *Front. Earth Sci.* **2018**, *12*, 325–338. [CrossRef]
98. Zhang, Y.; Zeng, Z.; Li, X.; Yin, X.; Wang, X.; Chen, S. High-potassium volcanic rocks from the O kinawa T rough: Implications for a cryptic potassium-rich and DUPAL -like source. *Geol. J.* **2017**, *53*, 1755–1766. [CrossRef]
99. Chen, Z.; Zeng, Z.; Yin, X.; Wang, X.; Zhang, Y.; Chen, S.; Shu, Y.; Guo, K.; Li, X. Petrogenesis of highly fractionated rhyolites in the southwestern Okinawa Trough: Constraints from whole-rock geochemistry data and Sr-Nd-Pb-O isotopes. *Geol. J.* **2018**, *54*, 316–332. [CrossRef]
100. Li, X.; Zeng, Z.; Chen, S.; Ma, Y.; Yang, H.; Zhang, Y.; Chen, Z. Geochemical and Sr-Nd-Pb isotopic compositions of volcanic rocks from the Iheya Ridge, the middle Okinawa Trough: Implications for petrogenesis and a mantle source. *Acta Oceanol. Sin.* **2018**, *37*, 73–88. [CrossRef]
101. Chen, Z.; Zeng, Z.; Qi, H.; Wang, X.; Yin, X.; Chen, S.; Yang, W.; Ma, Y.; Zhang, Y. Amphibole perspective to unravel the rhyolite as a potential fertile source for the Yonaguni Knoll IV hydrothermal system in the southwestern Okinawa Trough. *Geol. J.* **2020**, *55*, 4279–4301. [CrossRef]
102. Vidal, P.; Clauer, N. Pb and Sr isotopic systematics of some basalts and sulfides from the East Pacific Rise at 21°N (project RITA). *Earth Planet. Sci. Lett.* **1981**, *55*, 237–246. [CrossRef]
103. Chen, H.J. U, Th, and Pb isotopes in hot springs on the Juan de Fuca Ridge. *J. Geophys. Res. Solid Earth* **1987**, *92*, 11411–11415. [CrossRef]
104. Hegner, E.; Tatsumoto, M. Pb, Sr, and Nd isotopes in basalts and sulfides from the Juan de Fuca Ridge. *J. Geophys. Res. Space Phys.* **1987**, *92*, 11380–11386. [CrossRef]
105. Hinkley, T.K.; Tatsumoto, M. Metals and isotopes in Juan de Fuca Ridge hydrothermal fluids and their associated solid materials. *J. Geophys. Res. Space Phys.* **1987**, *92*, 11400–11410. [CrossRef]
106. Fouquet, Y.; Marcoux, E. Lead isotope systematics in Pacific hydrothermal sulfide deposits. *J. Geophys. Res. Space Phys.* **1995**, *100*, 6025–6040. [CrossRef]
107. Charlou, J.; Donval, J.; Fouquet, Y.; Jean-Baptiste, P.; Holm, N. Geochemistry of high H2 and CH4 vent fluids issuing from ultramafic rocks at the Rainbow hydrothermal field (36°14′N, MAR). *Chem. Geol.* **2002**, *191*, 345–359. [CrossRef]
108. Yao, H.-Q.; Zhou, H.-Y.; Peng, X.-T.; Bao, S.-X.; Wu, Z.-J.; Li, J.-T.; Sun, Z.-L.; Chen, Z.-Q.; Li, J.-W.; Chen, G.-Q. Metal sources of black smoker chimneys, Endeavour Segment, Juan de Fuca Ridge: Pb isotope constraints. *Appl. Geochem.* **2009**, *24*, 1971–1977. [CrossRef]
109. Stavinga, D.; Jamieson, H.; Layton-Matthews, D.; Paradis, S.; Falck, H. Geochemical and mineralogical controls on metal(loid) mobility in the oxide zone of the Prairie Creek Deposit, NWT. *Geochem. Explor. Environ. Anal.* **2017**, *17*, 21–33. [CrossRef]
110. Turekian, K.K. *Oceans*; Prentice-Hall: Engelwood Cliffs, NJ, USA, 1968.
111. Fallon, E.K.; Petersen, S.; Brooker, R.; Scott, T.B. Oxidative dissolution of hydrothermal mixed-sulphide ore: An assessment of current knowledge in relation to seafloor massive sulphide mining. *Ore Geol. Rev.* **2017**, *86*, 309–337. [CrossRef]

Journal of
Marine Science and Engineering

Article

Strontium, Hydrogen and Oxygen Behavior in Vent Fluids and Plumes from the Kueishantao Hydrothermal Field Offshore Northeast Taiwan: Constrained by Fluid Processes

Zhigang Zeng [1,2,3,*], Xiaoyuan Wang [1], Xuebo Yin [1], Shuai Chen [1,4], Haiyan Qi [1] and Chen-Tung Arthur Chen [5]

[1] Seafloor Hydrothermal Activity Laboratory, CAS Key Laboratory of Marine Geology and Environment, Institute of Oceanology, Chinese Academy of Sciences, Qingdao 266071, China; wangxiaoyuan@qdio.ac.cn (X.W.); re_hero@163.com (X.Y.); chenshuai@qdio.ac.cn (S.C.); qihaiyan@qdio.ac.cn (H.Q.)

[2] Laboratory for Marine Mineral Resources, Qingdao National Laboratory for Marine Science and Technology, Qingdao 266237, China

[3] University of Chinese Academy of Sciences, Beijing 101408, China

[4] Center for Ocean Mega-Science, Chinese Academy of Sciences, Qingdao 266071, China

[5] Department of Oceanography, National Sun Yat-Sen University, Kaohsiung 80424, Taiwan, China; ctchen@mail.nsysu.edu.tw

* Correspondence: zgzeng@ms.qdio.ac.cn

Abstract: Strontium (Sr), hydrogen (H) and oxygen (O) in vent fluids are important for understanding the water–rock interaction and hydrothermal flux in hydrothermal systems. We have analyzed the Sr, H and O isotopic compositions of seawater, vent fluid and hydrothermal plume samples in the Kueishantao hydrothermal field, as well as their calcium (Ca), total sulfur (S), Sr, arsenic (As), stibium (Sb), chlorine (Cl) and manganese (Mn) concentrations for understanding the origin and processes of fluids. The results suggest that most As, Sb and Mn are leached from andesitic rocks into the fluids, and most Ca and Cl remained in the deep reaction zone during the fluid–andesitic rock interaction. The ranges of $^{87}Sr/^{86}Sr$, $\delta D_{V\text{-SMOW}}$ and $\delta^{18}O_{V\text{-SMOW}}$ values in the yellow spring, white spring and plumes are small. The $^{87}Sr/^{86}Sr$, $\delta D_{V\text{-SMOW}}$ and $\delta^{18}O_{V\text{-SMOW}}$ values of fluids and plumes are like those of ambient seawater, indicating that the Sr, H and O of vent fluids and hydrothermal plumes are derived primarily from seawater. This suggests that the interaction of andesite and subseafloor fluid is of short duration and results in the majority of As, Sb and Mn being released into fluids, while most Ca and Cl remained in the deep reaction zone. In addition, there was no significant variation of Sr, H and O isotopic compositions in the upwelling fluid, keeping the similar isotopic compositions of seawater. There are obvious correlations among the pH values, As and Sb concentrations, and H isotopic compositions of the vent fluids and hydrothermal plumes, implying that the As and Sb concentrations and H isotopic compositions can trace the dispersion of plumes in the ambient seawater. According to the Sr concentrations and $^{87}Sr/^{86}Sr$ values, the water/rock ratios are 3076~8124, which is consistent with the idea that the interaction between fluid and andesite at the subseafloor is of short duration. The hydrothermal flux of Sr discharged from the yellow spring into the seawater is between 2.06×10^4 and 2.26×10^4 mol/yr, and the white spring discharges $1.18 \times 10^4 \sim 1.26 \times 10^4$ mol/yr Sr if just andesites appear in the reaction zone.

Keywords: strontium-hydrogen-oxygen isotopes; vent fluid; hydrothermal plume; Kueishantao hydrothermal field

Citation: Zeng, Z.; Wang, X.; Yin, X.; Chen, S.; Qi, H.; Chen, C.-T.A. Strontium, Hydrogen and Oxygen Behavior in Vent Fluids and Plumes from the Kueishantao Hydrothermal Field Offshore Northeast Taiwan: Constrained by Fluid Processes. *J. Mar. Sci. Eng.* **2022**, *10*, 845. https://doi.org/10.3390/jmse10070845

Academic Editor: Luca Cavallaro

Received: 7 May 2022
Accepted: 10 June 2022
Published: 21 June 2022

Publisher's Note: MDPI stays neutral with regard to jurisdictional claims in published maps and institutional affiliations.

1. Introduction

Extensive strontium (Sr), hydrogen (H), and oxygen (O) isotopic and chemical exchange depending on fluid temperatures, hydrothermal alteration and the water/rock ratios can be caused by hydrothermal circulation [1–5]. The intensity of the water–rock interaction has been recorded by these isotopes in the hydrothermal fluids, and they are particularly useful in the research of fluid origin and hydrothermal processes in subseafloor hydrothermal systems [1,6,7]. The mobile Sr, H and O isotopes are good markers of host rock composition and subseafloor alteration processes in hydrothermal systems [2,4,8,9]. Sr, H and O in hydrothermal fluids come from different sources, relying on the hydrothermal circulation system, seawater and fluid mixing, host rocks and magma [2,4,10,11]. Their isotopic ratios are used to identify different sources (rock, seawater and sediment) [2,4,12] and mixing versus conductive cooling or heating processes and may be valuable for constraining the phase separation conditions in seafloor hydrothermal systems.

However, it has been implied that the high-temperature hydrothermal circulation is insufficient to balance the other Sr input fluxes, as an increase of ~5.4×10^{-5} Myr^{-1} [13] has been observed in oceanic ^{87}Sr/^{86}Sr over the past few million years [14–16]. Many explanations have been put forward to explain the missing unradiogenic ^{87}Sr/^{86}Sr flux, such as hydrothermal systems occurring on the ridge flank [16], low-temperature alteration processes and the precipitation of carbonate [17,18]. The reactions happening in the hydrothermal systems are of great importance and are believed to be the main source of unradiogenic ^{87}Sr/^{86}Sr in seawater; however, the Sr flux into the oceans from high- and particularly low-temperature hydrothermal systems remain poorly quantified [14,17].

During subcritical phase separation, H isotope fractionation has been shown to depend on both temperature and salinity [19]. Furthermore, H isotope ratios of fluids are relatively insensitive to reactions with the host rock because H is abundant in the fluid relative to the mineral content. However, the H isotope fractionation data of brine and vapor coexisting in the two-phase supercritical region of seawater is completely lacking. The H isotope fractionation related to the phase separation of NaCl-H$_2$O fluid has only been obtained up to 350 °C [19]. Although the temperature of the vent fluid is generally close to 350 °C [20], phase separation occurs in the deep reaction zones at temperatures significantly higher than this. Many hydrothermal experiments studying the fractionation of seawater phase separation at 450 °C had obtained H isotopic fractionation factors, and these data were then used to explain the phase separation and segregation in deep-sea hydrothermal systems [2].

The first O isotope studies of black smoker hydrothermal fluids were published by the East Pacific Rise Study Group [21] and Welhan and Craig [22], which showed that the δ^{18}O values are close to that of ambient seawater; however, they increase by 1–2‰ owing to the reaction between basalt and seawater at elevated temperatures. Furthermore, Teagle et al. [23] used Sr and O isotope systematics to confirm that anhydrite on the Trans-Atlantic Geotraverse (TAG) Mound forms through the mixing between fluids and seawater, which is conductively heated to 100–180 °C before mixing; thus, anhydrite forms by a mixture of heated seawater and vent fluid at 230–320 °C. According to Bach and Humphris [24], the ^{87}Sr/^{86}Sr and δ^{18}O of the vent fluids on the oceanic ridge show a global correction with the rate of spreading (or the rate of magma supply). They believe that at low spreading rates (or low magma supply rates), there is significantly less Sr from seawater and higher δ^{18}O from seawater in vent fluids. This is due to a greater exchange of Sr and O with the crust during hydrothermal fluid circulation, possibly because of longer reaction paths under slow-spreading ridges. Using the datasets of hydrothermal vent fluids associated with ridges, linear regression equations and R2 values show that the δ^{18}O proposed by Bach and Humphris [24] is not statistically related to the spreading rate.

Field and experimental studies, as well as isotopic exchange computations [25–31], have clearly shown that both O and H isotope values increase due to water/rock interactions with the igneous crust. Due to the decrease of water/rock mass ratios caused by the evolution of hydrothermal fluids, the O and H isotope values of end-member vent fluids follow the calculated seawater/basalt reaction vector [25]. For slower spreading ridges, higher δD values are consistent with longer fluid paths and more fluid-rock interactions. However, δ^{18}O values would be even more responsive to fluid–rock interactions and have insignificant variation with the rate of spreading. With the increase of the hydrothermal vent fluid database, the short-term time variation in each field becomes more and more important.

The chemical characteristics of deep-sea hydrothermal fluids from mid-ocean ridges and back-arc basins, including Sr, H and O isotopic compositions, have been described by [20,32–35]. However, little is known about the chemical and Sr, H and O isotopic compositions of shallow hydrothermal vents fluids. The quantification of global hydrothermal flux into the ocean is also biased due to the lack of budget caused by the interaction of seawater and andesite. To better understand the fluid–andesitic rock interaction, the fluid and plume samples in the Kueishantao shallow hydrothermal field were collected and analyzed. The main purposes of this study are (1) to investigate the geochemistry characteristics and evolution of hydrothermal fluids and their plumes, (2) to study the rock–fluid interactions recorded by fluids and plumes, and (3) to assess the hydrothermal flux of Sr in the shallow hydrothermal field.

2. Geologic Setting

The Kueishantao shallow hydrothermal field (121°55′ E, 24°50′ N) is in the southeast of Kueishantao, northeastern Taiwan, and near the southern Okinawa Trough. The hydrothermal field covers an area of about 0.5 square kilometers (Figure 1). Kueishantao's last major eruption occurred about 7000 years ago [36], and the area around the hydrothermal vents is characterized by andesitic lava and pyroclastic flows. The andesitic magma is thought to be produced by a MORB-type magma with an assimilated 30% local continental crust [37]. The difference in Fe-Cu-Zn isotopic compositions between the Kueishantao andesites and the MORBs and the continental crust might indicate entrainment of carbonate sediment components into the andesitic magma [38].

The Kueishantao hydrothermal field has more than 30 hydrothermal vents at water depths of 10–30 m. The hydrothermal products in the form of chimneys, mounds and balls are mainly composed of natural sulfur. On 12 August 2000, a large yellow chimney approximately 6 m high was discovered at a water depth of 20 m. The geochemical study of the native sulfur chimneys and balls indicated that trace elements are derived primarily from the andesite and partly from seawater, and the interaction between the subseafloor fluid and the andesite has a short duration [39,40].

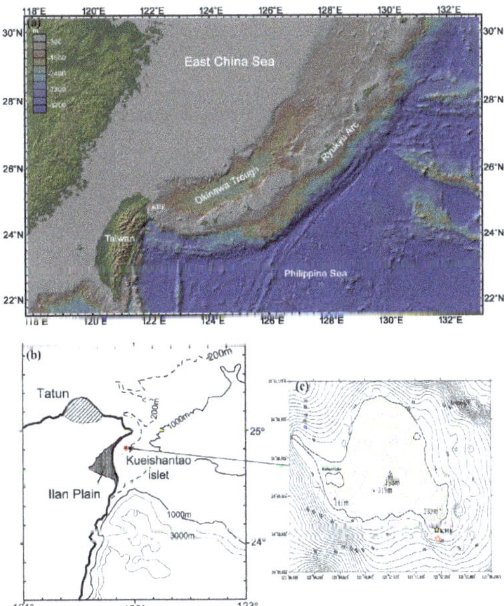

Figure 1. (**a**) The bathymetric map of Taiwan and the Okinawa Trough (based on [41]). (**b**) The tectonic map with the location of Kueishantao island (from [42]). (**c**) The location of the yellow spring

(yellow star, 108 °C) and white spring (red star, 51 °C) in the Kueishantao hydrothermal field (based on [41]).

There are two types of vents, "yellow spring" (78~116 °C) and "white spring" (30~65 °C) [42]. The yellow spring fluids are characterized by very low pH (~1.52) and a wide range of chemical compositions. The white spring fluids have a relatively low content of methane, iron and copper [42]. The temperature of the vent fluids varied with tides and reached the maximum temperature approximately 3.5 h after each high tide [42–44]. The Ca, Mg and K are significantly correlated with each other, indicating the same source of seawater [45]. The gas component is mainly CO_2, followed by N_2, CH_4 and H_2S [46].

3. Sampling and Methods

3.1. Sample Collection

On 31 May 2011, the samples of hydrothermal fluids and plumes were collected from the Kueishantao hydrothermal field by divers (Table 1, Figure 1). Four liter Pyrex bottles were used to sample the hydrothermal fluids at and in the vent. Two-valve polyethylene tubes were also employed to sample the fluids in order to assess the reliability of the data gained by the Pyrex bottles. The hydrothermal plume was sampled in 1 L Nalgene polypropylene bottles at approximately 2, 5 and 7 m below sea level above the yellow spring vent (water depth, 7.2 m) and approximately 5, 10, 13 and 15 m below sea level above the white spring vent (water depth, 15.1 m). Shallow seawater was sampled at a depth of 10 m near Kueiwei to eliminate the effect of hydrothermal activity. The fluid fluxes and temperatures were determined in situ. Kuo [43] described in detail the methods of fluid collection and the measurement of temperature and flux in situ.

Table 1. Location of the yellow and white springs in the Kueishantao hydrothermal field as well as the fluid temperature and flux [45].

Spring Type	Latitude (°N)	Longitude (°E)	Depth (m)	Fluid Temperature (°C)	Fluid Flux (m³/h)
Yellow spring	24.8349	121.96194	7.2	108	35.1
White spring	24.83412	121.96196	15.1	51	19.3

3.2. Analytical Methods

In the lab on the shore, the pH of the liquid was analyzed by a portable pH meter with a resolution of 0.01 (JENCO 6010, San Diego, CA, USA). Before measurement, calibrate the pH meter with buffer solutions of pH 4.00 and 6.86 (i.e., 0.05 mol/L potassium hydrogen phthalate (25 °C, pH = 4.00) and 0.025 mol/L mixed phosphate (25 °C, pH = 6.86) buffers). All liquids were filtered into 1 L Nalgene polypropylene bottles, which were previously soaked in 1:1 HNO_3 for 48 h, washed with distilled water, and then dried.

Calcium (Ca), total sulfur (S) and Sr were analyzed by the inductively coupled plasma optical emission spectrometer at the Shandong Institute of Geophysical and Geochemical Exploration, and the precision was <±5%. Arsenic (As) and stibium (Sb) were determined using atomic fluorescence spectrophotometer at the Qingdao Institute of Marine Geology, China Geological Survey Bureau. Manganese (Mn) was analyzed by inductively coupled plasma mass spectrometer, and chlorine (Cl) was determined using ion chromatography at the Institute of Oceanology, Chinese Academy of Sciences. The detailed chemical separation and measurement procedures of As, Sb, Mn and Cl were described in previous studies [47]. Seawater standard NASS-5 was used. The analytical precision of As, Sb and Mn was better than 5%, and that of Cl was ±1%.

The $^{87}Sr/^{86}Sr$ was measured using thermal ionization mass spectrometry (TIMS; Triton) at the State Key Laboratory of Geological Processes and Mineral Resources, China University of Geosciences. The procedures of chemical separation and determination were described in detail in previous studies [48,49]. For Sr, the procedural blanks were less than 50 pg. The determination of the NBS987 Sr isotopic standard gained over one year yielded an average value of $^{87}Sr/^{86}Sr$ = 0.710266 ± 0.000009 (2σ, n = 38). Rock standard AGV-2 was used for $^{87}Sr/^{86}Sr$. The analytical precision was <0.006%.

H and O isotopic ratios were measured by the stable isotope ratio mass spectrometer (MAT-253, Thermo Fisher, USA) at the Institute of Mineral Resources, Chinese Academy of Geological Sciences. Fluid samples were prepared for H isotope analyses using the Zn-reduction method and for O isotope analyses using the CO_2 equilibration technique. Values of H and O isotopes are calibrated relative to the V-SMOW standard. The analytical precision of the δD and $\delta^{18}O$ are better than 1.0‰ and 0.1‰, respectively.

4. Results

We report the element (Ca, total S, Sr, As, Sb, Cl and Mn) concentrations and $^{87}Sr/^{86}Sr$, δD and $\delta^{18}O$ compositions of fluid and plume samples collected from the Kueishantao hydrothermal field. At both yellow and white spring sites, the hydrothermal fluids are released directly from the andesite host rock due to an absence of sediment cover. Figure 1 displays the sample locations and element concentrations, and the $^{87}Sr/^{86}Sr$, δD and $\delta^{18}O$ values are listed in Table 2.

Table 2. pH, Ca, total S, Sr, As, Sb, Cl and Mn concentrations and $^{87}Sr/^{86}Sr$, $\delta D_{V\text{-SMOW}}$ and $\delta^{18}O_{V\text{-SMOW}}$ values of fluid and plume samples collected from the Kueishantao hydrothermal field.

Sample	Depth (m)	pH	Ca (mg/L)	Total S (mg/L)	Sr (mg/L)	As (µg/L)	Sb (µg/L)	Cl (mg/L)	Mn (µg/L)	$^{87}Sr/^{86}Sr$	$2\sigma(\pm)$	$\delta D_{V\text{-SMOW}}$ (‰)	$\delta^{18}O_{V\text{-SMOW}}$ (‰)
Ambient seawater	10	8.02	372.6	835	6.62	1.15	0.20	19,352	0.91	0.709177	0.000008	2	0.1
YSP, 0 m	0	6.15	381.3	829	6.37	5.59	0.29	20,814	10.8	0.709147	0.000008	2	0.1
YSP, −2 m	2	6.12	385.0	840	6.36	3.81	0.23	20,422	3.93	0.709156	0.000007	3	0.2
YSP, −5 m	5	5.60	382.2	833	6.34	6.01	0.28	20,810	3.45	0.709168	0.000006	5	0.2
YSF, out, bottle	7.2	2.81	371.2	810	5.87	22.4	0.77	18,052	4.55	-	-	1	0.1
YSF, in, bottle	>7.2	2.29	378.9	812	6.12	46.4	1.22	18,254	3.58	-	-	1	0.1
YSF, out, tube	7.2	-	345.0	838	6.32	-	0.03	17,380	18.5	-	-	-	-
YSF, in, tube	>7.2		359.1	1069	6.43	0.29	-	18,323	14.2	-	-	-	-
WSP, 0 m	0	6.14	381.3	832	6.35	5.65	0.27	19,068	6.24	0.709202	0.000006	0	0.1
WSP, −2 m	2	6.12	383.0	833	6.35	4.08	0.29	18,925	3.48	0.709186	0.000005	0	0.1
WSP, −5 m	5	5.91	379.3	823	6.30	6.46	0.32	20,205	4.34	0.709171	0.000006	2	0.1
WSP, −10 m	10	5.51	376.2	823	6.27	14.5	0.48	20,454	10.4	0.709178	0.000005	0	0.2
WSF, out, bottle	15.1	5.11	380.5	826	6.42	10.5	0.36	16,529	7.48	0.709158	0.000004	−4	0.2

Table 2. *Cont.*

Sample	Depth (m)	pH	Ca (mg/L)	Total S (mg/L)	Sr (mg/L)	As (μg/L)	Sb (μg/L)	Cl (mg/L)	Mn (μg/L)	^{87}Sr/^{86}Sr	2σ(±)	δD$_{V\text{-SMOW}}$ (‰)	δ^{18}O$_{V\text{-SMOW}}$ (‰)
WSF, in, bottle	15.1	4.67	369.4	963	6.12	15.8	0.36	16,951	23.3	-	-	−2	0.2
WSF, out, tube	15.1	5.10	383.2	802	6.56	27.6	0.47	17,811	14.4	0.709154	0.000007	1	0.2
WSF, in, tube	15.1	4.67	372.5	805	6.23	21.1	0.39	17,309	13.2	0.709164	0.000006	1	0.3

Sample	Range of Sr (ppm)	Average of Sr (ppm)
Kueishantao andesite	175–264	206 ($n = 41$)

"-" no detect; YSP represents yellow spring plume; YSF represents yellow spring fluid; WSP represents white spring plume; WSF represents white spring fluid.

The Ca, total S, Sr As, Sb, Cl and Mn concentrations in the yellow spring varied from 381.3 mg/L, 892 mg/L, 6.37 mg/L, 5.59 μg/L, 0.29 μg/L, 20,814 mg/L and 10.8 μg/L at 0 mbsl (meters below sea level) to 371.2 mg/L, 810 mg/L, 5.87 mg/L, 22.4 μg/L, 0.77 μg/L, 18,052 mg/L and 4.55 μg/L at 7.2 mbsl (Figure 2), respectively. The Ca, total S, Sr As, Sb, Cl and Mn concentrations in the white spring varied from 381.3 mg/L, 832 mg/L, 6.3 mg/L, 5.65 μg/L, 0.27 μg/L, 19,068 mg/L and 6.24 μg/L at 0 mbsl to 369.4 mg/L, 963 mg/L, 6.12 mg/L, 15.8 μg/L, 0.36 μg/L, 16,951 mg/L and 23.3 μg/L at 15.1 mbsl (Figure 2), respectively.

Compared with the Ca, Sr As and Cl concentrations in the yellow spring hydrothermal plumes (381.3 to 385.0 mg/L, 6.34 to 6.37 mg/L, 3.81 to 6.01 μg/L and 20,422 to 20,814 mg/L), those in the white spring hydrothermal plumes have a slightly wider range of variation (376.2 to 383.0 mg/L, 6.27 to 6.35 mg/L, 4.08 to 14.5 μg/L and 18,925 to 20,454 mg/L) than in (Figure 3). Most of the total S and Sr concentrations for both hydrothermal plumes are slightly lower than that of shallow seawater (835 and 6.62 mg/L) (Figure 2). The Ca, Sr and Cl concentrations of most hydrothermal fluids are clearly lower than that of shallow seawater and consistent with those (Cl 15,811–19,888 mg/L) of [39,42].

The δD$_{V\text{-SMOW}}$ and δ^{18}O$_{V\text{-SMOW}}$ values in the yellow spring range from 1‰ and 0.1‰ in fluid to 5‰ and 0.2‰ in the plume, and the ^{87}Sr/^{86}Sr, δD$_{V\text{-SMOW}}$ and δ^{18}O$_{V\text{-SMOW}}$ values in the white spring range from 0.709154, −4‰, and 0.2‰ in fluid to 0.709202, 2‰ and 0.2‰ in the plume (Table 2). The ^{87}Sr/^{86}Sr, δD$_{V\text{-SMOW}}$ and δ^{18}O$_{V\text{-SMOW}}$ values of the two hydrothermal plumes were similar (Table 2), and near the 0.709177, 2‰ and 0.1‰ values of ambient seawater (Figure 3). The ^{87}Sr/^{86}Sr values of the white spring hydrothermal fluids are distinctly lower than those of shallow seawater and the white spring hydrothermal plume (Figure 3).

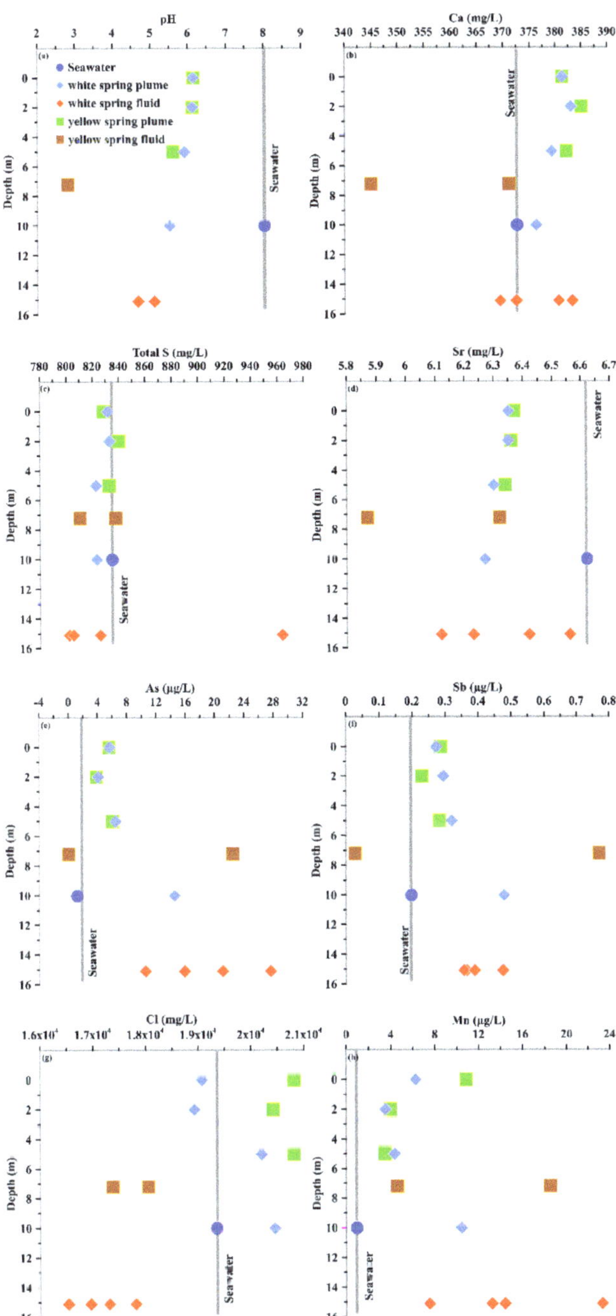

Figure 2. Changes in pH (**a**), Ca (**b**), total S (**c**), Sr (**d**), As (**e**), Sb (**f**), Cl (**g**) and Mn (**h**) concentrations from hydrothermal fluid to plume at the yellow and white spring sites.

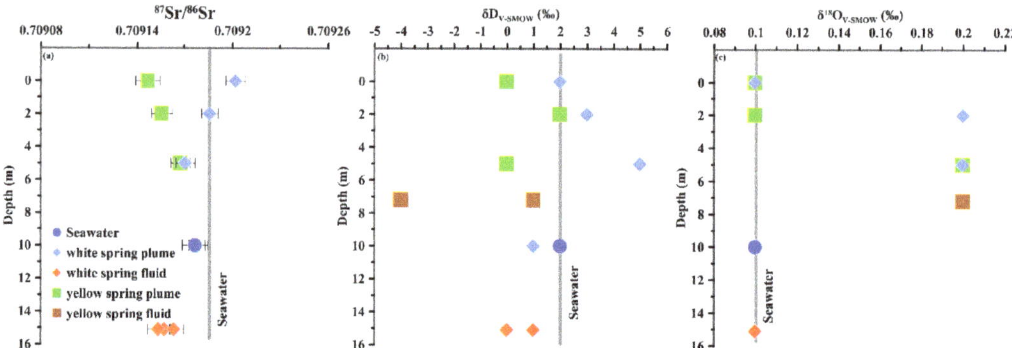

Figure 3. Variation of $^{87}Sr/^{86}Sr$ (**a**), δD (**b**) and $\delta^{18}O$ (**c**) values from hydrothermal fluid to plume at the yellow and white spring sites.

The hydrothermal fluids have a pH range of 2.29 to 5.11, and those of the plumes are 5.51~6.15. The pH values of the hydrothermal fluids and plumes are distinctly lower than that of shallow seawater (8.02), and from the hydrothermal fluids to the hydrothermal plumes, the pH value increases gradually as the water depth decreases (Figure 2). The As, Sb and Mn concentrations and $\delta D_{V\text{-}SMOW}$ values of the hydrothermal fluids and plumes display a good positive and negative correlation with pH values (Figure 4).

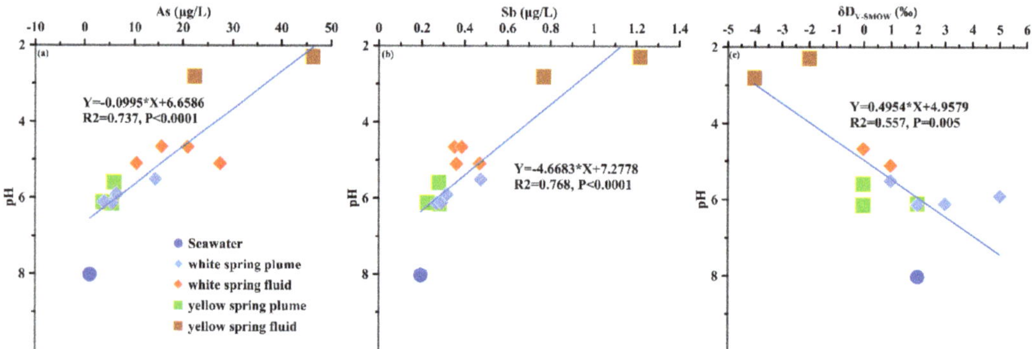

Figure 4. pH vs. As concentration (**a**), pH vs. Sb concentration (**b**) and pH vs. δD (**c**) values for fluids and plumes in the yellow and white springs.

From vent fluids to hydrothermal plumes, the pH values, As and Sb concentrations and H isotope compositions have a good correlation, and the concentrations of As and Sb decrease with the increase of pH value (Figure 4), which is like the boron (B) concentrations and isotopic compositions [42], indicating that the As and Sb concentrations and H isotopic compositions can be used to trace the variation of chemical characteristics during the hydrothermal plumes that disperse in the seawater.

Moreover, from the surface to the water depth of 5 m, the Cl and total S concentrations and $\delta^{18}O_{V\text{-}SMOW}$ values in the hydrothermal plumes are similar (Figure 2), which is due to the mixing of the hydrothermal plumes with shallow seawater (Figure 4B,C).

5. Discussion

5.1. Sources of Strontium, Hydrogen and Oxygen

Because Sr, H and O isotopes in seawater and andesite are different, the Sr, H and O in hydrothermal fluids might be derived from two components: seawater and andesite. In the Kueishantao hydrothermal field, the $^{87}Sr/^{86}Sr$, $\delta D_{V\text{-}SMOW}$ and $\delta^{18}O_{V\text{-}SMOW}$ values in ambient seawater are 0.709177, 2‰ and 0.1‰ with a Sr concentration of 6.62 mg/L, and an average Sr content in the andesite is 206 ppm (Table 2). The andesite in the Kueishantao hydrothermal field have a large range of $^{87}Sr/^{86}Sr$ (from 0.70577 to 0.70688) and $\delta^{18}O_{SMOW}$ (from 7.1‰ to 7.9‰) values, with Sr concentrations of 221–317 ppm [37].

However, the hydrothermal plumes have similar values of $^{87}Sr/^{86}Sr$, $\delta D_{V\text{-}SMOW}$ and $\delta^{18}O_{V\text{-}SMOW}$ to those of the shallow seawater (Table 2) (Figure 3), and the $^{87}Sr/^{86}Sr$ ratios of the hydrothermal fluids and plumes in the Kueishantao hydrothermal field were significantly higher than those (from 0.70382 to 0.7056) of hydrothermal fluids from the Eastern Lau spreading center [35], suggesting that the hydrothermal plumes and fluids are derived primarily from seawater.

A simple two-end-member mixing model using an equation:

$$M_{mix} = X \times M_{seawater} + (1 - X) \times M_{andesite} \tag{1}$$

where X is the amount of the seawater component; M is the $^{87}Sr/^{86}Sr$ and $\delta^{18}O$; M_{mix}, $M_{seawater}$ ($^{87}Sr/^{86}Sr_{seawater}$ 0.709177, $\delta^{18}O_{seawater}$ 0.1‰) and $M_{andesite}$ ($^{87}Sr/^{86}Sr_{andesite}$ from 0.70577 to 0.70688, $\delta^{18}O_{andesite}$ from 7.1‰ to 7.9‰) are the Sr and O isotopic compositions of hydrothermal plumes and fluids, seawater and andesite, respectively. The model shows that approximately 99% and 97% of the Sr and O are derived from seawater (Table 3).

Table 3. Calculated water/rock ratios, Sr and O source contributions and Sr fluxes in the Kueishantao hydrothermal field.

Spring Fluid	Water/Rock Ratio		Strontium Source Contributions (%)				Strontium Flux (mole/yr)	Oxygen Source Contributions (%)			
	$^{87}Sr/^{86}Sr$ andesite = 0.70577	$^{87}Sr/^{86}Sr$ andesite = 0.70688	Andesite [1]	Seawater [2]	Andesite [3]	Seawater [2]		Andesite [4]	Seawater [5]	Andesite [6]	Seawater [5]
YSF, out, bottle	-	-	-	-	-	-	2.06×10^4	1.43	98.57	1.28	98.72
YSF, in, bottle	-	-	-	-	-	-	2.15×10^4	1.43	98.57	1.28	98.72
YSF, out, tube	-	-	-	-	-	-	2.22×10^4	1.43	98.57	1.28	98.72
YSF, in, tube	-	-	-	-	-	-	2.26×10^4	2.86	97.14	2.56	97.44
WSF, out, bottle	5548	3730	0.56	99.41	0.83	99.17	1.24×10^4	0.00	100	0.00	100
WSF, in, bottle	-	-	-	-	-	-	1.18×10^4	-	-	-	-
WSF, out, tube	4578	3076	0.68	99.32	1.00	99.00	1.26×10^4	-	-	-	-
WSF, in, tube	8124	5467	0.38	99.62	0.57	99.43	1.20×10^4	0.00	100	0.00	100

"-" no data. Sr of spring fluid from andesite and seawater. Sr source contribution (X) can be calculated by using the simple two endmember mixing method: X = (($^{87}Sr/^{86}Sr_{spring}$ fluid − $^{87}Sr/^{86}Sr_{seawater}$)/($^{87}Sr/^{86}Sr_{andesite}$ − $^{87}Sr/^{86}Sr_{seawater}$) × 100); YSF represents yellow spring fluid; WSF represents white spring fluid; [1] represents $^{87}Sr/^{86}Sr_{andesite}$ = 0.70577 from [37]; [2] represents $^{87}Sr/^{86}Sr_{seawater}$ = 0.709177; [3] represents $^{87}Sr/^{86}Sr_{andesite}$ = 0.70688 from [37]. O source contribution (Y) can be calculated by using simple two end member mixing method: Y = (($\delta^{18}O_{spring}$ fluid − $\delta^{18}O_{seawater}$)/($\delta^{18}O_{andesite}$ − $\delta^{18}O_{seawater}$) × 100); [4] represents $\delta^{18}O_{andesite}$ = 7.1‰ from [40]; [5] represents $\delta^{18}O_{seawater}$ = 0.1; [6] means $\delta^{18}O_{andesite}$ = 7.9‰ from [37].

5.2. Seawater-Rock Interaction

Sr, H and O behave conservatively, as their behavior during hydrothermal alteration demonstrates [6]. The geochemistry of Sr, H and O in submarine hydrothermal systems has been used to constrain subseafloor fluid-rock interaction processes [23,30]. When high temperature fluids reacted with fresh basalt in mid-ocean ridges [50,51], Sr is believed to be extracted from basalt quantitatively during hydrothermal reactions without subsequent re-equilibration with secondary mineral phases [6,23,30], while O and H may be primarily derived from seawater. The relative concentrations of Sr, As, Sb and Mn in the hydrothermal fluids could reflect those in the volcanic rocks where they were leached [52,53].

Past studies of seafloor hydrothermal fields imply that hydrothermal fluids often have similar $^{87}Sr/^{86}Sr$, δD and $\delta^{18}O$ values to those of the nearby rocks ($^{87}Sr/^{86}Sr$) and seawater (δD and $\delta^{18}O$), indicating the leaching of Sr from the host rocks or mixing of H and O from seawater [6,23,30]. Therefore, the Sr, H and O isotopic compositions of hydrothermal fluids can be used to clarify different source rocks.

In both springs, the $^{87}Sr/^{86}Sr$ values of the hydrothermal fluids are obviously higher than that in hydrothermal fluids at the mid-ocean ridge (the East Pacific Rise near 9°N 0.70381–0.70790, [7]) and back-arc basin (Lau basin 0.70382–0.7056, [20]) (Figure 3), and the $^{87}Sr/^{86}Sr$, $\delta D_{V\text{-SMOW}}$ and $\delta^{18}O_{V\text{-SMOW}}$ values of Kueishantao vent fluids and hydrothermal plumes are like those of the ambient seawater, indicating a short-term interaction between subseafloor fluids and andesite.

Most As, Sb and Mn concentrations in hydrothermal fluids are much higher than those in sea level hydrothermal plumes, which have lower temperatures. However, the high As, Sb and Mn concentrations cannot be produced solely by local seawater contributions. The apparently higher concentrations of As, Sb and Mn in the hydrothermal fluids may show an extra source from the andesite basement. The seawater-andesite interaction will result in higher exchangeable As, Sb and Mn concentrations. The high As, Sb and Mn concentrations of most hydrothermal fluids (above seawater, Table 2) partly reflect the constraints by the andesite, which reacts with source fluids. These As-, Sb- and Mn-enriched fluids in both springs have low pH values and can be interpreted by the interaction between seawater and andesite, which explains the difference between Kueishantao and mid-ocean ridge hydrothermal systems. The $^{87}Sr/^{86}Sr$ values in the Kueishantao hydrothermal fluids are obviously higher than those in the hydrothermal fluids on the Valu Fa ridge in the Lau basin (Figure 3), indicating a short interaction between the andesite basement and seawater.

In the Kueishantao hydrothermal field, the concentrations of As, Sb and Mn in the hydrothermal plume and fluid are enriched relative to those in the seawater. This is different from the basalt-hosted hydrothermal systems in which As, Sb and Mn losses are caused by the entrainment of seawater during the formation of altered minerals at low temperatures, and subsequently, the mobilization of As, Sb and Mn in the high-temperature reaction zone [52,53], usually resulting in the slight enrichment of As, Sb and Mn in high-temperature hydrothermal fluids. The concentrations of As, Sb and Mn in the Kueishantao hydrothermal system are like those in the back-arc basins. At these two locations, the weak water–rock interaction results in the smaller removal of As, Sb and Mn in the hydrothermal fluid circulation. However, the unique enrichment of As, Sb and Mn in the hydrothermal fluids is a feature of hydrothermal systems with andesite basements because trace As, Sb and Mn have been leached out by seawater.

We calculated the water/rock ratios of the fluid-rock reaction as follows:

$$W/R = ((^{87}Sr/^{86}Sr_h - {}^{87}Sr/^{86}Sr_{andesite}) \times Sr_{andesite})/((^{87}Sr/^{86}Sr_{sw} - {}^{87}Sr/^{86}Sr_h) \times Sr_{sw}) \qquad (2)$$

where $^{87}Sr/^{86}Sr_h$, $^{87}Sr/^{86}Sr_{andesite}$ and $^{87}Sr/^{86}Sr_{sw}$ (0.709177) are the isotopic ratios in hydrothermal fluids, andesite and seawater, respectively, and $Sr_{andesite}$ (206 ppm) and Sr_{sw} (6.62 mg/L) are the Sr concentrations of andesite and seawater, respectively. The water/rock ratios for $^{87}Sr/^{86}Sr$ andesite = 0.70577 and 0.70688 are listed in Table 3, ranging from 4578 to 8124 and 3076 to 5467, respectively. The significant difference between these results and those (approximately 1.6) from fluids from the DSDP Hole 504B, Costa Rica Rift reported by Kawahata et al. [6] is ascribed to the shorter fluid pathways and seawater–rock interaction. However, the Sr/Ca ratio of KHF hydrothermal fluids and plumes (0.0165–0.0171, $n = 15$) is like that (0.0178) of ambient seawater and significantly higher than that (0.0042–0.0068, $n = 35$) of KHF andesites, which indicates the precipitation of anhydrite is negligible in the subseafloor recharge zone. This suggests that the fluids in the Kueishantao hydrothermal field have weaker water and rock interactions and shorter reaction paths, and the isotopic fractionation seems to decrease accompanying the increasing water/rock ratio, resulting in the $^{87}Sr/^{86}Sr$, δD and $\delta^{18}O$ values of upwelling fluids in the Kueishantao hydrothermal field like those of ambient seawater.

5.3. From Vent Fluid to Plume

The Ca and Cl concentrations of vent fluids are less than those of hydrothermal plumes and ambient seawater (Table 2), and the Ca and Cl concentrations of hydrothermal plumes are higher than those of shallow seawater. This suggests that the mixing of hydrothermal fluids and seawater leads to the enrichment of Ca and Cl in the hydrothermal plumes, and most Ca and Cl are retained in the deep reaction zone during the fluid-andesite interaction. Thus, the resulting Ca and Cl concentrations of upwelling and vent fluids are less than that of shallow seawater.

The As, Sb and Mn concentrations of most vent fluids are higher than those of hydrothermal plumes and shallow seawater (Table 2), and the As, Sb and Mn concentrations of most vent fluids are higher than those of hydrothermal plumes. This suggests that As, Sb and Mn in the hydrothermal fluids are leached from andesite during the fluid–rock interaction, and the mixing of the hydrothermal fluid and shallow seawater leads to the leaching of the As, Sb and Mn in the fluids into hydrothermal plumes.

In a plot of pH versus As, Sb and $\delta D_{V\text{-SMOW}}$, the linear correlation (Figure 4) indicates that the hydrothermal plumes are a two-component mixture of seawater and fluid. Therefore, the As, Sb and $\delta D_{V\text{-SMOW}}$ values of the Kueishantao hydrothermal fluids and plumes can be interpreted as the result of mixing seawater As, Sb and $\delta D_{V\text{-SMOW}}$ with fluid As and Sb extracted from the andesitic rocks.

5.4. Strontium Flux

The hydrothermal circulation under the sea floor leads to extensive chemical and isotopic exchange due to changes in fluid temperature. For Sr, the reaction direction changes from uptake by the rock at low temperature to release from the rock at high temperature [51]. However, the Sr and Cl are clearly like those of the hydrothermal fluids in the North Cleft segment on the Juan de Fuca Ridge [54] and those vented from the Plume vent at the South Cleft [55] (Figure 2). The Sr concentrations and $^{87}Sr/^{86}Sr$, $\delta D_{V\text{-SMOW}}$ and $\delta^{18}O_{V\text{-SMOW}}$ values of the white spring (51 °C) are like those of the yellow spring (108 °C), indicating that the exchange magnitude of Sr, H and O between the andesite and seawater is consistent within these two types of springs.

For calculating the Sr flux, it was presumed that the yellow and white springs have stable flow rates and that andesite is the only host rock in the reaction zone. Using measured Sr concentrations in hydrothermal fluids, the Sr flux calculated for the white spring vent into the ocean is between 2.06×10^4 and 2.26×10^4 mol/yr, and the Sr flux for the yellow spring vent is between 1.18×10^4 and 1.26×10^4 mol/yr (Table 3). Therefore, the Sr flux from the yellow spring is slightly higher than that from the white spring. Assuming that more than 30 vents occur in the Kueishantao hydrothermal field [42], the hydrothermal Sr flux is about 6×10^5 mol/yr with $F_{sr}{}^{sw}$ (fraction of Sr from seawater) values from 0.990 to 0.996, much higher than the $F_{sr}{}^{sw}$ values of 0.083—0.318 at hydrothermal vent fields on the Lau basin, Mariana Trough, Mid-Atlantic Ridge, Juan de Fuca Ridge and East Pacific Rise [24]. The difference also shows that there is a short lifetime of subseafloor fluid–rock interaction and/or short reaction paths at the Kueishantao hydrothermal field.

6. Conclusions

The element concentrations and $^{87}Sr/^{86}Sr$, $\delta D_{V\text{-SMOW}}$ and $\delta^{18}O_{V\text{-SMOW}}$ values of the Kueishantao hydrothermal fluids and plumes have smaller ranges of variation than those of fluids released from back-arc basins and mid-ocean ridges. The $^{87}Sr/^{86}Sr$ values and Sr concentrations of fluids range from 0.709154 to 0.709164 and 5.87 to 6.56 mg/L, respectively. For the plumes, the $^{87}Sr/^{86}Sr$ values range from 0.709147 to 0.709202, and the Sr concentrations range from 6.27 to 6.37 mg/L. From hydrothermal fluid to hydrothermal plume, the As, Sb and Mn concentrations decrease with the reduction of water depth, whereas Cl and Ca concentrations tend to increase. The As, Sb and Mn concentrations of the hydrothermal fluids and plumes of both yellow and white springs are higher than those of shallow seawater (As, Sb and Mn are 1.15, 0.20 and 0.91 mg/L), and the $^{87}Sr/^{86}Sr$ values of most vent fluids of the white spring are obviously less than that of shallow seawater. The variations of As, Sb and Mn concentration and $\delta D_{V\text{-}SMOW}$ values of the hydrothermal plumes can be interpreted by the mixing of vent fluids with seawater. Using As, Sb and H isotopes of hydrothermal plumes can describe their diffusive processes.

Sr, H and O in Kueishantao vent fluids are primarily derived from the ambient seawater. The As, Sb, Mn, Ca and Cl concentrations of hydrothermal fluid are influenced slightly by weak interactions between seawater and rock in the deep reaction zones, which is characterized by leaching of As, Sb and Mn from andesite. The hydrothermal flux of Sr in the Kueishantao hydrothermal field is between 1.18×10^4 and 2.26×10^4 mol/yr. Compared with the fluxes in other hydrothermal fields, the low flux might be caused by lower formation temperatures, higher seawater/andesite ratios (3076–8124), the shorter duration of seawater–andesite interaction and/or the shorter fluid paths.

However, most As, Sb, and Mn in the Kueishantao hydrothermal fluids are leached from andesitic rocks into the fluids. Most Ca and Cl remained in the reaction zone of the subseafloor hydrothermal field during seawater and rock interactions. At present, the sampling numbers of hydrothermal fluids, plumes and andesite are not enough. In the future, it will be possible to understand the material and heat flux of hydrothermal fluids and plumes formed by subseafloor magma chambers to seawater and their effect on seawater and ecologic environments by carrying out long-term continual observations.

Author Contributions: Conceptualization, data curation, formal analysis, funding acquisition, writing and editing, Z.Z.; methodology, formal analysis, data curation, writing—review and editing, X.W.; methodology, writing—review and editing, X.Y. and S.C.; data curation, methodology, validation, H.Q.; writing—review and editing, C.-T.A.C. All authors have read and agreed to the published version of the manuscript.

Funding: This work was supported by the NSFC Major Research Plan on West-Pacific Earth System Multi-spheric Interactions (grant number 91958213), the Strategic Priority Research Program of the Chinese Academy of Sciences (grant number XDB42020402), the Special Fund for the Taishan Scholar Program of Shandong Province (grant number ts201511061), the National Key Basic Research Program of China (grant number 2013CB429700), and the Science and Technology Development Program of Shandong Province (grant number 2013GRC31502).

Institutional Review Board Statement: Not applicable.

Informed Consent Statement: Not applicable.

Data Availability Statement: All the data that support the findings of this study are given in the main text.

Acknowledgments: We thank Bing-Jye Wang and Sea-watch Company for sampling the hydrothermal fluids and plumes.

Conflicts of Interest: The authors declare no conflict of interest.

References

1. Shanks, W.C., III; Seyfried, W.E., Jr. Stable isotope studies of vent fluids and chimney minerals, southern Juan de Fuca Ridge: Sodium metasomatism and seawater sulfate reduction. *J. Geophys. Res.* **1987**, *92*, 11387–11399. [CrossRef]
2. Berndt, M.E.; Seal, R.R.; Shanks, W.C.; Seyfried, W.E., Jr. Hydrogen isotope systematics of phase separation in submarine hydrothermal systems: Experimental calibration and theoretical models. *Geochim. Cosmochim. Acta* **1996**, *60*, 1595–1604. [CrossRef]
3. Liakhovitch, V.; Quick, J.E.; Gregory, R.T. Hydrogen and Oxygen Isotope Constraints on Hydrothermal Alteration of the Trinity Peridotite, Klamath Mountains, California. *Int. Geol. Rev.* **2005**, *47*, 203–214. [CrossRef]
4. Gena, K.R.; Chiba, H.; Mizuta, T.; Matsubaya, O. Hydrogen, oxygen and sulfur isotope studies of seafloor hydrothermal system at the Desmos caldera, Manus back-arc basin, Papua New Guinea: An analogue of terrestrial acid hot crater-lake. *Resour. Geol.* **2006**, *56*, 183–190. [CrossRef]
5. Voigt, M.; Pearce, C.R.; Baldermann, A.; Oelkers, E.H. Stable and radiogenic strontium isotope fractionation during hydrothermal seawater-basalt interaction. *Geochim. Cosmochim. Acta* **2018**, *240*, 131–151. [CrossRef]
6. Kawahata, H.; Kusakabe, M.; Kikuchi, Y. Strontium, oxygen, and hydrogen isotope geochemistry of hydrothermally altered and weathered rocks in DSDP Hole 504B, Costa Rica Rift. *Earth Planet. Sci. Lett.* **1987**, *85*, 343–355. [CrossRef]

7. Ravizza, G.; Blusztajn, J.; Von Damm, K.L.; Bray, A.M.; Bach, W.; Hart, S.R. Sr isotope variations in vent fluids from 9°46′–9°54′ N East Pacific Rise: Evidence of a non-zero-Mg fluid component. *Geochim. Cosmochim. Acta* **2001**, *65*, 729–739. [CrossRef]

8. Dekov, V.M.; Cuadros, J.; Shanks, W.C.; Koski, R.A. Deposition of talc-kerolite-smectite-smectite at seafloor hydrothermal vent fields: Evidence from mineralogical, geochemical and oxygen isotope studies. *Chem. Geol.* **2008**, *247*, 171–194. [CrossRef]

9. Wilckens, F.K.; Reeves, E.P.; Bach, W.; Seewald, J.S.; Kasemann, S.A. Application of B, Mg, Li and Sr isotopes in acid-sulfate vent fluids and volcanic rocks as tracers for fluid-rock interaction in back-arc hydrothermal systems. *Geochem. Geophys. Geosyst.* **2019**, *20*, 5849–5866. [CrossRef]

10. Palmer, M.R. Controls over the chloride concentration of submarine hydrothermal vent fluids: Evidence from Sr/Ca and ^{87}Sr/^{86}Sr ratios. *Earth Planet. Sci. Lett.* **1992**, *109*, 37–46. [CrossRef]

11. Webber, A.P.; Roberts, S.; Burgess, R.; Boyce, A.J. Fluid mixing and thermal regimes beneath the PACMANUS hydrothermal field, Papua New Guinea: Helium and oxygen isotope data. *Earth Planet. Sci. Lett.* **2011**, *304*, 93–102. [CrossRef]

12. Jean-Baptise, P.; Dapoigny, A.; Stievenard, M.; Charlou, J.L.; Fouquet, Y.; Donval, J.P.; Auzende, J.M. Helium and oxygen isotope analyses of hydrothermal fluids from the East Pacific Rise between 17° S and 19° S. *Geo-Mar. Lett.* **1997**, *17*, 213–219. [CrossRef]

13. Hodell, D.A.; Mead, G.A.; Mueller, P.A. Variation in the strontium isotopic composition of seawater (8 Ma to present): Implications for chemical weathering rates and dissolved fluxes to the oceans. *Chem. Geol. Isot. Geosci. Sect.* **1990**, *80*, 291–307. [CrossRef]

14. Davis, A.C.; Bickle, M.J.; Teagle, D.A.H. Imbalance in the oceanic strontium budget. *Earth Planet. Sci. Lett.* **2003**, *211*, 173–187. [CrossRef]

15. Palmer, M.R.; Edmond, J.M. The strontium isotope budget of the modern ocean. *Earth Planet. Sci. Lett.* **1989**, *92*, 11–26. [CrossRef]

16. Butterfield, D.A.; Nelson, B.K.; Wheat, C.G.; Mottl, M.J.; Roe, K.K. Evidence for basaltic Sr in midocean ridge-flank hydrothermal systems and implications for the global oceanic Sr isotope balance. *Geochim. Cosmochim. Acta* **2001**, *65*, 4141–4153. [CrossRef]

17. Coogan, L.A.; Dosso, S.E. Alteration of ocean crust provides a strong temperature dependent feedback on the geological carbon cycle and is a primary driver of the Sr isotopic composition of seawater. *Earth Planet. Sci. Lett.* **2015**, *415*, 38–46. [CrossRef]

18. Antonelli, M.A.; Pester, N.J.; Brown, S.T.; DePaolo, D.J. Effect of paleoseawater composition on hydrothermal exchange in midocean ridges. *Proc. Natl. Acad. Sci. USA* **2017**, *114*, 12413–12418. [CrossRef]

19. Horita, J.; Cole, D.R.; Weslowski, D.J. The activity composition relationship of oxygen and hydrogen isotopes in aqueous salt solutions: III. Vapor-liquid water equilibration of NaCl solutions to 350 °C. *Geochim. Cosmochim. Acta* **1995**, *6*, 1139–1152. [CrossRef]

20. Campbell, A.C.; Palmer, M.R.; Klinkhammer, G.P.; Bowers, T.S.; Edmond, J.M.; Lawrence, J.R.; Casey, J.F.; Thompson, G.; Humphris, S.; Rona, P.; et al. Chemistry of hot springs on the Mid-Atlantic Ridge. *Nature* **1988**, *335*, 514–519. [CrossRef]

21. East-Pacific-Rise-Study-Group. Crustal processes of the mid-ocean ridge. *Science* **1981**, *213*, 31–40. [CrossRef] [PubMed]

22. Welhan, J.A.; Craig, H. Methane, hydrogen and helium in hydrothermal fluids at 21° N on the East Pacific Rise. In *Hydrothermal Processes at Seafloor Spreading Centers*; NATO conference series: IV, Marine Sciences; Rona, P.A., Bostrom, K., Laubier, L., Smith, K.L., Jr., Eds.; Plenum Press: New York, NY, USA, 1983; pp. 391–409.

23. Teagle, D.A.H.; Alt, J.C.; Chiba, H.; Humphris, S.E.; Halliday, A.N. Strontium and oxygen isotopic constraints on fluid mixing, alteration and mineralization in the TAG hydrothermal deposit. *Chem. Geol.* **1998**, *149*, 1–24. [CrossRef]

24. Bach, W.; Humphris, S.E. Relationship between the Sr and O isotope compositions of hydrothermal fluids and the spreading and magma-supply rates at oceanic spreading centers. *Geology* **1999**, *27*, 1067–1070. [CrossRef]

25. Shanks, W.C., III; Bohlke, J.K.; Seal, R.R., II. Stable isotopes in Mid-Ocean Ridge Hydrothermal Systems: Interactions between fluids, minerals and organisms. In *Physical, Chemical, Biological, and Geological Interactions within Hydrothermal Systems*; Geophysical Monograph Series; Humphris, R.A., Zierenberg, R.A., Mullineaux, L.S., Thomson, R.E., Eds.; American Geophysical Union: Washington, DC, USA, 1995; Volume 91, pp. 194–221.

26. Muehlenbachs, K.; Clayton, R.N. Oxygen isotope geochemistry of submarine greenstones. *Can. J. Earth Sci.* **1972**, *9*, 471–478. [CrossRef]

27. Stakes, D.S.; O'Neil, J.R. Mineralogy and stable isotope geochemistry of hydrothermally altered oceanic rocks. *Earth Planet. Sci. Lett.* **1982**, *57*, 285–304. [CrossRef]

28. Bowers, T.S.; Taylor, H.P., Jr. An Integrated chemical and stable isotope model of the origin of mid-ocean ridge hot spring systems. *J. Geophys. Res.* **1985**, *90*, 12583–12606. [CrossRef]

29. Cole, D.R.; Mottl, M.J.; Ohmoto, H. Isotopic exchange in mineral-fluid systems: II. Oxygen and hydrogen isotopic investigation of the experimental basalt-seawater system. *Geochim. Cosmochim. Acta* **1987**, *51*, 1523–1538. [CrossRef]

30. Bowers, T.S. Stable isotope signatures of water-rock interaction in mid-ocean ridge hydrothermal systems: Sulfur, oxygen and hydrogen. *J. Geophys. Res.* **1989**, *94*, 5775–5786. [CrossRef]

31. Böhlke, J.K.; Shanks, W.C., III. Stable isotope study of hydrothermal vents at Escanaba Trough, NE Pacific: Observed and calculated effects of sediment-seawater interaction. In *Geologic, Hydrothermal, and Biologic Studies at Escanaba Trough, Gorda Ridge, Offshore Northern California, Gorda Ridge*; Morton, J.L., Zierenberg, R.A., Reiss, C.A., Eds.; U.S. Geological Survey Bull: Reston, VA, USA, 2022; Volume 1994, pp. 223–239.

32. Fouquet, Y.; Von Stackelberg, U.; Charlou, J.L.; Donval, J.P.; Erzinger, J.; Foucher, J.P.; Herzig, P.; Mühe, R.; Soakai, S.; Wiedicke, M.; et al. Hydrothermal activity and metallogenesis in the Lau back-arc basin. *Nature* **1991**, *349*, 778–781. [CrossRef]

33. James, R.H.; Elderfield, H.; Palmer, M.R. The chemistry of hydrothermal fluids from the Broken Spur site, 29°N Mid-Atlantic Ridge. *Geochim. Cosmochim. Acta* **1995**, *59*, 651–659. [CrossRef]

34. Schmidt, K.; Garbe-Schönberg, D.; Koschinsky, A.; Strauss, H.; Jost, C.L.; Klevenz, V.; Königer, P. Fluid elemental and stable isotope composition of the Nibelungen hydrothermal field (8°18′ S, Mid-Atlantic Ridge): Constraints on fluid-rock interaction in heterogeneous lithosphere. *Chem. Geol.* **2011**, *280*, 1–18. [CrossRef]

35. Mottl, M.J.; Seewald, J.S.; Wheat, C.G.; Tivey, M.K.; Michael, P.J.; Proskurowski, G.; McCollom, T.M.; Reeves, E.; Sharkey, J.; You, C.-F.; et al. Chemistry of hot springs along the Eastern Lau Spreading Center. *Geochim. Cosmochim. Acta* **2011**, *75*, 1013–1038. [CrossRef]

36. Chen, Y.-G.; Wu, W.-S.; Chen, C.-H.; Liu, T.-K. A date for volcanic eruption inferred from a siltstone xenolith. *Quat. Sci. Rev.* **2001**, *20*, 869–873. [CrossRef]

37. Chen, C.-H.; Lee, T.; Shieh, Y.-N.; Chen, C.-H.; Hsu, W.-Y. Magmatism at the onset of back-arc basin spreading in the Okinawa Trough. *J. Volcanol. Geotherm. Res.* **1995**, *69*, 313–322.

38. Zeng, Z.; Li, X.; Chen, S.; Zhang, Y.; Chen, Z.; Chen, C.-T.A. Iron-Copper-Zinc Isotopic Compositions of Andesites from the Kueishantao Hydrothermal Field off Northeastern Taiwan. *Sustainability* **2022**, *14*, 359. [CrossRef]

39. Zeng, Z.G.; Liu, C.H.; Chen, C.-T.A.; Yin, X.B.; Chen, D.G.; Wang, X.Y.; Wang, X.M.; Zhang, G.L. Origin of a native sulfur chimney in the Kueishantao hydrothermal field, offshore northeast Taiwan. *Sci. China (Ser. D Earth Sci.)* **2007**, *50*, 1746–1753. [CrossRef]

40. Zeng, Z.G.; Chen, C.-T.A.; Yin, X.B.; Zhang, X.Y.; Wang, X.Y.; Zhang, G.L.; Wang, X.M.; Chen, D.G. Origin of native sulfur ball from the Kueishantao hydrothermal field offshore northeast Taiwan: Evidence from trace and rare earth element composition. *J. Asian Earth Sci.* **2011**, *40*, 661–671. [CrossRef]

41. Zeng, Z.G.; Wang, X.Y.; Chen, C.-T.A.; Yin, X.B.; Chen, S.; Ma, Y.Q.; Xiao, Y.K. Boron isotope compositions of fluids and plumes from the Kueishantao hydrothermal field off northeastern Taiwan: Implications for fluid origin and hydrothermal processes. *Mar. Chem.* **2013**, *157*, 59–66. [CrossRef]

42. Chen, C.T.A.; Zeng, Z.G.; Kuo, F.W.; Yang, T.F.; Wang, B.J.; Tu, Y.Y. Tide-influenced acidic hydrothermal system offshore NE Taiwan. *Chem. Geol.* **2005**, *224*, 69–81. [CrossRef]

43. Kuo, F.W. Preliminary Investigation of Shallow Hydrothermal Vents on Kueishantao Islet of Northeastern Taiwan. Master's Thesis, Institute of Marine Geology and Chemistry, National Sun Yat-Sen University, Kaohsiung, Taiwan, China, 2001; p. 81. (In Chinese).

44. Chen, C.T.A.; Wang, B.J.; Huang, J.F.; Lou, J.Y.; Kuo, F.W.; Tu, Y.Y.; Tsai, H.S. Investigation into extremely acidic hydrothermal fluids off Kueishantao islet, Taiwan. *Acta Oceanol. Sin.* **2005**, *24*, 125–133.

45. Chen, X.-G.; Lyu, S.-S.; Garbe-Schönberg, D.; Lebrato, M.; Li, X.; Zhang, H.-Y.; Zhang, P.-P.; Chen, C.-T.A.; Ye, Y. Heavy metals from Kueishantao shallow-sea hydrothermal vents, offshore northeast Taiwan. *J. Mar. Syst.* **2018**, *180*, 211–219. [CrossRef]

46. Chen, X.G.; Zhang, H.Y.; Li, X.H.; Chen, C.-T.A.; Yang, T.F.; Ye, Y. The chemical and isotopic compositions of gas discharge from shallow-water hydrothermal vents at Kueishantao, offshore northeast Taiwan. *Geochem. J.* **2016**, *50*, 341–355. [CrossRef]

47. Zeng, Z.; Wang, X.; Qi, H.; Zhu, B. Arsenic and Antimony in Hydrothermal Plumes from the Eastern Manus Basin, Papua New Guinea. *Geofluids* **2018**, *2018*, 6079586. [CrossRef]

48. Chen, F.; Li, X.-H.; Wang, X.-L.; Li, Q.-L.; Siebel, W. Zircon age and Nd-Hf isotopic composition of the Yunnan Tethyan belt, southwestern China. *Int. J. Earth Sci.* **2007**, *96*, 1179–1194. [CrossRef]

49. Chen, F.; Zhu, X.-Y.; Wang, W.; Wang, F.; Pham, T.-H.; Siebel, W. Single-grain detrital muscovite Rb-Sr isotopic composition as an indicator of provenance for the Carboniferous sedimentary rocks in northern Dabie, China. *Geochem. J.* **2009**, *43*, 257–273. [CrossRef]

50. Von Damm, K.L.; Edmond, J.M.; Grant, B.; Measures, C.I.; Walden, B.; Weiss, R.F. Chemistry of submarine hydrothermal solutions at 21° N, East Pacific Rise. *Geochim. Cosmochim. Acta* **1985**, *49*, 2197–2220. [CrossRef]

51. Spivack, A.J.; Edmond, J.M. Boron isotope exchange between seawater and the oceanic crust. *Geochim. Cosmochim. Acta* **1987**, *51*, 1033–1043. [CrossRef]

52. Alt, J.C.; Honnorez, J.; Laverne, C.; Emmermann, R. Hydrothermal alteration of 1 km section through the upper oceanic crust, deep sea drilling project Hole 504B: Mineralogy, chemistry, and evolution of seawater-basalt interaction. *J. Geophys. Res.* **1986**, *91*, 10309–10335. [CrossRef]

53. Patten, C.G.C.; Pitcairn, I.K.; Teagle, D.A.H.; Harris, M. Mobility of Au and related elements during the hydrothermal alteration of the oceanic crust: Implications for the sources of metals in VMS deposits. *Miner. Depos.* **2015**, *51*, 179–200. [CrossRef]

54. Butterfield, D.A.; Massoth, G.J. Geochemistry of north Cleft segment vent fluids: Temporal changes in chlorinity and their possible relation to recent volcanism. *J. Geophys. Res.* **1994**, *99*, 4951–4968. [CrossRef]

55. Von Damm, K.L. Seafloor hydrothermal activity: Black smoker chemistry and chimneys. *Annu. Rev. Earth Planet. Sci. Lett.* **1990**, *18*, 173–204. [CrossRef]

Journal of
Marine Science and Engineering

Article

The Geochemical Features and Genesis of Ferromanganese Deposits from Caiwei Guyot, Northwestern Pacific Ocean

Linzhang Wang [1,2] and Zhigang Zeng [1,2,3,*]

[1]   Seafloor Hydrothermal Activity Laboratory, CAS Key Laboratory of Marine Geology and Environment, Institute of Oceanology, Chinese Academy of Sciences, Qingdao 266071, China
[2]   University of Chinese Academy of Sciences, Beijing 100049, China
[3]   Laboratory for Marine Mineral Resources, Qingdao National Laboratory for Marine Science and Technology, Qingdao 266071, China
*   Correspondence: zgzeng@ms.qdio.ac.cn

Abstract: The ferromanganese deposit is a type of marine mineral resource rich in Mn, Fe, Co, Ni, and Cu. Its growth process is generally multi-stage, and the guyot environment and seawater geochemical characteristics have a great impact on the growth process. Here, we use a scanning electron microscope, X-ray diffraction (XRD), inductively coupled plasma optical emission spectrometer (ICP-OES), X-ray fluorescence (XRF), and inductively coupled plasma mass spectrometry (ICP-MS) to test and analyze the texture morphology, microstructure, mineralogical features, geochemical features of ferromanganese crusts deposits at different distribution locations on Caiwei Guyot. The ferromanganese deposits of Caiwei Guyot are ferromanganese nodules on the slope and board ferromanganese crusts on the mountaintop edge, which are both of hydrgenetic origin. Hydrgenetic origin reflects that the metal source is oxic seawater. Global palaeo-ocean events control the geochemistry compositions and growth process of ferromanganese crusts and the nodule. Ferromanganese crusts that formed from the late Cretaceous on the mountaintop edge have a rough surface with black botryoidal shapes, showing an environment with strong hydrodynamic conditions, while the ferromanganese nodule that formed from the Miocene on the slope has an oolitic surface as a result of water depth. What is more, nanoscale or micron-scale diagenesis may occur during the growth process, affecting microstructure, mineralogical and geochemical features.

Keywords: Caiwei Guyot; ferromanganese crust; ferromanganese nodule; geochemistry

Citation: Wang, L.; Zeng, Z. The Geochemical Features and Genesis of Ferromanganese Deposits from Caiwei Guyot, Northwestern Pacific Ocean. *J. Mar. Sci. Eng.* **2022**, *10*, 1275. https://doi.org/10.3390/jmse10091275

Academic Editor: Assimina Antonarakou

Received: 22 July 2022
Accepted: 5 September 2022
Published: 9 September 2022

Publisher's Note: MDPI stays neutral with regard to jurisdictional claims in published maps and institutional affiliations.

1. Introduction

Ferromanganese deposits are common deep-sea mineral resources, which mainly exist in the form of ferromanganese crust and ferromanganese nodules. Ferromanganese crusts primarily grow on rock surfaces located without sediment on the seamount's ridges, and plateaus at water depths of 400–7000 m [1,2], while ferromanganese nodules mainly grow on the surface of sediment-covered abyssal plains and the slope of seamounts at water depths of approximately 3500 to 6500 m [1]. They are rich in metals such as Co, Ni, Cu, Pb, and Zn, rare earth elements (REEs), and platinum group elements (PGEs), which have high economic value and resource potential [1]. Ferromanganese deposits in the ocean are primarily divided into three types: hydrogenetic, diagenetic and hydrothermal [3]. The metal ions of hydrogenetic ferromanganese deposits are sourced from seawater and form ferromanganese oxides by colloidal precipitation under oxidation conditions. The growth rates of hydrogenetic ferromanganese deposits are extremely slow (1–10 mm/Ma) [3,4]. The metal ions of diagenetic ferromanganese deposits are derived from pore water in sediments or sediment water under suboxidation conditions [5]. The metal ions of hydrothermal ferromanganese deposits are sourced from the mixture of medium–low temperature hydrothermal fluid with seawater [6,7]. In addition, the formation of ferromanganese deposits is often hydrogenetic–diagenetic and hydrogenetic–hydrothermal mixed types [8–10]. Ferromanganese deposits are regarded as a hydrogenetic type [11–13] formed in paleo-ocean and paleo-environments, recording the evolution of ocean and climate for the past 60–100 Ma, and are important in large amounts of paleo-ocean and paleo-environment information [2,14]. In recent years, several studies have shown that microorganisms are also involved in the formation of marine ferromanganese oxides [15–17]. After metals are adsorbed to ferromanganese oxides, the precipitation of Co can be influenced by bacterial activity. Microorganisms as catalysts not only promote the formation of ferromanganese deposits, but also the enrichment of metals [17].

Ferromanganese deposits are distributed in the Pacific, Atlantic, Arctic, and Indian oceans, though most are in the Pacific Ocean [1]. Scholars have also performed a significant amount of research on these deposits [9,18–21]. The formation and mineral and geochemistry compositions of hydrogenetic ferromanganese deposits are the most common in the Pacific [1]. The seamounts of the Western and Central Pacific are the main distribution areas of ferromanganese deposits. Caiwei Guyot, the location of samples collected in this study, is located in the Magellan Seamounts to the east of the Mariana Trench.

In this paper, a comprehensive study on the ferromanganese deposits (ferromanganese crusts on the mountaintop edge and the nodule on the slope) of Caiwei Guyot has been carried out based on detailed mineralogical and geochemical analysis including major elements, trace elements and rare earth elements. The growth rates and ages of ferromanganese deposits are estimated by Co-Chronometer. Finally, we propose the formation model of ferromanganese deposits by combining them with the paleoceanographic events in the Northwest Pacific since the late Cretaceous.

2. Geological Setting and Oceanography

The Caiwei seamounts are located along the northeast edge of the eastern Mariana basin in the western Pacific. They are distributed in a NW chain in the Magellan seamounts, which consists of the Vlinder, Loah, ITA Mai Tai, other seamounts, with a northeastern trend. The Caiwei seamounts include two relatively independent guyots: the larger main guyot is Caiwei Guyot, and the smaller subsidiary guyot is Caiqi Guyot. The distance between the two guyots is 10.2 km. Substrate rocks are mainly composed of tholeiitic basaltic pillow lavas from the early Cretaceous, subalkaline basalt from the early Cretaceous to Paleogene, alkaline basalt and pyroclastic rocks from the Neogene; sedimentary caprocks are mainly composed of bioherm limestone and mudstone from the early Cretaceous to Paleocene, foraminiferal limestone from the Eocene to the early Miocene and calcium and argillaceous sediments such as foraminiferal sand and soft mud from the Tertiary to the Holocene [22–27]. The age of the guyot is 80–120 Ma. The thickness of the overlying sedimentary layer is approximately 25–100 m. The average slope of the Caiwei Guyot is 20°–30° [28]. There are collapsed flanks and landslide topography on the Caiwei Guyot [28,29]. The existing survey results show that ferromanganese deposits are primarily distributed on the mountaintop edge and slope [29]. The samples used in this study were collected from the mountaintop edge and the slope of Caiwei Guyot (Figure 1c).

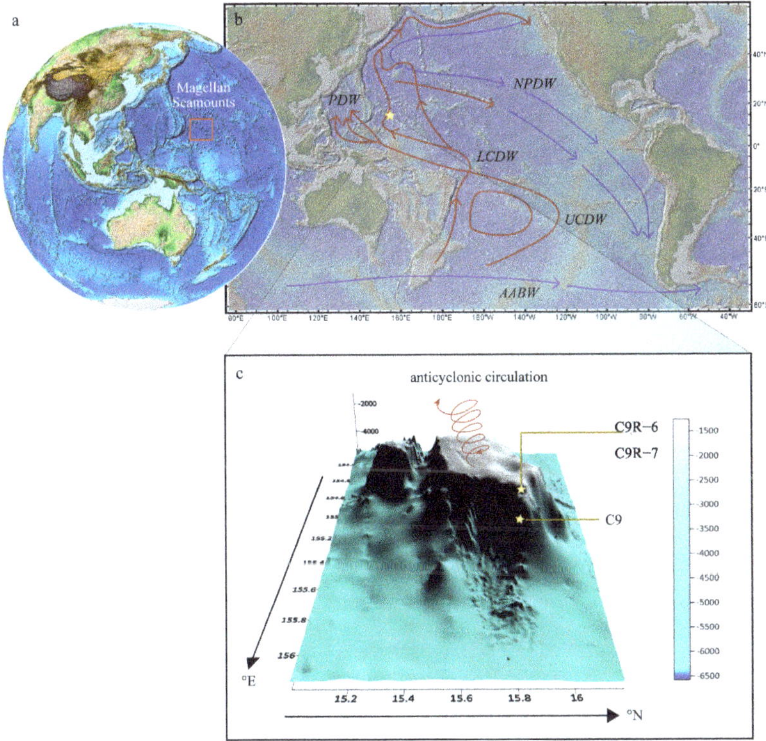

Figure 1. (**a**) The location of the study area. (**b**) The location of the study area and the modern Pacific Ocean circulation system, the yellow star represents Caiwei Guyot. The red and blue lines represent the lower circumpolar deep water (LCDW), the upper circumpolar deep water (UCDW), the Pacific deep water (PDW), the north Pacific deep water (NPDW), and the Antarctic bottom water (AABW) (modified by [30]). (**c**) Topographic map of Caiwei Guyot (data are quoted from https://www.gebco.net on 15 July 2022), the yellow star represents HOBAB5-C9, HOBAB5-C9R-6, HOBAB5-C9R-7 the red line represents clockwise anticyclone circulation (Taylor Column).

Caiwei Guyot is located in the Magellan seamounts of the eastern Mariana Basin (Figure 1a). The deep-water system in the western Pacific Ocean was primarily controlled by CDW (circumpolar deep water) and AABW (Antarctic bottom water) in the Southern Ocean [30]. CDW can be divided into upper circumpolar deep water (UCDW) with maximum temperature and lower circumpolar deep water (LCDW) with maximum salinity [31]. LCDW is mixed by AABW and NPDW (north Pacific deep water) and can upwell the depth of 2000 m–3000 m [32]. Due to the complex submarine topography of the low latitude western Pacific, after LCDW enters the Pacific, it is divided into eastern branch current and western branch current [33–38]. The eastern branch partly enters the Central Pacific Basin with a volume transport of 3.7 Sv (1 Sv − 10^6 m^3 s^{-1}) [39]. Additionally, the other part enters Northwest Pacific Basin through the Wake Island Passage [36,37]. The western branch current passes through the Magellan Seamounts and the Mid-Pacific Seamounts with a volume transport of 2.1 Sv [36]. UCDW and LCDW become the older NPDW by the mixing of deep water [36], and NPDW enters the western Pacific and the Philippine sea again [30] (Figure 1b). LCDW can erode and sweep sediments on Caiwei Guyot, resulting in substrate exposure that provides an ideal place for the growth of ferromanganese crusts and sediment discontinuity [28,40,41]. LCDW has high oxidation, creating an oxidation environment for the guyot, promoting the reaction of metal ions, and forming the precipitation of oxides and hydroxides [42]. The topography of Caiwei Guyot can promote the strengthening of internal waves to a certain extent, which can promote the mixing of deep water [43–45]. Internal waves (clockwise anticyclonic circulation called Taylor Column phenomenon) in Caiwei Guyot [40] (Figure 1c) can also communicate oxygen minimum zone (OMZ) with oxygen-enriched and iron-enriched deep water [27]. The metal ions such as Mn^{2+} and Co^{2+} in OMZ can be oxidized better [27].

3. Samples and Methods

The three ferromanganese deposit samples are collected by TV grab and Remote Operated Vehicle (ROV) on the Caiwei Guyot in Magellan seamounts during the HOBAB5 cruise expedition in 2018 (Figure 1a,b). The station information is shown in Table 1 and Figure 1. C9R-6 and C9R-7 are collected by ROV, and C9 is collected by TV grab.

Table 1. Location of samples from Caiwei Guyot.

Cruise ID	Water Depth	Latitude (N)	Latitude (E)
HOBAB5-C9	3116 m	15°51′10.883″	155°35′15.620″
HOBAB5-C9R-6	1650.1 m	15°51′08.672″	155°29′17.412″
HOBAB5-C9R-7	1608.4 m	15°51.16997′	155°29.20822′

The ferromanganese deposit samples were washed with clean water and dried, and photos were taken on a large scale. Then, the samples were divided into two parts with a cutter, cleaned with ultrapure water, and put into the oven for 60 min. After being dried, one is used to make probe pieces for microscopic observation, and the other is divided into many layers due to the internal structure along the growth direction as shown in Figure 2. Each layer is independent and continuous. Grind them to 200 mesh with agate mortar, and then place them into the oven for 60 min. After being dried, the samples were transferred to the dryer.

a C9R-6

Figure 2. *Cont.*

b C9R-6

c C9R-7

Figure 2. *Cont.*

d C9R-7

e C9

Figure 2. *Cont.*

f C9

Figure 2. (**a**) Internal texture of C9R-6, which is composed of obvious changes of layers that are are mostly composed of dense ferromanganese oxide layers with low porosity and loose ferromanganese oxide layers with high porosity filled with siliceous debris, but the layers are not nearly parallel; (**b**) botryoidal shapes whose diameter is 1–2 cm on the surface; (**c**) internal texture of C9R-7, which is composed of obvious changes of nearly parallel layers that are mostly composed of dense ferromanganese oxide layers with low porosity and loose ferromanganese oxide layers with high porosity filled with siliceous debris; (**d**) botryoidal shapes whose diameter is 1–2 cm on the surface; (**e**) internal texture of C9, which is composed of a substrate as the nucleus inside, and outer layers composed of ferromanganese oxides; (**f**) smooth oolitic surface structure where siliceous debris can be seen in the cracks.

3.1. Mineralogy

The microscopic analysis of the samples was conducted in the Key Laboratory of Marine Geology and Environment, Institute of Oceanography, Chinese Academy of Sciences. The samples were carbon-coated and observed by a VECE3 TESCAN scanning electron microscope combined with an Oxford EDS Oxford X-ray spectrometer (UK). The scanning electron microscope instrument used 20 kV high voltage, the emission current was 1.4–1.9 nA, and the working distance was 15 mm. The X-ray spectrometer used 20 kV excitation voltage. The mineral composition was determined by Rigaku D/MAX 2500 PC 18 kw X-ray diffraction (XRD) in the Key Laboratory of Marine Mineral Resources of Ministry of Land and Resources, Guangzhou Marine Geological Survey. The analytical conditions for XRD were: Cu Kα radiation at 40 kV and 300 mA, graphite monochromator. The speed of continuous scanning was 2.5° (2θ)/min, the step was 0.01° (2θ) and scans ranged from 5 to 75° (2θ). Ambient temperature was 25 ± 2 °C.

3.2. Geochemistry

The composition of major element of ferromanganese nodule was carried out by 5300 DV plasma emission spectrometer (ICP-OES) in the Analytical Laboratory, Beijing Research Institute of Uranium Geology and AB104L based on "Methods for chemical analysis of silicate rocks—Part 32: Determination of 20 components including aluminium oxide etc.-Mixed acid digestion-inductively coupled plasma atomic emission spectrometer", while the composition of major element of ferromanganese crusts was carried out by Axios-mAX wavelength dispersive X-ray fluorescence spectrometer (XRF) in the Analytical Laboratory, Beijing Research Institute of Uranium Geology based on GB/T 14506.28-2010 "Methods for chemical analysis of silicate rocks—Part 28: Determination of 16 major and minor elements content" due to different powder quality. The precision of the analysis was better than that of $\pm5\%$. The compositions of trace and rare earth elements were analyzed by ELAN 9000 inductively coupled plasma mass spectrometry (ICP-MS) in the Key Laboratory of Marine Geology and Environment, Institute of Oceanography, Chinese Academy of Sciences. Firstly, the sample was dried in the oven at 105 °C for 2 h, then approximately 40 mg powdered samples were put into Teflon digestion beakers, then 0.5 mL of HF, 0.5 mL of HNO_3, and 1.5 mL of HCl were added into Teflon digestion beakers. Then, the powder was sealed and heated in an electric oven and heated at 150 °C for 12 h. After cooling, the cover of the beakers was opened, and the beaker was heated on a hot plate to dryness. After the acid in the beakers was evaporated to dryness, 1 mL of HNO_3 was added, and then the acid was evaporated to dryness without sealing. After steaming dry, 1 mL of HNO_3 and 1 mL of H_2O were added to the beakers. Then, the beakers were again sealed and heated in an electric oven at 150 °C to dissolve the residue. Finally, H_2O was added to 40 g before the compositions of trace and rare earth elements were measured by ICP-MS. The digestion process of standard samples, blank samples, and parallel samples are the same as above. Re element was used as the internal standard, and the external standard samples included (GBW07315; GBW07316; BCR-2; BHVO-2; Nod-A-1, Nod-P-1, GBW07295, GBW07296). The precision of the analysis was better than that of $\pm5\%$. The test accuracy was better than 5%.

The pattern diagram of rare earth elements is made according to concentrations normalized to the Post Archean Australian Shale (PAAS) values. $Ce_{SN}/Ce_{SN}{}^{*} = 2Ce_{SN}/(La_{SN} + Pr_{SN})$, $Eu_{SN}/Eu_{SN}{}^{*} = 2Eu_{SN}/(Sm_{SN} + Gd_{SN})$, where SN means concentrations normalized to the PAAS values [46].

The following formula is used to calculate the growth rate of the ferromanganese crusts: $GR = 0.68/(Co_n)^{1.67}$ [47], where $Co_n = Co \times 50/(Fe + Mn)$ with Co, Fe, and Mn expressed as weight percent (wt.%). Pearson correlation coefficient is used to measure the linear relationship between two variables. Statistical significance will be at either 99% or 95% (CL). Q-mode factor analysis is used to analyze the relationship between each element and the influencing factors in order to determine which elements are in different factors. Here, we regard each factor as different mineral phases in ferromanganese crusts, and the elements in the factor exist in mineral phases.

4. Results

4.1. Texture and Microstructure

4.1.1. Textural Morphology

C9R-6 is a board ferromanganese crust that is about 10 cm long (Figure 2a,b). There are black botryoidal shapes whose diameter is 1–2 cm on the surface (Figure 2b). Obvious changes of layers can be seen inside, but the layers are not nearly parallel (Figure 2a). C9R-6 is divided into six layers due to the internal structure (Figure 2a): the layer composed of outermost black ferromanganese oxides is divided into C9R-6-1 and C9R-6-2; the layer composed of black ferromanganese oxides consolidated with yellow detrital minerals is divided into C9R-6-3, C9R-6-4, and C9R-6-5, which have high porosity (Figure 2a). The ferromanganese oxides at the bottom are C9R-6-6.

C9R-7 is also a board ferromanganese crust with a length of about 10 cm (Figure 2c,d). The crust has a rough surface on which black botryoidal shapes develop (Figure 2d). The diameter of each botryoidal ball is about 1–2 cm (Figure 2d). Obvious changes of nearly parallel layers can be seen inside (Figure 2c), which are mostly composed of dense ferromanganese oxide layers with low porosity and loose ferromanganese oxide layers with high porosity filled with siliceous debris. The layers grow orderly and have obvious rhythmic characteristics. C9R-7 is divided into six layers due to the internal structure (Figure 2c).

C9 is a ferromanganese nodule whose diameter is about 4 cm (Figure 2e,f), and the surface is an oolitic structure (Figure 2f). White siliceous debris can be seen in the cracks on the surface (Figure 2f). There is a substrate as the nucleus inside, and ferromanganese oxides precipitate as the outer layer (Figure 2e). In the early growth stage, the bottom structure is dark, dense, and relatively hard, while in the late growth stage, the top structure is relatively loose with high porosity (Figure 2e). Siliceous debris can be seen in the gaps (Figure 2e). It is divided into one layer every 2 mm along the growth direction with an average of five layers. The top layer is stage II (C9-1 and C9-2), and the bottom layer is stage I (C9-3, C9-4, and C9-5).

4.1.2. Microstructure

The microstructure of ferromanganese deposits is shown in Figures 3 and 4. Columnar structure, laminar structure, and mottled structure are mainly developed in ferromanganese deposits on Caiwei Guyot (Figures 3 and 4).

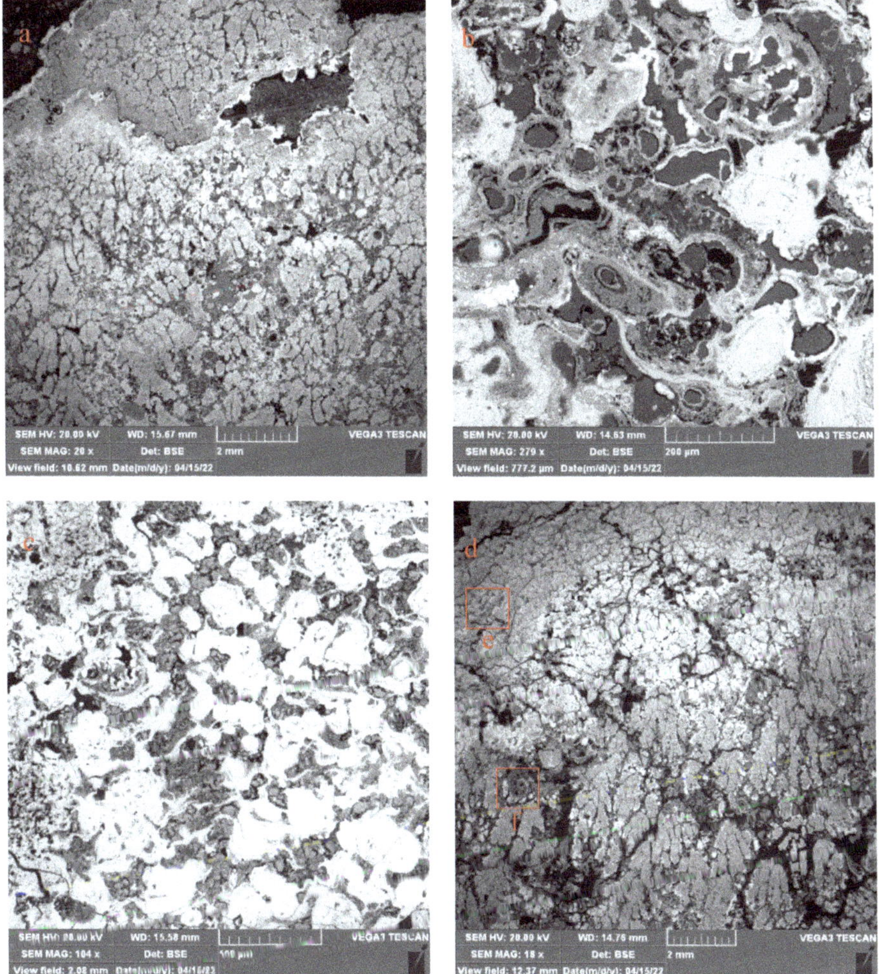

Figure 3. *Cont.*

Figure 3. (**a**) Disorderly arranged columns form a shape similar to dendrite; (**b**) banded Si-Fe-rich cements; (**c**) mottled structure composed of siliceous debris and ferromanganese oxides; (**d**) disorderly arranged columns developed into columnar structure; (**e**) laminar structure composed of ferromanganese oxides in Figure 3d; (**f**) foraminifera entered in gaps between columns in Figure 3d, and Mn-Fe rich oxides precipitated around it over time.

Along the growth direction of C9R-6, disorderly arranged columns can be seen along the growth direction. Branches appear in the growth process of the columns, forming a shape similar to the dendrite (Figure 3a,b). Banded Si-Fe-rich cement exists in the gaps (Figure 3b). Different from C9R-6, there are many types of structures in C9R-7. Along the growth direction, a mottled structure composed of siliceous debris and ferromanganese oxides can be seen (Figure 3c,e). Then, disorderly arranged columns developed into a columnar structure with low porosity and gradually formed a dendritic shape like C9R-6 (Figure 3d). In the later stage of the growth process, the columnar structure is rarely seen and gradually transits to a laminar structure (Figure 3d,e). Organic remnants such as foraminifera surrounded by Mn-Fe-rich oxides precipitated with seawater and were transformed by hydrogenesis and diagenesis (Figure 3f).

Along the growth direction of C9, at the junction of the nucleus and ferromanganese oxides, it is a mainly columnar structure composed of upright columns in stage I (Figure 4a,b). The columns connect end to end, whose diameter ranges from 100 μm to 300 μm (Figure 4a,b). There are gaps between the columns in the later stage of the growth process. There are mainly three groups in gaps between columns in stage II: siliceous debris; fan-shaped or circular layered Mn-rich oxides (Figure 4d–f); banded Si-Fe-rich cement (Figure 4d–f). The entry of clastic materials interrupted the precipitation of ferromanganese oxides and further promoted the formation of columnar structures. Columns connected from head to tail gradually become forked dendritic columns, and finally, the mottled structure gradually formed in stage II (Figure 4a,b).

4.2. Mineralogy

The XRD diffraction patterns of each layer of ferromanganese deposits are shown in Figure 5. The diffraction peak of XRD diffraction patterns is not sharp, and the peak intensity are widened and dispersed because of low crystallinity of minerals. X-ray diffraction data for samples of ferromanganese crusts and ferromanganese nodule show that Mn-rich oxide minerals are dominant minerals. Quartz, albite, anorthite, and phillipsite exist as accessory minerals. Fe-rich oxides with poor crystallinity are amorphous, and no obvious diffraction peak is found. Mn-rich oxide minerals are mainly Fe-rich vernadite (1.42 Å and 2.45 Å) and and 10 Å manganese minerals (todorokite). Fe-rich vernadite exists in all the layers of ferromanganese crusts and nodule. Todorokite exists in the stage II in the ferromanganese nodule. In addition, todorokite exists in all layers of C9R-6, which just exists in the younger layers (C9R-7-1 and C9R-7-2) of C9R-7. Phillipsite exists in the layers of stage I in the ferromanganese nodule, older layer of C9R-6 and younger layers in C9R-7.

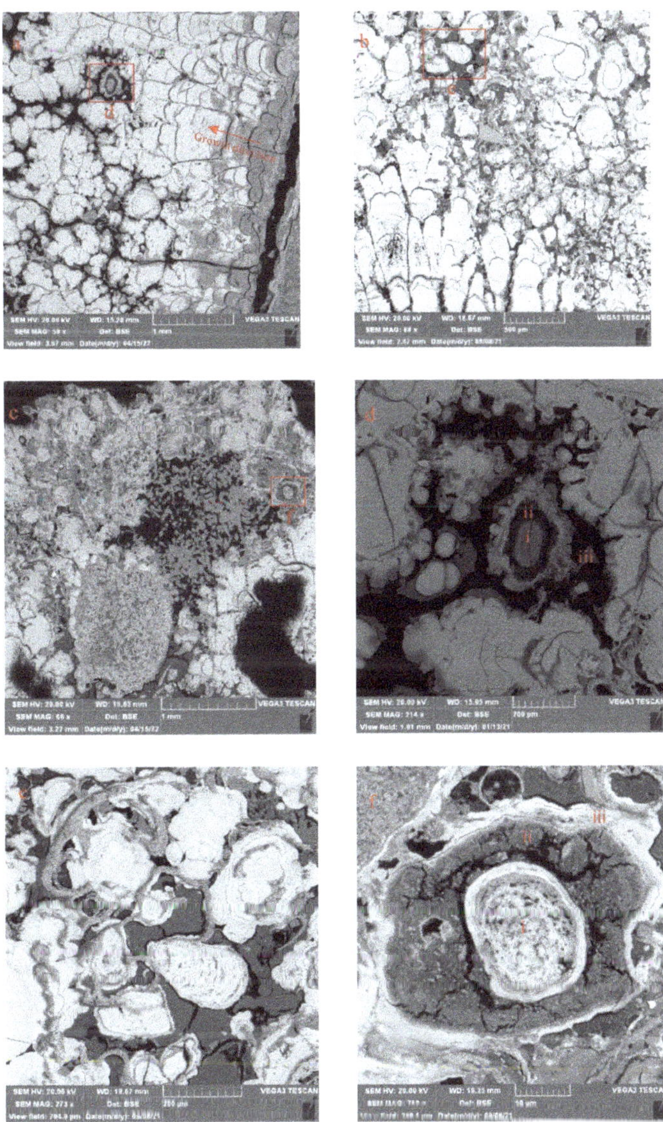

Figure 4. (**a**) Columnar structure composed of upright columns along the growth direction at the junction of the nucleus and ferromanganese oxides; (**b**) columnar structure gradually transit to mottled structure in stage II; (**c**) mottled structure composed of siliceous debris and Mn-rich oxides in stage II; (**d**) groups in gaps between columns in Figure 4a,i. Mn-rich oxides, ii. Ca-carbonate, iii. Si-Fe-rich cement; (**e**) Mn-rich oxides connected by banded Si-Fe-rich cement in Figure 4b; (**f**) groups in gaps between columns in Figure 4c,i. Mn-rich oxides, ii. organic matter where Mn- rich oxides scatter; iii. Si-Fe-rich cement.

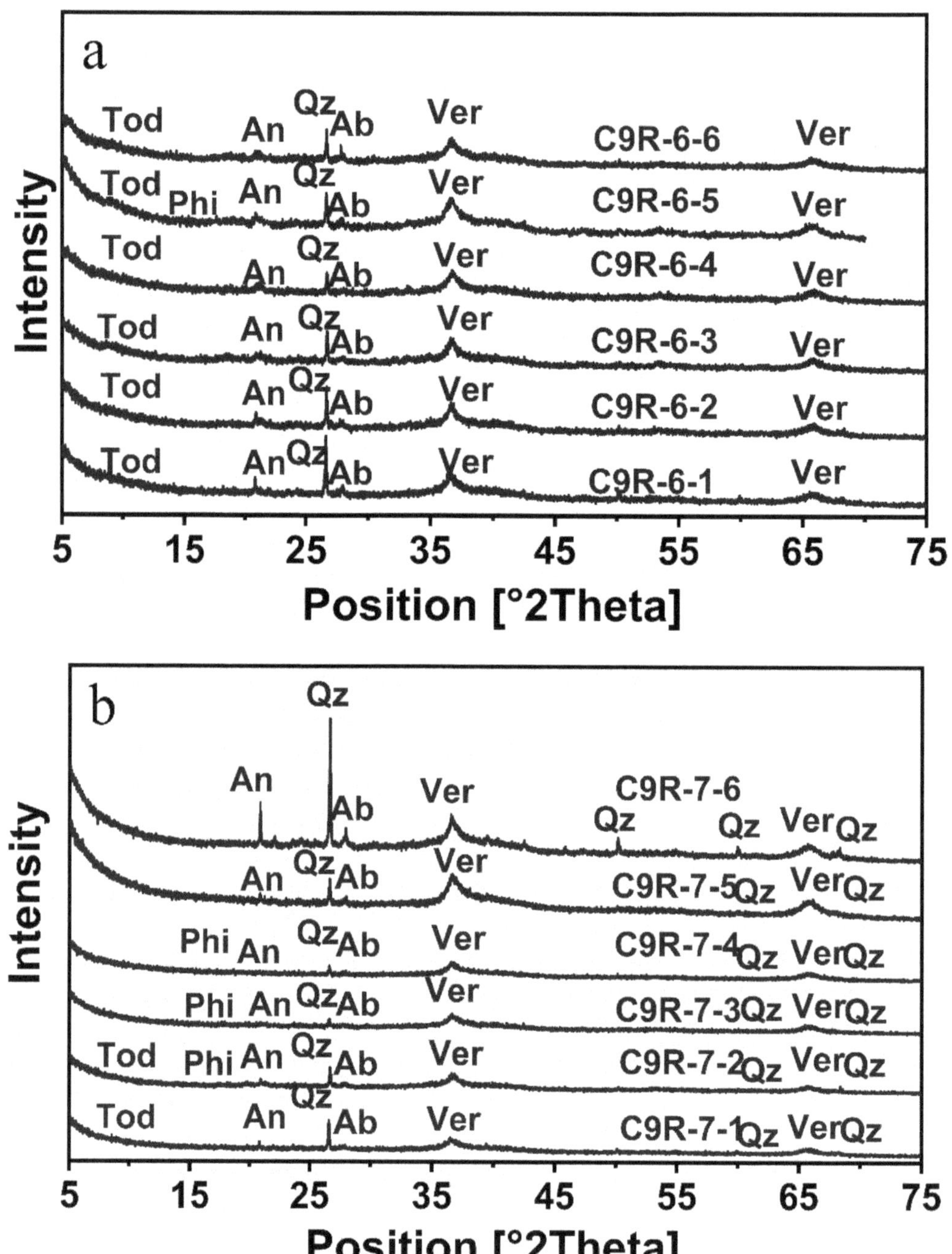

Figure 5. *Cont.*

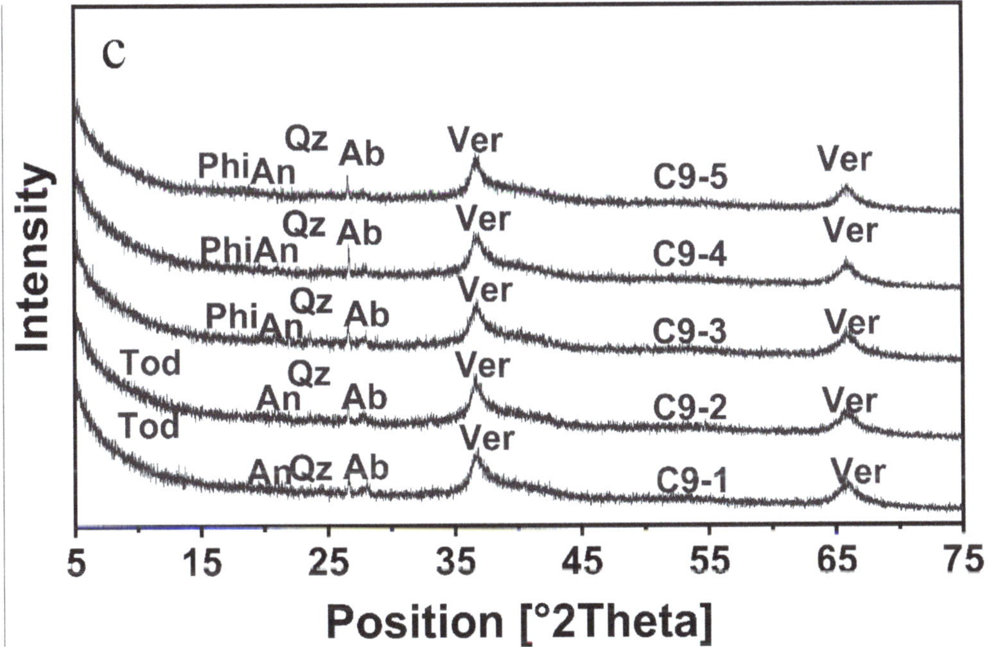

Figure 5. (**a–c**) The XRD patterns for ferromanganese deposits on the slope of Caiwei Guyot; Ver = Fe-vernadite, Ab = Al, Tod = todorokite, Qz = Quartz, Ab= Albite, Anorthite = An, Phi = Phillipsite.

4.3. Geochemistry

4.3.1. Geochemistry Compositions of the Ferromanganese Crusts

The chemical compositions of major and trace elements of each layer are shown in Table 2. The contents of Mn are 14.7–25.2% with an average value of 21.0%. The contents of Fe are 11.8–19.4% with an average value of 17.3%. The contents of Mn and Fe gradually decrease from the bottom to the top in C9R-7, while the contents of Mn and Fe in C9R-6 almost have no change during the growth process (Figure 6). The Mn/Fe ratios range from 1.06 to 1.45. The contents of Al are 0.66–2.22% with an average of 1.24%, and the contents of Si are 3.40–5.83% with an average of 4.84%. The contents of Al and Si in C9R-7 increase from the bottom to the top. The Si/Al ratios range from 2.21 to 5.28.

Among the trace elements, the contents of elements with economic value are extremely high, in which the contents of Co are 3841–8327 ppm with an average of 6163 ppm. The contents of Ni are 3000–5759 ppm with an average of 4593 ppm. The contents of Cu are 357–1664 ppm with an average of 915 ppm, and the contents in C9R-7 are lower than that in C9R-6 (nearly two-fold). The contents of Zn are 437–672 ppm with an average of 558 ppm. The contents of Co, Ni, Cu and Zn gradually decrease from the bottom to the top in C9R-7 in general (Figure 6).

The contents of REY are 1410–2117 ppm with an average of 1747 ppm. The contents of Ce are the highest in REY with an average of 803 ppm, ranging from 553–998 ppm, and accounting for nearly 50% in REY. The contents of LREE are 1079–1652 ppm with an average of 1380 ppm. The contents of HREE are 138–238 ppm with an average of 180 ppm. LREE/HREE ratios range from 6.42 to 8.81. The contents of Y range from 147 ppm to 227 ppm with an average of 186ppm. It can be seen that each layer of samples shows a positive Ce anomaly and negative Y anomaly in the PAAS standardization diagram of REY (Figure 7).

Table 2. Geochemistry compositions of the ferromanganese crusts.

Sample	Mn(wt%))	Fe	Al	Si	Mn/Fe	Si/Al	Co/(Cu + Ni)	Co/(Fe + Mn)	Ca	P	Mg	K	Na	Ti	Growth Rate	Li (ppm)	Be	Sc	V
C9R-6-1	21.6	18.2	1.20	5.77	1.18	4.81	1.38	175	2.20	0.42	1.27	0.61	2.41	0.96	0.85	3.88	4.59	2.02	441
C9R-6-2	20.5	19.4	1.50	5.75	1.06	3.83	0.98	136	2.13	0.43	1.32	0.70	2.45	1.05	1.30	6.67	4.89	3.22	455
C9R-6-3	20.8	19.0	1.50	5.08	1.09	3.39	0.75	139	2.32	0.51	1.38	0.65	2.00	1.09	1.24	11.35	4.81	4.05	460
C9R-6-4	23.4	17.4	1.10	3.69	1.35	3.35	0.85	137	2.66	0.49	1.28	0.61	1.84	1.32	1.28	4.98	5.63	3.80	527
C9R-6-5	20.4	19.3	1.34	5.18	1.06	3.87	1.07	171	2.02	0.43	1.36	0.54	1.76	1.09	0.89	7.77	4.58	3.67	481
C9R-6-6	21.0	18.2	1.44	5.44	1.15	3.78	0.89	150	2.15	0.40	1.37	0.53	1.76	1.10	1.10	9.69	3.63	2.32	474
C9R-7-1	15.9	11.8	1.20	5.83	1.35	4.86	0.97	139	1.92	0.78	1.18	0.40	1.92	0.91	1.25	7.67	5.46	9.94	555
C9R-7-2	14.7	12.5	2.22	4.90	1.17	2.21	1.68	208	1.86	0.72	1.32	0.63	1.79	1.05	0.64	3.08	4.70	7.43	595
C9R-7-3	22.0	18.5	0.980	4.84	1.19	4.96	1.36	147	2.32	0.46	1.23	0.54	2.41	0.97	1.13	2.38	4.92	2.69	577
C9R-7-4	22.0	18.4	1.05	4.53	1.20	4.31	1.15	147	2.30	0.42	1.23	0.522	1.99	1.11	1.14	3.48	5.47	5.43	607
C9R-7-5	24.8	17.5	0.700	3.68	1.41	5.28	1.40	191	2.47	0.44	1.28	0.47	2.20	0.95	0.73	1.54	5.18	1.51	595
C9R-7-6	25.2	17.4	0.660	3.40	1.45	5.12	1.40	196	2.50	0.43	1.24	0.49	2.17	0.95	0.70	0.87	4.11	0.97	543

Sample	Co	Ni	Cu	Zn	Ga	Rb	Sr	Zr	Nb	Mo	Cd	Cs	Ba	Hf	Ta	W	Tl	Pb
C9R-6-1	6962	4300	741	437	6.46	7.47	1026	507	55.1	318	5.16	0.75	1102	10.2	0.74	85.7	47.3	1740
C9R-6-2	5417	4430	1071	528	8.51	9.89	1017	626	74.5	284	5.45	1.15	1335	14.5	1.00	83.4	44.8	1578
C9R-6-3	5556	5759	1664	646	9.53	9.58	1037	664	78.8	284	6.70	1.08	1493	16.3	1.06	82.8	42.4	1727
C9R-6-4	5589	5094	1485	672	8.75	8.09	1255	758	79.2	353	6.78	0.85	1894	16.8	1.34	111	48.7	1556
C9R-6-5	6769	5346	1002	582	9.18	9.17	1127	629	75.8	281	6.16	1.01	1351	14.2	0.95	78.9	36.0	1800
C9R-6-6	5872	5469	1152	600	9.71	9.89	1090	618	65.2	290	6.60	1.08	1366	13.6	0.91	82.5	36.0	1661
C9R-7-1	3841	3250	704	535	6.86	15.8	1254	628	62.3	336	3.27	1.61	1424	11.3	0.57	67.7	84.4	1412
C9R-7-2	5639	3000	357	452	5.33	7.29	1315	527	46.3	393	3.96	0.58	1095	7.3	0.48	69.9	66.5	1670
C9R-7-3	5961	3816	566	556	9.15	8.20	1480	683	61.8	376	6.68	0.76	1519	12.5	0.79	102	23.7	1715
C9R-7-4	5932	4375	792	650	10.1	9.85	1596	807	76.6	384	6.96	0.96	1813	15.9	0.93	109	29.4	1745
C9R-7-5	8091	5075	709	558	8.52	5.51	1518	642	66.6	459	7.19	0.42	1463	11.11	0.80	127	30.2	1957
C9R-7-6	8327	5201	731	484	7.91	4.32	1371	567	59.1	429	6.72	0.31	1416	9.90	0.72	127	33.4	1999

Sample	Bi	Th	U	La	Ce	Pr	Nd	Sm	Eu	Gd	Tb	Dy	Y	Ho	Er	Tm	Yb	Lu
C9R-6-1	27.9	19.2	11.9	255	717	41.4	180	38.1	9.97	44.8	7.27	40.4	147	9.61	26.0	4.28	25.7	4.21
C9R-6-2	34.7	13.5	10.8	235	752	39.0	172	34.9	8.84	44.1	6.92	39.9	166	9.69	27.0	4.40	27.0	4.60
C9R-6-3	38.7	11.4	11.1	248	846	38.5	169	35.2	8.68	44.8	6.74	38.6	182	9.39	26.3	4.37	26.5	4.51
C9R-6-4	48.2	8.51	11.7	246	998	42.4	185	38.3	10.6	48.3	7.15	41.7	202	10.2	28.0	4.57	27.9	4.70
C9R-6-5	30.2	17.9	12.0	268	840	43.7	194	40.2	11.0	47.9	7.69	43.6	178	10.4	28.5	4.74	28.1	4.63
C9R-6-6	37.8	16.5	11.2	238	861	42.2	184	38.2	9.87	46.0	7.28	40.8	164	9.75	26.8	4.59	27.2	4.63
C9R-7-1	15.2	11.9	10.7	242	553	44.1	191	39.8	9.77	46.2	7.28	41.8	192	9.75	25.3	4.02	25.5	3.92
C9R-7-2	16.3	21.2	12.0	239	643	46.6	200	43.0	10.3	48.3	7.61	43.3	187	9.85	25.2	4.02	25.0	3.80
C9R-7-3	38.1	24.7	12.4	280	777	63.8	274	57.8	13.6	65.5	10.4	57.5	210	12.9	35.7	5.86	34.9	5.70
C9R-7-4	45.5	20.9	13.2	302	923	66.4	285	61.3	14.8	68.7	10.8	58.5	227	13.6	37.4	6.16	36.9	5.83
C9R-7-5	41.0	19.2	14.7	303	854	58.1	252	53.6	13.8	60.7	9.90	54.7	200	12.7	34.3	5.62	33.4	5.57
C9R-7-6	40.0	17.0	15.3	309	877	51.1	227	48.1	12.4	55.2	8.79	50.1	180	11.9	32.4	5.37	31.2	5.06

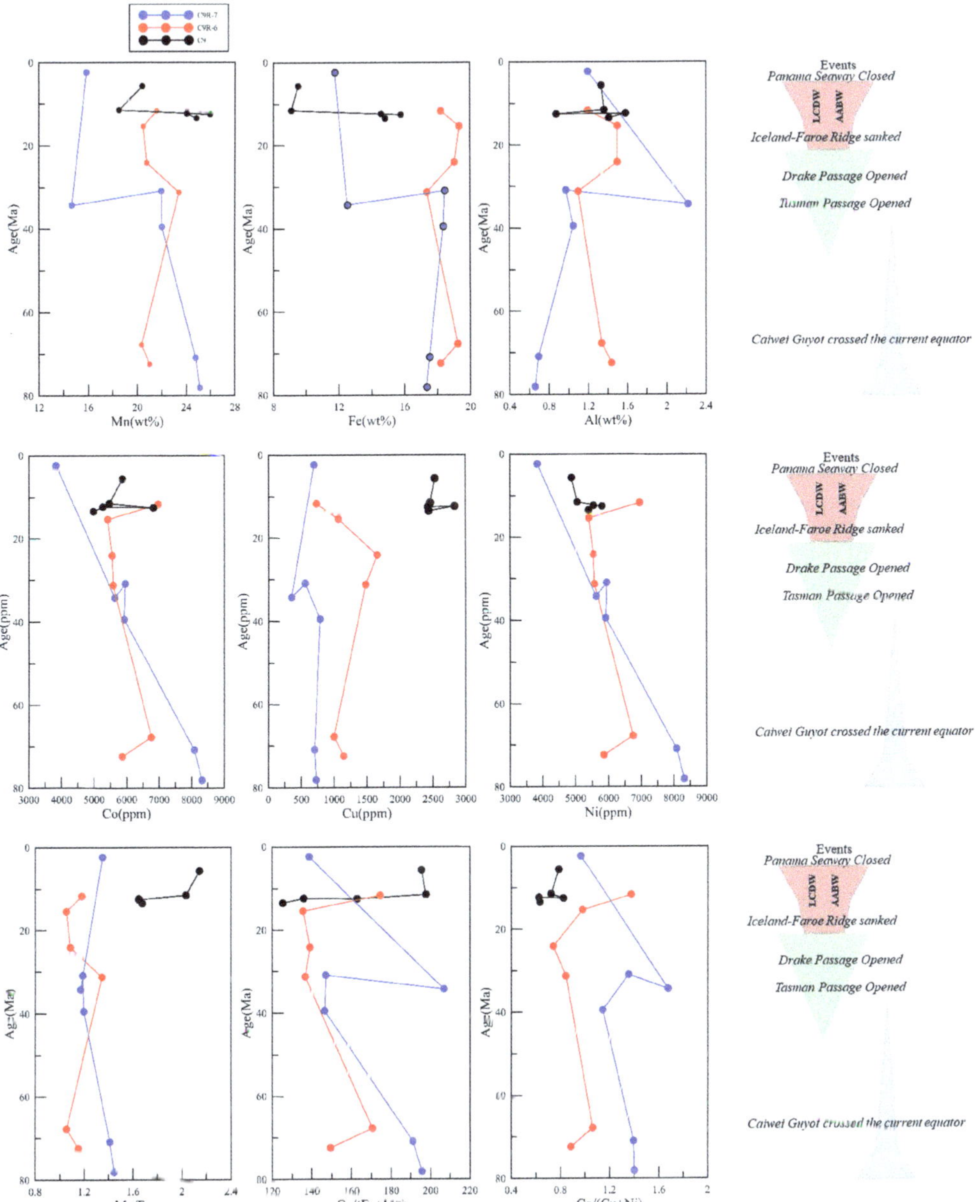

Figure 6. Geochemical profile of elements through ferromanganese deposits on Cawei Guyot. Major paleoceanographic and tectonic events are identified on the right. Blue color represents paleoceanographic and tectonic events "Caiwei Guyot crossed the current equator at about 70Ma", green color represents paleoceanographic and tectonic events "Drake Passage and Tasmanian Passage opened"; red color represents paleoceanographic and tectonic events "Iceland-Faroe Ridge sanked, LCDW and AABW strengthened, and Panama Seaway closed".

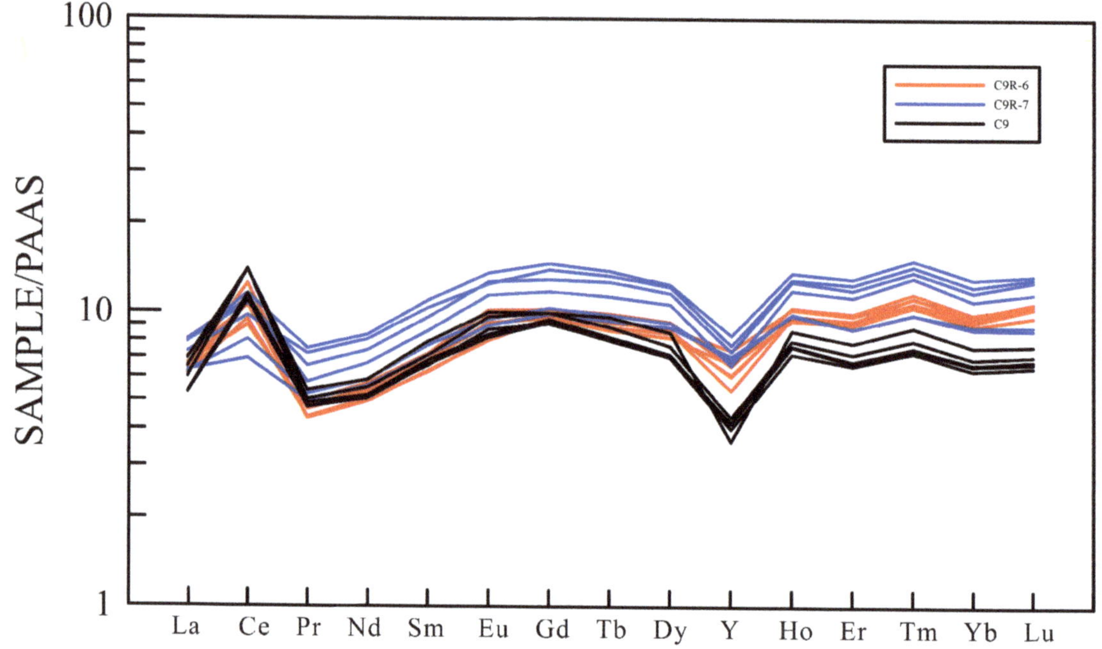

Figure 7. PAAS shale-normalized rare earth element plots for ferromanganese deposits of Caiwei Guyot.

4.3.2. Geochemistry Compositions of the Ferromanganese Nodule

The chemical compositions of major and trace elements of each layer are shown in Table 3. Among the major elements, the contents of Mn are 18.5–26.0% with an average value of 22.8%, which gradually decreases from stage I to stage II (Figure 6). The contents of Fe are 9.10–15.8% with an average value of 12.8%, which gradually decreases from stage I to stage II. The Mn/Fe ratios range from 1.65 to 2.14, which gradually increase from stage I to stage II (Figure 6). The contents of Al are 0.88–1.58% with an average value of 1.31%, and the contents of Ca are 2.04–2.89% with an average value of 2.45%.

Table 3. Geochemistry compositions of the ferromanganese nodule.

Sample	Mn (wt%)	Fe	Mn/Fe	Al	Ca	P	Co/(Cu + Ni)	Co/(Fe + Mn)	K	Ti	Mg	Na	Li (ppm)	Be	Sc	V	Co	Ni
C9-1	20.4	9.52	2.14	1.33	2.13	0.46	0.79	195.56	0.48	1.42	1.26	1.71	3.21	4.52	5.52	493	5851	4871
C9-2	18.5	9.10	2.03	1.36	2.04	0.51	0.73	197.73	0.50	1.34	1.20	1.62	3.56	4.88	6.38	491	5457	5047
C9-3	26.0	15.8	1.65	0.880	2.60	0.32	0.83	163.05	0.58	1.32	1.29	1.99	2.00	3.86	3.42	388	6806	5807
C9-4	24.0	14.6	1.65	1.58	2.58	0.35	0.63	136.06	0.87	1.15	1.24	1.88	3.93	5.05	3.90	366	5255	5545
C9-5	24.8	14.8	1.68	1.41	2.89	0.45	0.64	125.38	0.73	1.17	1.27	1.91	3.66	4.96	3.77	370	4973	5390

Sample	Cu	Zn	Ga	Rb	Sr	Zr	Nb	Mo	Cd	Cs	Ba	Hf	Ta	W	Tl	Pb
C9-1	2532.	639	6.97	4.93	1195	481	44.8	519	7.28	0.23	1392	6.24	0.73	72.4	103	1359
C9-2	2461	698	7.60	5.79	1168	503	47.9	548	6.98	0.26	1470	7.54	0.86	74.3	115	1400
C9-3	2427	472	6.60	3.70	898	447	39.9	343	5.21	0.19	1491	10.4	1.0	83.7	95.2	1333
C9-4	2833	487	8.06	6.29	802	477	42.3	318	5.25	0.25	1575	12.1	0.95	77.2	78.4	1204
C9-5	2439	461	8.48	5.17	801	464	38.9	315	4.96	0.20	1589	11.4	0.94	77.7	73.3	1243

Sample	U	La	Ce	Pr	Nd	Sm	Eu	Gd	Tb	Dy	Y	Ho	Er	Tm	Yb	Lu
C9-1	10.2	203	875	42.0	173	37.0	8.88	43.1	6.22	33.1	114	7.62	19.2	3.04	18.7	2.92
C9-2	10.0	203	915	43.3	177	38.1	9.08	44.1	6.39	33.8	113	7.64	19.5	3.08	18.9	2.97
C9-3	10.5	265	1110	47.8	199	44.1	10.8	46.4	7.47	40.1	98.4	8.61	22.5	3.62	21.6	3.35
C9-4	7.45	229	1113	42.0	176	36.6	9.46	42.7	6.26	33.0	109	7.18	18.9	2.98	18.0	2.83
C9-5	7.76	249	1113	44.7	188	39.6	10.4	46.4	6.91	36.0	119	7.95	20.6	3.26	19.7	3.09

Among the trace elements, the contents of elements with economic value are significantly high. The contents of Co are 4972–6806 ppm with an average value of 5668 ppm. The contents of Co gradually increase to the maximum value in stage I, and gradually decrease in stage II (Figure 6). The contents of Cu are 2428–2833 ppm with an average value of 2538 ppm, which is much higher than ferromanganese crusts, and the contents basically do not change (Figure 6). The contents of Zn are 461–698 ppm with an average value of 550 ppm, and the contents gradually increase from stage I to stage II. The contents of Ni are 4871–5807 ppm with an average value of 5332 ppm, and gradually increase to the maximum value in stage I and gradually decrease in stage II (Figure 6).

The contents of REY (REE + Y) are 1587–1929 ppm with an average value of 1781 ppm. The contents of Ce content are the highest with an average of 1025 ppm, ranging from 875 ppm to 1114 ppm and accounting for more than 50%. The contents of LREE are 1338–1677 ppm with an average of 1531 ppm. The contents of HREE are 240–263 ppm with an average of 251 ppm. The contents of Y are 98.4–119 ppm with an average of 111 ppm. LREE/HREE ratios range from 5.39 to 6.69, which shows LREE enrichment. It can be seen that each layer of the sample shows positive Ce anomaly and negative Y anomaly in the PAAS standardization diagram of REY (Figure 7).

4.4. Growth Rates and Age Model

According to the Co empirical formula above, the growth rates of C9R-6 are 0.85–1.30 mm/Ma (Table 2), and the average growth rate is 1.11 mm/Ma. The ages of each layer of C9R-6 are estimated to be 11.75 Ma, 15.42 Ma, 24.14 Ma, 31.25 Ma, 67.77 Ma, and 72.42 Ma, which are from the late Cretaceous to the middle Miocene. The growth rates of C9R-7 range from 0.64 mm/Ma to 1.25 mm/Ma (Table 3), and the average growth rate is 0.93 mm/Ma. The age of each layer of C9R-7 is estimated to be 2.40 Ma, 34.29 Ma, 30.91 Ma, 39.48 Ma, 70.90 Ma, and 78.12 Ma, which is from the late Cretaceous to the early Pleistocene. The growth rates of C9 are 0.69–1.48 mm/Ma (Table 3), and the average growth rate is 1.02 mm/Ma, which is much lower than the growth rates of ferromanganese nodules in the Peru Basin [48]. The ages of each layer of C9 are estimated to be 5.67 Ma, 11.54 Ma, 12.55 Ma, 12.37 Ma, and 13.48 Ma, which grew from the Miocene.

4.5. Element Correlations and Factor Analysis

Element correlations of ferromanganese crusts on the mountaintop edge of Caiwei Guyot are shown in Table S1 in Supplementary Materials. Mn shows a significantly positive correlation with Ca, Co, Cd, W, Bi, Ce, and Lu (99% CL) and shows a positive correlation with Fe, Ni, Pb, U, La, Er, Tm, and Yb (95% CL); Mn is negatively correlated with Al, P, Sc, and Tl (99%CL) and shows a negative correlation with Si and Rb (95% CL). Fe shows a positive correlation with Ni, Ga, Cd, and Bi (99% CL) and shows a positive correlation with Mn, Nb, Hf, Ta, and Ce (95% CL); Fe shows a negative correlation with P, Sc and Tl (99% CL). Mn/Fe shows a positive correlation with Mo (99% CL) and shows a positive correlation with W and U (95% CL), Mn/Fe is negatively correlated with Ga and Nb (95%CL); Cu is positively correlated with Zn and Ni (99% CL); the growth rates are positively correlated with Li, Cu, Zn, Rb, Nb and Ta (95% CL) and negatively correlated with Co (99% CL). Q-mode factor analysis of ferromanganese crusts produces four factors that account for 89.6% of the variance, with Factors 1–4 accounting for 35.5%, 23.0%, 18.5%, and 12.6% of the variance, respectively (Table S2 in Supplementary Materials). Factor 1 is interpreted to represent aluminosilicate minerals: Al, Mg, K, Li, V, Cu, Si, Zr, Mo, W, Tl, U, and REY (except Ce). Factor 2 is interpreted to represent residual biogenic: Fe, Ca, Ti, Ni, Cu, Zn, Ga, Zr, Nb, Cd, Ba, Hf, Ta, Bi, Th and Ce. Factor 3 is interpreted to represent Mn-rich oxides: Mn, Si, Ca, Li, Sc, Co, Rb, U, La and Ce. Factor 4 is interpreted to represent Fe-rich oxyhydroxide: Fe, P, Be, Sc and V.

Element correlations of the ferromanganese nodule on the slope of Caiwei Guyot are shown in Table S3 in Supplementary Materials. Mn is significantly positively correlated with Fe and Na (99% CL) and negatively correlated with Sc and Zn (99% CL), Mn is positively correlated with Ca, Ni, Bi, La, and Ce (95% CL) and negatively correlated with V, Sr, Zr, Nb, Mo, and Cd (95% CL). Fe is significantly positively correlated with Ca, Na, Hf, Bi, and Ce (99% CL) and significantly negatively correlated with Sc, V, Zn, Mo, and Cd (99% CL). Fe is positively correlated with Ni and La (95% CL) and negatively correlated with Sr and Nb (95% CL). Co has no significant correlation with Fe and Mn, while Co is negatively correlated with Al and Li (95% CL) and significantly negatively correlated with Na and Be (99% CL), indicating that the entry of debris affects the adsorption of Co. Mn/Fe is extremely positively correlated with Sr, V, Mo, and Cd (99% CL) and positively correlated with the Zn and growth rates (95% CL). Mn/Fe shows a negative correlation with Fe, Hf, Bi, and Ce (99%CL) and a negative correlation with Mn, Ca, Na, Ni, Ta, and La (95% CL). The growth rates are extremely positively correlated with V and Sr (99% CL) and are positively correlated with Mn/Fe, Ti, Zn, Mo, Cd, Tl, and W (95% CL). The growth rates show an extremely negative correlation with Ba, Hf, and Bi (99% CL) and a negative correlation with K and Ce (95% CL). Q-mode factor analysis of the ferromanganese nodule produces three factors that account for 95.7% of the variance, with Factors 1–3 accounting for 44.9%, 39.8%, and 11.0% of the variance, respectively (Table S4 in Supplementary Materials). Factor 1 is interpreted to represent ferromanganese oxides: Mn, Fe, Ca, P, K, Ti, Na, Sc, V, Ni, Zn, Ga, Sr, Zr, Nb, Mo, Cd, Ba, Hf, Ta, W, Tl, Pb, Bi, U, La, Ce, and Eu. Factor 2 is interpreted to represent aluminosilicate minerals: Al, Mg, Li, Be, Co, Cu, Rb, Zr, Cs, W, Th, and REY (except Ce and Y). Factor 3 is interpreted to represent fluorocarbon phosphate: P, Li, Be, Co, Ga, and Y.

5. Discussion

Guyot is a complex environment and is controlled by geology, biological activities, geochemical properties of seawater, and physical oceanography, which further affects the characteristics of ferromanganese deposits. Ferromanganese deposits on Caiwei Guyot can be divided into ferromanganese crusts and ferromanganese nodules. In the next section, we focus on the texture morphology, microstructure, mineralogical features, and geochemical features of ferromanganese crusts and nodules. Meanwhile, mineralogical and geochemical analyses are essential for discussing growth processes and environments.

5.1. Texture and Surface

Ferromanganese crusts and nodules have different textures and surfaces due to different oceanographic and geological conditions [49,50]. The surface of the ferromanganese nodule is an oolitic structure with cracks (Figure 2f), while the surfaces of board ferromanganese crusts have a surface with black botryoidal shapes (Figure 2b,d), showing smoother surface texture at a greater depth [50]. These black botryoidal shapes are composed of ferromanganese oxides. Accumulation of ferromanganese oxides reflects the hydrodynamic conditions of the growth process. A more dynamic environment will form a botryoidal structure [2]. The speed of the bottom current (>7 cm/s) on the mountaintop edge and the hydrodynamic conditions were strong because of the Taylor Column phenomenon on Caiwei Guyot [40], while the speed of the current was low and the hydrodynamic conditions are weak on the slope of Caiwei Guyot [40]. In addition, the bottom current with a high-speed bottom on the mountaintop edge sweeps the sediment and exposes the hard substrate [40,41], which is also conducive to the rapid accumulation of ferromanganese oxide. The same phenomenon occurs in ferromanganese crusts on Takuyo-Daigo Seamount, NW Pacific, and Tropic Seamount, Atlantic [41,50], whose botryoidal shapes are bigger at shallower water depths. What is more, the speed of the bottom current and hydrodynamic intensity are in a state of equilibrium where calcareous or siliceous sediments can be cleaned, but botryoidal shapes cannot be eroded [41]. Board ferromanganese crusts generally grow in situ on the mountain top, and finally aggregate for mineralization on the hard substrate. What is more, due to the erosion of the bottom current, the increase in slope gradient, and other reasons, collapse often occurs on Caiwei Guyot, which is common in the Western Pacific Seamounts [29,51]. Ferromanganese nodule grows in an environment with relatively weaker hydrodynamic force and may be buried by sediments brought by the collapse of Caiwei Guyot.

5.2. Genesis Type of Ferromanganese Crusts and the Nodule

Mn-(Cu + Co + Ni)-Fe ternary diagram is used to divide ferromanganese deposits in the ocean into three types: hydrogenetic, diagenetic and hydrothermal [52]. (Fe + Mn)/4-100 × (Zr + Y + Ce) − 10 × (Cu + Ni + Co) ternary diagram [53] and $Ce_{SN}/Ce_{SN}{}^*$-Nd and $Ce_{SN}/Ce_{SN}{}^*$-Y_{SN}/Ho_{SN} [3] are also used to classify the type. Each layer in board ferromanganese crusts and layers of stage I in the ferromanganese nodule fall into the hydrogenentic area (Figures 8 and 9), showing the hydrogenentic origin and the enrichment of Cu, Co and Ni. While layers of stage II in the ferromanganese nodule are closer to diagenetic origin (Figure 8a). The authors also used the Mn/Fe ratio to distinguish the genesis type of ferromanganese crusts. It is generally believed that ferromanganese crust whose Mn/Fe is less than 2.5 is hydrogenetic origin [54]. On the contrary, it is diagenetic origin. Mn/Fe of each layer in board ferromanganese crusts of Caiwei Guyot is less than 2.5, which is hydrogenetic origin. Although the Mn/Fe ratios in stage I of the ferromanganese nodule is less than 2.5, greater than 2, which is much higher than those in board ferromanganese crusts, indicating that the nodule has a diagenetic trend in the growth process. Meanwhile, the enrichment of LREE reflects the hydrogenetic origin [55]. Positive Ce anomaly and negative Y anomaly are other evidence of hydrogenetic origin. REE generally has +3 valence, while Ce has Ce^{3+} and Ce^{4+}. Ce^{3+} in the marine environment is oxidized to Ce^{4+} to form CeO_2, which precipitates from seawater, resulting in a strong loss of Ce in seawater [56]. Therefore, the PAAS standardization diagrams of REY of ferromanganese crusts and the nodules show strong positive Ce anomalies after PAAS standardization, which are often considered to be the influence of seawater oxidation [57]. The ion radius and valence of the Y element are similar to other rare earth elements, but Y has no 4f electrons, so it is difficult to form a more stable complex. Therefore, its chemical behavior is different from Ho. When ferromanganese crusts and nodules formed, Y and Ho would differentiate, resulting in Y negative anomaly [58,59]. Fe-vernadite is found in each layer in the XRD patterns (Figure 5), which is a nanocrystalline turbostratic phyllomanganate phase of hydrogenetic origin. The crystal structure of vernadite is a layered structure, $(Mn^{4+}O_6)^{8-}$ octahedron is connected to each other through the edge. The octahedron layer is mainly Mn^{4+}, and the metal cations can replace the Mn^{4+} by isomorphism [60–62]. 10 Å manganese minerals (todorokite) are found in stage II of C9, C9R-6 and the younger layers of C9R-7 (C9R-7-1 and C9R-7-2) (Figure 5). 10 Å manganese minerals (todorokite) have tunnel structure. K^+, Mg^{2+}, Ba^{2+}, and water molecules usually occupy the sites in the todorokite tunnel. Meanwhile, Co, Ni, and Cu may mainly replace Mn at the corner of the tunnel [60–65] 10 Å manganese minerals (todorokite) are formed by the influence of pore water on the seawater-sediment interface under suboxic conditions and the formation of todorokite from vernadite is known from ferromanganese crusts and nodules when oxic conditions have been changed into suboxic conditions, which is considered to be a typical feature of diagenetic ferromanganese crust and nodule [2,66–69]. We believe that nanoscale and micron-scale diagenesis may occur in the internal layers, but bulk geochemical features are not obvious on the whole.

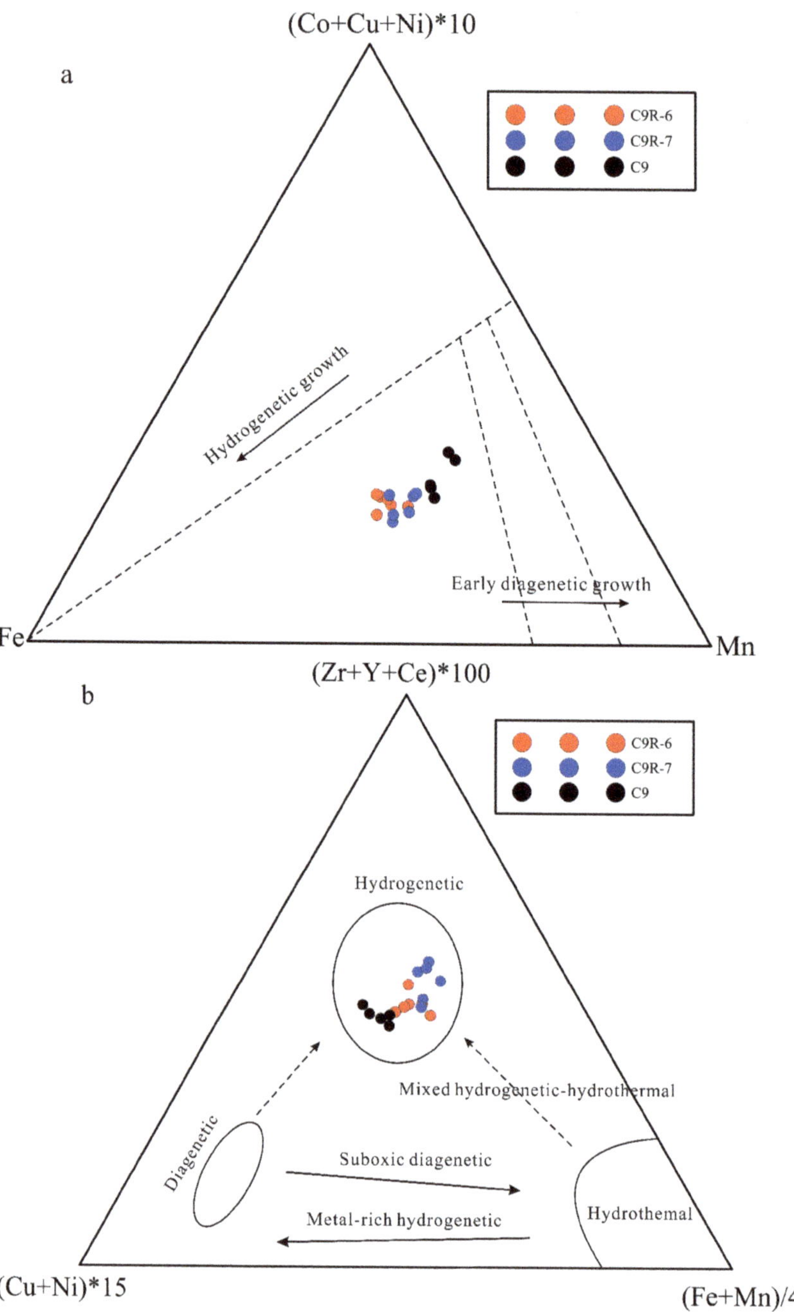

Figure 8. (**a**) Ternary diagram for the genetic classification of ferromanganese deposits. (**b**) Ternary discrimination diagram for the genetic classification of ferromanganese deposits.

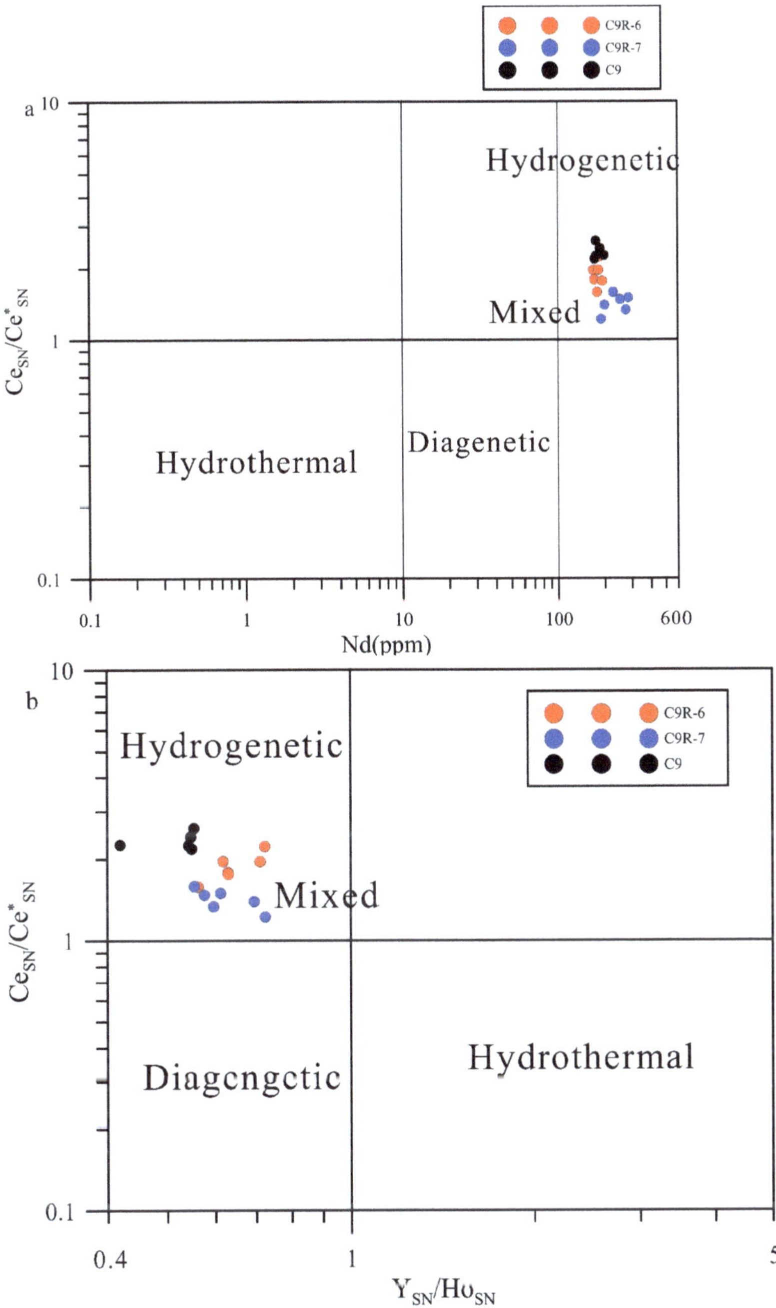

Figure 9. Discrimination graphs of ferromanganese deposits from Caiwei Guyot. (**a**) Ce_{SN}/Ce_{SN}^* ratios vs. Nd contents. (**b**) Ce_{SN}/Ce_{SN}^* ratios vs. Y_{SN}/Ho_{SN} ratios.

5.3. Genesis of Ferromanganese Deposits

Paleoceanographic events affecting ferromanganese crusts and nodule in the northwest Pacific since the late Cretaceous are consistent. According to the mineralogical and geochemical characteristics above, ferromanganese crusts on the mountaintop edge of

Caiwei Guyot are hydrogenetic origin, whose metal source is oxic seawater [70]. Based on the previous empirical formula calculation of Co, the oldest layers can reach the late Cretaceous, when AABW had not been formed. The formation of ferromanganese crusts may be mainly controlled by the middle ocean current and seamount authigenic current [42]. There are relatively more ore-forming elements in seawater, providing sufficient material source for the mineralization of the ferromanganese crusts. At this stage (78.12 Ma), ferromanganese oxides precipitated in form of plaque structure with occasional siliceous debris and biological sediments in the gaps, which formed a mottled structure (Figure 3c), indicating that the ferromanganese crusts precipitated in the strong hydrodynamic environment on the mountaintop edge of Caiwei Guyot [2]. The contents of Mn and Fe are high, as well as Co, which has the characteristics of obvious hydrogenetic origin. The content of Si and Al is less, suggesting that they are less affected by terrestrial materials. With the drift of the Pacific plate, Caiwei Guyot drifted to the northwest mainly in zonal motion and crossed the current equator about 60–70 Ma ago [71]. The guyot where the ferromanganese crusts are located migrated to the equatorial high productivity zone of the open ocean. As time goes by, due to the crossing of the equator, continental weathering is gradually strengthened. The global temperature was high during this period. Without AABW (78.12–31.25 Ma), the dissolved oxygen content of deep water was very low, and the content of Mn and Co gradually decreased, while the contents of Si and Al gradually increased. Co/(Cu + Ni) and Co/(Fe + Mn) ratios reach the minimum value at 31.25 Ma (Figure 6), indicating that oxidation conditions fell into a decline. Obvious changes in layers can be seen inside. At this time, the texture of ferromanganese crusts was loose with high porosity, the precipitation of ferromanganese oxides was prevented due to the entry of silicon debris, and the growth of columnar structures was inhibited. Phillipsite existed in this stage (Figure 5a,b), which is formed by the alteration of volcanic glass leading to the increase in pH value at the sediment-water interface [72–74]. Phillipsite usually appears in the form of authigenic minerals and exists in the gaps of the columnar structure, which are likely to be physically combined with the columnar growth process in the growth process.

From the late Oligocene to the early Miocene, the Drake Passage and Tasmanian Passage completely opened and the deep-water channel was formed [75], which contributed to the formation of the modern circulation system. AABW with high content of dissolved oxygen has an impact on the growth process of ferromanganese crusts on the Pacific Seamounts [76–78]. After entering the Northwest Pacific Ocean, LCDW (mixed by AABW and NPDW) with dissolved oxygen is the main source of dissolved oxygen in the Magellan seamount area. When a significant global cooling event occurs, the degree of environmental oxidation in the Pacific Seamounts will also increase. At this stage (31.25–21.14 Ma), due to the role of LCDW, the deep water in the ocean began to have a dissolved oxygen supply, and OMZ began to develop in the Pacific Ocean. The growth process of ferromanganese crusts is related to OMZ on seamounts [12]. The water depth of OMZ in the Caiwei Guyot is less than 1000 m [79]. OMZ can promote the dissolution of Mn in OMZ and ferromanganese crusts with high Mn content form below the OMZ [20,80]. Dissolved oxygen oxidizes Mn^{2+}, forming hydrated complex ions. Co, Ni, and Ce show a positive correlation with Mn, indicating that Ni, Cu, and Ce are distributed in the manganate octahedral layer of Fe-vernadatie [81]. Fe-vernadatie adsorbs oxidized Ce, as well as Co^{2+}, Ni^{2+}, and other metal cations. Higher contents of Co are the result of the oxidation of these elements on the surface of Fe-vernadite, indicating an extremely effective process of enrichment of trace elements [82]. Co/(Cu + Ni) and Co/(Fe + Mn) ratios increase with the formation of the modern circulation system, indicating the enhancement of oxidation conditions (Figure 6). The columnar structure formed in this stage (Figure 3). The formation of the southeast ice sheet of the Antarctic, colder climate and the sank of Iceland–Faroe Ridge at about 15 Ma promote the high-density deep water of the Arctic Ocean to flow into the Atlantic Ocean and other oceans [76,83], AABW is further enhanced, making the oxidation degree of Caiwei Guyot higher. At this stage (15.42 Ma–0), Co/(Cu + Ni) and Co/(Fe + Mn) ratios reach the maximum value in C9R-6-1 (Figure 6). Meanwhile, the enhancement of LCDW and the strong hydrodynamic force on the mountaintop edge promote the formation of a dense laminar structure with low porosity.

Different from the ferromanganese crusts on the mountaintop edge, the ages of the nodule on the slope are younger, whose growth process is affected by AADW that becomes LCDW when entering the Northwest Pacific Ocean. The enhancement of the LCDW oxidation degree has a great effect on the texture, microstructure, mineralogical, and geochemical features of the nodule. According to the mineralogical and geochemical characteristics, the ferromanganese nodule on the slope of Caiwei Guyot can be divided into two stages. The layers with hydrogenetic origin in stage I (13.48–11.54 Ma) are formed in the middle Miocene. Lower temperatures of deep water, the expansion of the Antarctic ice sheet, and the enhancement of deep-water circulation from the late Miocene [75,84] make the high-density seawater of the Arctic Ocean flow into the Atlantic Ocean and other oceans with the circulation system. As a result, LCDW was further strengthened. LCDW with dissolved oxygen flew through Caiwei Guyot, strengthening the oxidation of seawater. Fe^{2+} and Mn^{2+} in seawater accumulated to form Fe-rich and Mn-rich oxide colloids [80,85], which precipitated in form of a columnar structure connected end to end around the nucleus. This process is the same as that of ferromanganese crusts. The contents of Mn and Fe are relatively high, and the content of Co also gradually increases. Compared with Atlantic ferromanganese nodules, ferromanganese nodules on Caiwei Guyot enrich Mn, Co, and Ni, while losing Fe [86]. Co/(Cu + Ni) and Co/(Fe + Mn) ratios increase with the enhancement of oxidation of seawater (Figure 6). Fe-vernadite is the main Mn-rich oxide in stage I, and quartz, albite, anorthite, and phillipsite exist as accessory minerals.

The layers in stage II (11.54 Ma–0) are formed in the middle and late Miocene. The contents of Mn, Fe, and Co are reduced to varying degrees relative to stage I. High Mn/Fe ratios (>2) and 10 Å manganese minerals (todorokite) indicate nanoscale or micron-scale diagenesis occurred in the internal layers of stage II. Todorokite has a tunnel structure. Meanwhile, Co, Ni, and Cu may mainly replace Mn at the corner of the tunnel. Cu is enriched in diagenetic layers of the nodule. Meanwhile, the growth rates in stage II are higher than those in stage I. Previous studies have shown that the diagenetic process is mainly due to the following reasons: i, adsorbed metals are released due to particle dissolution and will be enriched in pore water and further have a reaction with manganese oxides [87,88]; ii. phillipsite formed by the alteration of volcanic glass leads to the increase in pH value, resulting in the additional oxidation of Mn^{2+} [72,74]; iii. the decomposition of organic matter and the dissolution of biological components related to biological activities promote the release of metal ions, and metal elements can be adsorbed into manganese oxides [87,89]. Metals of particle dissolution may have an impact on diagenesis at the micron level, but the specific impact is unknown. What is more, it is believed that 10 Å manganese minerals (todorokite) are considered to be the evidence of diagenesis [2,66–69], but there is no or very little phillipsite in some layers where 10 Å manganese minerals (todorokite) exist. We exclude that phillipsite may be the cause of internal diagenesis of the nodule. Biological activities on the slope in Caiwei Guyot are more active than those on the mountaintop [90]. Siliceous debris and organic matter precipitated in gaps with seawater in stage II. As a result, the texture is loose with high porosity. Siliceous debris, fan-shaped or circular layered Mn-rich oxides, and banded Si-Fe-rich cement are filled in the gaps (Figure 4). Though Co/(Cu + Ni) and Co/(Fe + Mn) ratios are higher than those of stage I (Figure 6), the internal columnar structure provides space for nano-scale or micron-scale diagenesis. The degradation of organic matter may produce a local (micron-scale) reducing microenvironment [55]. The change of redox potential promotes the reactivation of Fe from surrounding biological carbonates [55], forming banded Si-Fe-rich oxides, which are cemented and precipitated in gaps. Banded Si-Fe-rich oxides mostly appear in the middle and late stages of growth and are connected with columns or layered manganese-rich oxides dispersed in gaps (Figure 4). The mottled structure gradually formed in stage II (Figure 4b).

6. Conclusions

In this study, we analyze the texture morphology, microstructure, mineralogical features, and geochemical features of ferromanganese crusts on the mountaintop edge of Caiwei Guyot and the ferromanganese nodule on the slope of Caiwei Guyot. Global palaeo-ocean events control on the geochemistry compositions and growth process of ferromanganese crusts and the nodule.

Ferromanganese crusts on the mountaintop edge of Caiwei Guyot are board ferromanganese crusts that are of hydrogenetic origin with many black botryoidal shapes on the surface, suggesting board ferromanganese crusts grow in strong hydrodynamic conditions. Obvious changes in layers can be seen inside. Fe-vernadite is the main Mn-rich mineral. The oldest layers can reach the late Cretaceous age when AABW had not been formed. The formation of crusts in this process was related to the middle ocean current and seamount authigenic current. Ferromanganese oxides precipitated in form of plaque structure with occasional siliceous debris and biological sediments in the gaps, which formed a mottled structure. With the drift of the guyot, continental weathering is gradually strengthened, and the dissolved oxygen content of deep water was very low. Co/(Cu + Ni) and Co/(Fe + Mn) ratios reach the minimum value at 31.25 Ma. After the formation of the modern circulation system, LCDW (mixed by AABW and NPDW) with dissolved oxygen is the main source of dissolved oxygen in the guyot, which further strengthens oxidation. Global palaeo-ocean events control the geochemistry compositions and growth process of ferromanganese crusts. For example, the formation of the south ice sheet of the Arctic, colder climate, and the sank of Iceland–Faroe Ridge at about 15 Ma make Co/(Cu + Ni) and Co/(Fe + Mn) ratios continue to increase. 10 Å manganese minerals (todorokite) indicate that nanoscale or micron-scale diagenesis in the internal layers may occur.

The surface of the ferromanganese nodule is an oolitic structure, showing a smoother surface texture at a greater depth. The ferromanganese nodule on the slope grows in an environment with relatively weaker hydrodynamic conditions. The growth process is divided into two stages. Stage I is formed in the middle Miocene and is affected by LCDW. The columnar structure formed under the action of LCDW. Fe-vernadite, 10 Å manganese minerals (todorokite), quartz, albite, and anorthite can be seen. Stage I (13.48–11.54 Ma) of the ferromanganese nodule shows hydrogenetic origins due to geochemical features, indicating that the metal source is oxic seawater. As a result, the contents of Cu and Fe are high, and the growth rates are low. Stage II (11.54 Ma–0) is formed in the middle and late Miocene. Nanoscale or micron-scale diagenesis in the internal layers may occur. The texture is loose with high porosity, and siliceous debris and organic matter are carried in gaps between columns by seawater, creating a reducing microenvironment that promotes the formation of banded Si-Fe-rich oxides. The contents of Mn, Fe, and Co gradually decrease. High Mn/Fe ratios (>2) and 10 Å manganese minerals (todorokite) prove it. Meanwhile, the growth rates in stage II are higher than those in stage I. The mottled structure gradually formed in stage II. The content of Cu of the ferromanganese nodule is much higher than those of ferromanganese crusts.

Supplementary Materials: The following supporting information can be downloaded at: https://www.mdpi.com/article/10.3390/jmse10091275/s1, Table S1–S4. Table S1: Pearson correlation for major and trace elements of ferromanganese crusts; Table S2: Statistical factor analysis for major and trace elements of ferromanganese crusts; Table S3: Pearson correlation for major and trace elements of the ferromanganese nodule; Table S4: Statistical factor analysis for major and trace elements of the ferromanganese nodule.

Author Contributions: Conceptualization, L.W.; methodology, L.W.; software, L.W.; validation, L.W.; formal analysis, L.W.; investigation, L.W.; resources, Z.Z.; data curation, L.W.; writing—original draft preparation, L.W.; writing—review and editing, L.W. and Z.Z.; visualization, L.W.; supervision, Z.Z.; project administration, Z.Z.; funding acquisition, Z.Z. All authors have read and agreed to the published version of the manuscript.

J. Mar. Sci. Eng. **2022**, 10, 1275

Funding: This work was supported by the NSFC Major Research Plan on West-Pacific Earth System Multi-spheric Interactions [grant number 91958213], the Strategic Priority Research Program of the Chinese Academy of Sciences [grant number XDB42020402], the National Programme on Global Change and Air–Sea Interaction [grant number GASI-GEOGE-02], the International Partnership Program of the Chinese Academy of Sciences [grant number 133137KYSB20170003], the Special Fund for the Taishan Scholar Program of Shandong Province [grant number ts201511061], and the National Key Basic Research Program of China [grant number 2013CB429700].

Informed Consent Statement: Not applicable.

Data Availability Statement: All the data are given in the Supporting Information.

Acknowledgments: We would like to thank the crews of the R/V *Kexue* during the HOBAB 5 cruise for their help with sample collection. We are grateful for the valuable comments and suggestions from the anonymous reviewers and editors.

Conflicts of Interest: The authors declare no conflict of interest.

References

1. Hein, J.R.; Koschinsky, A. Deep-Ocean Ferromanganese Crusts and Nodules. In *Treatise on Geochemistry*, 2nd ed.; Holland, H., Turekian, K., Eds.; Elsevier: Amsterdam, The Netherlands, 2014; Volume 13, pp. 273–291.
2. Hein, J.R.; Koschinsky, A.; Bau, M.; Manheim, F.; Kang, J.-K.; Roberts, L. Cobalt-Rich Ferromanganese Crusts in the Pacific. In *Handbook of Marine Mineral Deposits*; Cronan, D., Ed.; CRC Press: London, UK, 2000.
3. Bau, M.; Schmidt, K.; Koschinsky, A.; Hein, J.; Kuhn, T.; Usui, A. Discriminating between different genetic types of marine ferro-manganese crusts and nodules based on rare earth elements and yttrium. *Chem. Geol.* **2014**, *381*, 1–9. [CrossRef]
4. Hein, J.R.; Mizell, K.; Koschinsky, A.; Conrad, T.A. Deep-ocean mineral deposits as a source of critical metals for high- and green-technology applications: Comparison with land-based resources. *Ore Geol. Rev.* **2013**, *51*, 1–14. [CrossRef]
5. Heller, C.; Kuhn, T.; Versteegh, G.J.M. The geochemical behavior of metals during early diagenetic alteration of buried manganese nodules. *Deep-Sea Res. Part I-Oceanogr. Res. Pap.* **2018**, *142*, 16–33. [CrossRef]
6. Fitzgerald, C.E.; Gillis, K.M. Hydrothermal manganese oxide deposits from Baby Bare seamount in the Northeast Pacific Ocean. *Mar. Geol.* **2006**, *225*, 145–156. [CrossRef]
7. Pelleter, E.; Fouquet, Y.; Etoubleau, J.; Cheron, S.; Labanieh, S.; Josso, P.; Langlade, J. Ni-Cu-Co-rich hydrothermal manganese mineralization in the Wallis and Futuna back-arc environment (SW Pacific). *Ore Geol. Rev.* **2017**, *87*, 126–146. [CrossRef]
8. Baturin, G.N.; Dobretsova, I.G.; Dubinchuk, V.T. Hydrothermal manganese mineralization in the Peterbourgskoye ore field (North Atlantic). *Oceanology* **2014**, *54*, 222–230. [CrossRef]
9. González, F.J.; Somoza, L.; Hein, J.R.; Medialdea, T.; León, R.; Urgorri, V.; Reyes, J.; Martín-Rubí, J.A. Phosphorites, Co-rich Mn nodules, and Fe-Mn crusts from Galicia Bank, NE Atlantic: Reflections of Cenozoic tectonics and paleoceanography. *Geochem. Geophys. Geosyst.* **2016**, *17*, 346–374. [CrossRef]
10. Hein, J.R.; Koschinsky, A.; Halbach, P.; Manheim, F.T.; Bau, M.; Kang, J.-K.; Lubick, N. Iron and manganese oxide mineralization in the Pacific. *Geol. Soc. Special Publ.* **1997**, *119*, 123–138. [CrossRef]
11. Halbach, P. Processes controlling the heavy metal distribution inPacific f erromanganese nodules and crusts. *Geol. Rundsch.* **1986**, *75*, 235–247. [CrossRef]
12. Hein, J.R.; Schwab, W.C.; Davis, A. Cobalt- and platinum-rich ferromanganese crusts and associated substrate rocks from the Marshall Islands. *Mar. Geol.* **1988**, *78*, 255–283. [CrossRef]
13. Koschinsky, A.; Halbach, P. Sequential leaching of marine ferromanganese precipitates: Genetic implications. *Geochim. Cosmochim. Acta* **1995**, *59*, 5113–5132. [CrossRef]
14. McMurtry, G.M.; VonderHaar, D.L.; Eisenhauer, A.; Mahoney, J.J.; Yeh, H.W. Cenozoic accumulation history of a Pacific ferromanganese crust. *Earth Planet. Sci. Lett.* **1994**, *125*, 105–118. [CrossRef]
15. Jiang, X.D.; Sun, X.M.; Guan, Y.; Gong, J.L.; Lu, Y.; Lu, R.F.; Wang, C. Biomineralisation of the ferromanganese crusts in the Western Pacific Ocean. *J. Asian Earth Sci.* **2017**, *136*, 58–67. [CrossRef]
16. Jiang, X.D.; Gong, J.L.; Ren, J.B.; Liu, Q.S.; Zhang, J.; Chou, Y.M. An interdependent relationship between microbial ecosystems and ferromanganese nodules from the Western Pacific Ocean. *Sediment. Geol.* **2020**, *398*, 105588. [CrossRef]
17. Sujith, P.P.; Gonsalves, M.J.B.D. Ferromanganese oxide deposits: Geochemical and microbiological perspectives of interactions of cobalt and nickel. *Ore Geol. Rev.* **2021**, *139*, 104458. [CrossRef]
18. Benites, M.; Hein, J.R.; Mizell, K.; Blackburn, T.; Jovane, L. Genesis and Evolution of Ferromanganese Crusts from the Summit of Rio Grande Rise, Southwest Atlantic Ocean. *Minerals* **2020**, *10*, 349. [CrossRef]
19. Chen, S.; Yin, X.; Wang, X.; Huang, X.; Ma, Y.; Guo, K.; Zeng, Z. The geochemistry and formation of ferromanganese oxides on the eastern flank of the Gagua Ridge. *Ore Geol. Rev.* **2018**, *95*, 118–130. [CrossRef]
20. Hein, J.R.; Conrad, T.; Mizell, K.; Banakar, V.K.; Frey, F.A.; Sager, W.W. Controls on ferromanganese crust composition and reconnaissance resource potential, Ninetyeast Ridge, Indian Ocean. *Deep-Sea Res. Part I Oceanogr. Res. Pap.* **2016**, *110*, 1–19. [CrossRef]

21. Hein, J.R.; Konstantinova, N.; Mikesell, M.; Mizell, K.; Fitzsimmons, J.N.; Lam, P.J.; Till, C.P. Arctic deep water ferromanganese-oxide deposits reflect the unique characteristics of the Arctic Ocean. *Geochem. Geophys. Geosyst.* **2017**, *18*, 3771–3800. [CrossRef]
22. Epp, D. Possible perturbations to hotspot traces and implications for the origin and structure of the Line Islands. *J. Geophys. Res.-Solid Earth* **1984**, *89*, 11273–11286. [CrossRef]
23. Lonsdale, P. Geography and history of the Louisville Hotspot Chain in the southwest Pacific. *J. Geophys. Res.* **1988**, *93*, 3078. [CrossRef]
24. Wessel, P.; Kroenke, L. A geometric technique for relocating hotspots and refining absolute plate motions. *Nature* **1997**, *387*, 365–369. [CrossRef]
25. He, G.; Zhao, Z.; Zhu, K. *Cobalt-Rich Crust Resources in the Western Pacific*; Geological Publishing House: Beijing, China, 2001. (In Chinese)
26. Koppers, A.P.; Staudigel, H.; Pringle, M.S.; Wijbrans, J.R. Short-lived and discontinuous intraplate volcanism in the South Pacific: Hot spots or extensional volcanism? *Geochem. Geophys. Geosyst.* **2003**, *4*, 1089. [CrossRef]
27. Ren, X.W. The Metallogenic System of Co-rich Manganese Crusts in Western Pacific. Master's Thesis, Institute of Oceanology, Chinese Academy of Sciences, Qingdao, China, 2005. (In Chinese).
28. Zhao, B.; Wei, Z.; Yang, Y.; He, G.; Zhang, H.; Ma, W. Sedimentary characteristics and the implications of cobalt-rich crusts resources at Caiwei Guyot in the Western Pacific Ocean. *Mar. Georesour. Geotechnol.* **2019**, *38*, 1037–1045. [CrossRef]
29. Yang, Y.; He, G.; Ma, J.; Yu, Z.; Yao, H.; Deng, X.; Liu, F.; Wei, Z. Acoustic quantitative analysis of ferromanganese nodules and cobalt-rich crusts distribution areas using EM122 multibeam backscatter data from deep-sea basin to seamount in Western Pacific Ocean. *Deep-Sea Res. Part I Oceanogr. Res. Pap.* **2020**, *161*, 103281. [CrossRef]
30. Kawabe, M.; Fujio, S. Pacific ocean circulation based on observation. *J. Oceanogr.* **2010**, *66*, 389–403. [CrossRef]
31. Gordon, A.L.; Visbeck, M.; Huber, B. Export of Weddell Sea deep and bottom water. *J. Geophys. Res.-Oceans* **2001**, *106*, 9005–9017. [CrossRef]
32. Kawabe, M.; Fujio, S.; Yanagimoto, D.; Tanaka, K. Water masses and currents of deep circulation southwest of the Shatsky Rise in the western North Pacific. *Deep-Sea Res. Part I Oceanogr. Res. Pap.* **2009**, *56*, 1675–1687. [CrossRef]
33. Johnson, H.P.; Hautala, S.L.; Bjorklund, T.A.; Zarnetske, M.R. Quantifying the North Pacific silica plume. *Geochem. Geophys. Geosyst.* **2006**, *7*, Q05011. [CrossRef]
34. Kawabe, M. Deep Water Properties and Circulation in the Western North Pacific. In *Elsevier Oceanography Series*; Elsevier: Amsterdam, The Netherlands, 1993; pp. 17–37.
35. Kawabe, M.; Taira, K. Flow distribution at 165° E in the Pacific Ocean. In *Biogeochemical Processes and Ocean Flux in the Western Pacific*; Sakai, H., Nozaki, Y., Eds.; Terra Scientific Publishing Company (TERRAPUB): Tokyo, Japan, 1995; pp. 629–649.
36. Kawabe, M.; Fujio, S.; Yanagimoto, D. Deep-water circulation at low latitudes in the western North Pacific. *Deep-Sea Res. Part I Oceanogr. Res. Pap.* **2003**, *50*, 631–656. [CrossRef]
37. Kawabe, M.; Yanagimoto, D.; Kitagawa, S.; Kuroda, Y. Variations of the deep western boundary current in Wake Island Passage. *Deep-Sea Res. Part I-Oceanogr. Res. Pap.* **2005**, *52*, 1121–1137. [CrossRef]
38. Kawabe, M.; Yanagimoto, D.; Kitagawa, S. Variations of deep western boundary currents in the Melanesian Basin in the western North Pacific. *Deep-Sea Res. Part I Oceanogr. Res. Pap.* **2006**, *53*, 942–959. [CrossRef]
39. Kato, F.; Kawabe, M. Volume transport and distribution of deep circulation at 165°W in the North Pacific. *Deep-Sea Res. Part I Oceanogr. Res. Pap.* **2009**, *56*, 2077–2087. [CrossRef]
40. Guo, B.; Wang, W.; Shu, Y.; He, G.; Zhang, D.; Deng, X.; Wang, J. Observed Deep Anticyclonic Cap Over Caiwei Guyot. *J. Geophys. Res. Ocean.* **2020**, *125*, e2020JC01625. [CrossRef]
41. Yeo, I.A.; Howarth, S.A.; Spearman, J.; Cooper, A.; Crossouard, N.; Taylor, J.; Murton, B.J. Distribution of and Hydrographic Controls on Ferromanganese Crusts: Tropic Seamount, Atlantic. *Ore Geol. Rev.* **2019**, *114*, 103131. [CrossRef]
42. Wu, G.; Zhou, H.; Zhang, H.; Pulyaeva, I.; Liu, J. Two main formation episodes of ferromanganese crusts in the Pacific Ocean. *Acta Geol. Sinica* **2006**, *80*, 12.
43. Rolf, G.L.; Todd, D.M. Topographically induce mixing around shallow seamount. *Science* **1997**, *276*, 1831–1833.
44. Lavelle, J.W.; Lozovatsky, I.D.; Smith, D.C. Tidally induced turbulent mixing at Irving Seamount-Modeling and measurements. *Geophys. Res. Lett.* **2004**, *31*, L10308. [CrossRef]
45. Rudnick, D.L.; Boyd, T.J.; Brainard, R.E.; Sanford, T.B. From tides to mixing along the Hawaiian ridge. *Science* **2003**, *301*, 355–357. [CrossRef]
46. McLennan, S.M. Rare earth elements in sedimentary rocks; influence of provenance and sedimentary processes. *Rev. Mineral. Geochem.* **1989**, *21*, 169–200.
47. Manheim, F.T.; Lane-Bostwick, C.M. Cobalt in ferromanganese crusts as a monitor of hydrothermal dischargeon the Pacificsea floor. *Nature* **1988**, *335*, 59–62. [CrossRef]
48. Von Stackelberg, U. Manganese Nodules of the Peru Basin. In *Handbook of Marine Mineral Deposits*; Cronan, D.S., Ed.; CRC Press: Boca Raton, FL, USA, 2000; pp. 197–238.
49. Glasby, G.P.; Mountain, B.; Vineesh, T.C.; Banakar, V.; Rajani, R.; Ren, X. Role of hydrology in the formation of Co-rich Mn crusts from the equatorial N Pacific, Equatorial S Indian Ocean and the NE Atlantic Ocean. *Resour. Geol.* **2020**, *60*, 165–177. [CrossRef]

50. Usui, A.; Nishi, K.; Sato, H.; Nakasato, Y.; Thornton, B.; Kashiwabara, T.; Urabe, T. Continuous growth of hydrogenetic ferromanganese crusts since 17 Myr ago on Takuyo-Daigo Seamount, NW Pacific, at water depths of 800–5500 m. *Ore Geol. Rev.* **2017**, *87*, 71–87. [CrossRef]
51. Li, Z.; Li, H.; Hein, J.R.; Dong, Y.; Wang, M.; Ren, X.; Chu, F. A possible link between seamount sector collapse and manganese nodule occurrence in the abyssal plains, NW Pacific Ocean. *Ore Geol. Rev.* **2021**, *138*, 104378. [CrossRef]
52. Halbach, P.; Hebisch, U.; Scherhag, C. Geochemical variations of ferromanganese nodules and crusts from different provinces of the Pacific Ocean and their genetic control. *Chem. Geol.* **1981**, *34*, 3–17. [CrossRef]
53. Josso, P.; Pelleter, E.; Pourret, O.; Fouquet, Y.; Etoubleau, J.; Cheron, S.; Bollinger, C. A new discrimination scheme for oceanic ferromanganese deposits using high field strength and rare earth elements. *Ore Geol. Rev.* **2017**, *87*, 3–15. [CrossRef]
54. Halbach, P.; Segl, M.; Puteanus, D.; Mangini, A. Co-fluxes and growth rates in ferromanganese deposits from central Pacific seamount areas. *Nature* **1983**, *304*, 716–719. [CrossRef]
55. Josso, P.; Rushton, J.; Lusty, P.; Matthews, A.; Chenery, S.; Holwell, D.; Murton, B. Late cretaceous and Cenozoic paleoceanography from north-East Atlantic ferromanganese crust microstratigraphy. *Mar. Geol.* **2020**, *422*, 106122. [CrossRef]
56. Piper, D.Z. Rare earth elements in ferromanganese nodules and other marine phases. *Geochim. Cosmochim. Acta* **1974**, *38*, 1007–1022. [CrossRef]
57. Amakawa, H.; Ingri, J.; Masuda, A.; Shimizu, H. Isotopic compositions of Ce, Nd and Sr in ferromanganese nodules from the Pacific and Atlan tic oceans, the Baltic and Barents Seas, and the Gulf of Bothni a. *Earth Planet. Sci. Lett.* **1991**, *105*, 554–565. [CrossRef]
58. Bau, M.; Koschinsky, A.; Dulski, P. Comparison of the partitioning behaviours of yttrium, rare earth elements, and titanium between hydrogenetic marine ferromanganese crusts and seawater. *Geochim. Cosmochim. Acta* **1996**, *60*, 1709–1725. [CrossRef]
59. Menendez, A.; James, R.H.; Roberts, S. Controls on the distribution of rare earth elements in deep-sea sediments in the North Atlantic Ocean. *Ore Geol. Rev.* **2017**, *87*, 100–113. [CrossRef]
60. Giovanoli, R. Vernadite is random-stacked birnessite-A discussion of the paper by F. V. Chukhrov et al.: Contributions to the mineralogy of authigenic manganese phases from marine manganese deposits. *Miner. Depos.* **1980**, *15*, 251–253.
61. Villalobos, M.; Lanson, B.; Manceau, A.; Toner, B.; Sposito, G. Structural model for the biogenic Mn oxide produced by Pseudomonas putida. *Am. Miner.* **2006**, *91*, 489–502. [CrossRef]
62. Grangeon, S.; Lanson, B.; Lanson, M.; Manceau, A. Crystal structure of Ni-sorbed synthetic vernadite: A powder X-ray diffraction study. *Mineral. Mag.* **2008**, *72*, 1279–1291. [CrossRef]
63. Feng, X.H.; Tan, W.F.; Liu, F.; Wang, J.B.; Ruan, H.D. Synthesis of todorokite at atmospheric pressure. *Chem. Mat.* **2004**, *16*, 4330–4336. [CrossRef]
64. Post, J.E.; Heaney, P.J.; Hanson, J. Synchrotron X-ray diffraction study of the structure and dehydration behavior of todorokite. *Am. Miner.* **2003**, *88*, 142–150. [CrossRef]
65. Kim, J.; Kwon, K.D. Tunnel cation and water structures of todorokite: Insights into metal partitioning and isotopic fractionation. *Geochim. Cosmochim. Acta* **2022**, in press. [CrossRef]
66. Koschinsky, A.; Stascheit, A.; Bau, M.; Halbach, P. Effects of phosphatization on the geochemical and mineralogical composition of marine ferromanganese crusts. *Geochim. Cosmochim. Acta* **1997**, *61*, 4079–4094. [CrossRef]
67. Marino, E.; González, F.J.; Kuhn, T.; Madureira, P.; Wegorzewski, A.V.; Mirao, J.; Lunar, R. Hydrogenetic, Diagenetic and Hydrothermal Processes Forming Ferromanganese Crusts in the Canary Island Seamounts and Their Influence in the Metal Recovery Rate with Hydrometallurgical Methods. *Minerals* **2019**, *9*, 439. [CrossRef]
68. Wegorzewski, A.V.; Grangeon, S.; Webb, S.M.; Heller, C.; Kuhn, T. Mineralogical transformations in polymetallic nodules and the change of Ni, Cu and Co crystal-chemistry upon burial in sediments. *Geochim. Cosmochim. Acta* **2022**, *282*, 19–37. [CrossRef]
69. Zhong, Y.; Liu, Q.S.; Chen, Z.; Gonzalez, F.J.; Hein, J.R.; Zhang, J.; Zhong, L.F. Tectonic and paleoceanographic conditions during the formation of ferromanganese nodules from the northern South China Sea based on the high-resolution geochemistry, mineralogy and isotopes. *Mar. Geol.* **2019**, *410*, 146–163. [CrossRef]
70. Groves, D.I.; Santosh, M.; Müller, D.; Zhang, L.; Deng, J.; Yang, L.Q.; Wang, Q.F. Mineral systems: Their advantages in terms of developing holistic genetic models and for target generation in global mineral exploration. *Geosyst. Geoenviron* **2022**, *1*, 100001. [CrossRef]
71. Smoot, N.C. Orthogonal Intersections of Megatrends in the Western Pacific Ocean Basin: A Case Study of the Mid-Pacific Mountains. *Geomorphology* **1999**, *30*, 323–356. [CrossRef]
72. Bischoff, J.L.; Piper, D.Z.; Leong, K. The aluminosilicate fraction of North Pacific manganese nodules. *Geochim. Cosmochim. Acta* **1981**, *45*, 2047–2063. [CrossRef]
73. Hem, J.D. Rates of manganese oxidation in aqueous systems. *Geochim. Cosmochim. Acta* **1981**, *45*, 1369–1374. [CrossRef]
74. Wise, W.S. Minerals. Zeolites. In *Encyclopedia of Geology*; Selley, R.C., Cocks, L.R.M., Plimer, I.R., Eds.; Elsevier: Oxford, UK, 2005; pp. 591–600.
75. Zachos, J.; Pagani, M.; Sloan, L.; Thomas, E.; Billups, K. Trends, rhythms, and aberrations in global climate 65 Ma to present. *Science* **2001**, *292*, 686–693. [CrossRef]
76. Wu, G.H.; Zhou, H.Y.; Zhang, H.S.; Ling, H.F.; Ma, W.L.; Zhao, H.Q.; Chen, J.L.; Liu, J.H. New index of ferromanganese crusts reflecting oceanic environmental oxidation. *Sci. China Earth Sci.* **2007**, *50*, 371–384. [CrossRef]

77. Aplin, A.; Michard, A.; Albarede, F. ^{143}Nd/^{144}Nd in Pacific ferromanganese encrustations and nodules. *Earth Planet. Sci. Lett.* **1986**, *81*, 7–14. [CrossRef]

78. Banakar, V.K.; Borole, D.V. Depth profiles of 230Th excess transition metals and mineralogy of ferromanganese crusts of the Central Indian basin and implications for paleoceanographic influence on crust genesis. *Chem. Geol.* **1991**, *94*, 33–44. [CrossRef]

79. Liu, Q.; Huo, Y.Y.; Wu, Y.H.; Bai, Y.; Yuan, Y.; Chen, M. Bacterial community on a guyot in the northwest Pacific Ocean influenced by physical dynamics and environmental variables. *J. Geophys. Res.-Biogeosci.* **2019**, *124*, 2883–2897. [CrossRef]

80. Halbach, P.; Puteanus, D. The influence of the carbonate dissolution rate on the growth and composition of Co-rich ferromanganese crusts from Central Pacific seamount areas. *Earth Planet. Sci. Lett.* **1984**, *68*, 73–87. [CrossRef]

81. Byrne, R.H. Inorganic speciation of dissolved elements in seawater: The influence of pH on concentration ratios. *Geochem. Trans.* **2002**, *3*, 11–16. [CrossRef]

82. Hein, J.R.; Koschinsky, A.; Kuhn, T. Deep-ocean polymetallic nodules as a resource for critical materials. *Nat. Rev. Earth Environ.* **2020**, *1*, 158–169. [CrossRef]

83. Schnitker, D. North Atlantic oceanography as possible cause of Antarctic glaciation and eutrophication. *Nature* **1980**, *284*, 615–616. [CrossRef]

84. Kennett, J.P. Paleo-oceanography: Global ocean evolution. *Rev. Geophys. Space. Phys.* **1983**, *21*, 1258–1274.

85. Halbach, P.E.; Jahn, A.; Cherkashov, G. Marine co-rich ferromanganese crust deposits: Description and formation, occurrences and distribution, estimated worldwide resources. In *Deep-Sea Mining: Resource Potential, Technical and Environmental Considerations*; Sharma, R., Ed.; Springer: Cham, Germany, 2007; pp. 1–535.

86. Marino, E.; González, F.J.; Somoza, L.; Lunar, R.; Ortega, L.; Vázquez, J.T.; Reyes, J.; Bellido, E. Trategic and rare elements in Cretaceous-Cenozoic cobalt-rich ferromanganese crusts from seamounts in the Canary Island Seamount Province (northeastern tropical Atlantic). *Ore Geol. Rev.* **2017**, *87*, 41–61. [CrossRef]

87. Dymond, J.; Lyle, M.; Finney, B.; Piper, D.Z.; Murphy, K.; Conard, R.; Pisoas, N. Ferromanganese nodules from the MANOP Sites H, S, and R-Control of mineralogical and chemical composition by multiple accretionary processes. *Geochim. Cosmochim. Acta* **1984**, *48*, 931–949. [CrossRef]

88. Lyle, M.; Heath, G.R.; Robbins, J.M. Transport and release of transition elements during early diagenesis: Sequential leaching of sediments from MANOP sites M and H, Part, I. pH5 acetic acid leach. *Geochim. Cosmochim. Acta* **1984**, *48*, 1705–1715. [CrossRef]

89. Wegorzewski, A.V.; Kuhn, T. The influence of suboxic diagenesis on the formation of manganese nodules in the Clarion Clipperton nodule belt of the Pacific Ocean. *Mar. Geol.* **2014**, *357*, 123–138. [CrossRef]

90. Yang, Z.; Qian, Q.; Chen, M.; Zhang, R.; Yang, W.; Zheng, M.; Qiu, Y. Enhanced but highly variable bioturbation around seamounts in the northwest Pacific. *Deep-Sea Res. Part I Oceanogr. Res. Pap.* **2019**, *156*, 103190. [CrossRef]

Journal of
Marine Science and Engineering

Brief Report

Timescales of Magma Mixing Beneath the Iheya Ridge, Okinawa Trough: Implications for the Stability of Sub-Seafloor Magmatic Systems

Yuxiang Zhang [1,2,3,*], **Zuxing Chen** [1,2,3] **and Zhigang Zeng** [1,2,3,*]

1   Seafloor Hydrothermal Activity Laboratory, CAS Key Laboratory of Marine Geology and Environment, Institute of Oceanology, Chinese Academy of Sciences, Qingdao 266071, China
2   Laboratory for Marine Mineral Resources, Qingdao National Laboratory for Marine Science and Technology, Qingdao 266071, China
3   Center for Ocean Mega-Science, Chinese Academy of Sciences, Qingdao 266071, China
*   Correspondence: yxzhang@qdio.ac.cn (Y.Z.); zgzeng@qdio.ac.cn (Z.Z.)

Abstract: Submarine volcanic eruptions can be destructive for marine environments and resources. Magma mixing is considered to be an important trigger for volcanic eruptions. Determining the magma residence time from mixing to eruption is conducive to assessing the stability of magmatic systems, especially beneath the seafloor where in situ volcano monitoring is inaccessible. Here, we estimated the timescale of magma mixing beneath the Iheya Ridge, Okinawa Trough, which is characterized by pervasive magma mixing. We focused on andesitic and rhyolitic magma generated by basalt–rhyolite mixing and rhyolite–rhyolite mixing, respectively. By taking advantage of the Mg diffusion chronometry, we showed that the andesitic magma resided in the magma chamber for very short time (~0.1–0.3 years), whereas the residence time of the rhyolitic magma was much longer (~80–120 years). The different times might be in part related to the different rheology of the mixed magmas. The short residence time of the andesitic magma suggested efficient magma mixing that allowed the andesites to be erupted, which may explain the appearance of scarce andesites in basalt–rhyolite dominant settings. However, the rapid mixing and eruption of magma is a disadvantage for the development and preservation of seafloor hydrothermal resources. Therefore, we suggest that the stability of sub-seafloor magma systems must be evaluated during the assessment of seafloor sulfide resources and mining prospects.

Keywords: magma residence time; magma mixing; diffusion chronometry; magma eruption; seafloor hydrothermal activity

Citation: Zhang, Y.; Chen, Z.; Zeng, Z. Timescales of Magma Mixing Beneath the Iheya Ridge, Okinawa Trough: Implications for the Stability of Sub-Seafloor Magmatic Systems. *J. Mar. Sci. Eng.* **2023**, *11*, 375. https://doi.org/10.3390/jmse11020375

Academic Editor: János Kovács

Received: 25 December 2022
Revised: 29 January 2023
Accepted: 30 January 2023
Published: 8 February 2023

1. Introduction

Sub-seafloor magma systems are huge energy reservoirs, which are not only the most important heat source for seafloor hydrothermal activities but can also supply volatiles and/or metals to hydrothermal systems [1–3]. However, submarine volcanic eruptions can be extremely destructive. For example, they can trigger serious natural disasters such as earthquakes and tsunamis [4,5]. In addition, huge amounts of heat and toxic gases emitted from the seafloor could have catastrophic impacts on marine environments and ecosystems [3,6]. Furthermore, they can also damage previously formed hydrothermal systems, which is a potential threat that should not be ignored in the assessment of seafloor sulfide resources and mining feasibility. Therefore, it is essential to evaluate the stability of sub-seafloor magma systems and assess the residence times of magmas prior to eruption.

While it is difficult to monitor magmatic activity beneath the seafloor in situ, the instabilities of a deep magma reservoir may be partly reflected by complex magmatic processes such as magma mixing, which can be recorded by the mineralogy of erupted lavas [7–12]. The timescales of magmatic processes can be estimated by diffusive re-equilibration within crystals [13–21].

The Okinawa Trough (OT) is a young back-arc basin that is characterized by complex volcanic systems, widespread seafloor hydrothermal systems, and extremely high heat flow values [22–24]. Widespread magma mixing is suggested to have occurred in the shallow magma chambers of the OT [25,26]. However, the timescales of magma mixing have not been well constrained. In this study, we focused on the magma system beneath the Iheya Ridge, where basalt–rhyolite mixing and rhyolite–rhyolite mixing have been recorded by the chemical zonation of minerals [26], and estimated the residence times of magmas from mixing to eruption using diffusion chronometry.

2. Geological Setting and Samples

The OT is located in the eastern margin of the East China Sea Shelf (Figure 1). It is a newly developed back-arc basin, and its initial extension can be traced back to the middle-to-late Miocene in response to the subduction of the Philippine Sea plate [27]. The OT is generally divided into three segments from the north to the south, namely the NOT, MOT, and SOT. Several central grabens, which represent the active back-arc spreading centers, have developed in the MOT and SOT. The SOT and MOT are also characterized by widespread hydrothermal activities and complex volcanic systems [23,28].

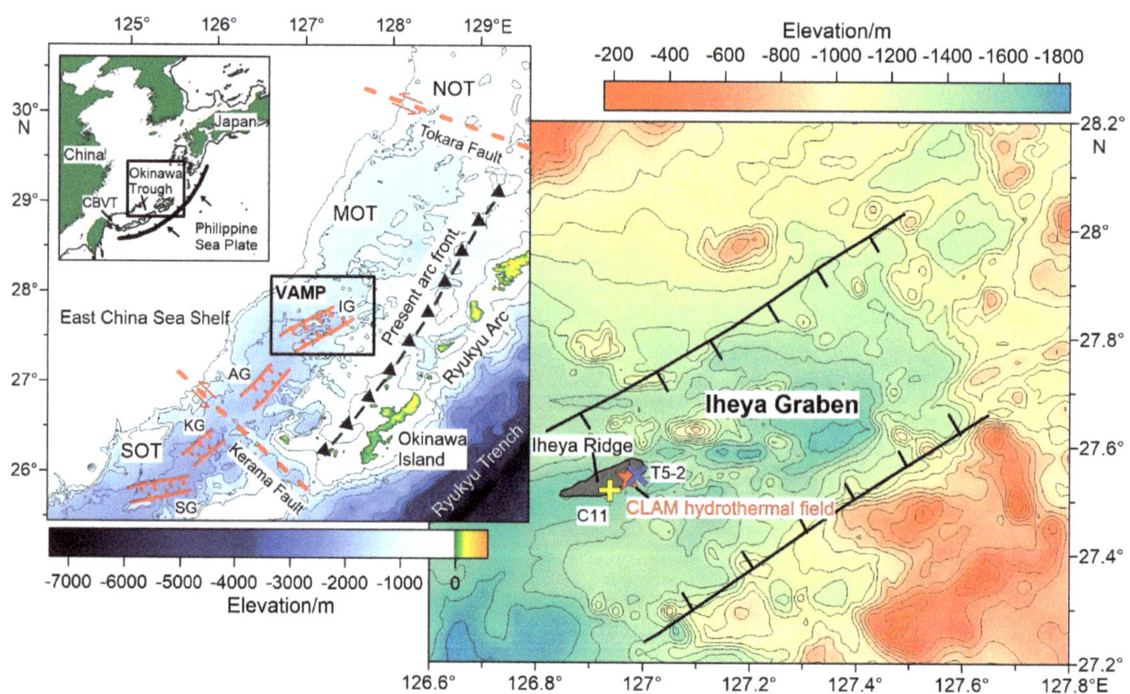

Figure 1. Bathymetric map of the Iheya Graben showing the locations of the Iheya Ridge (after [26]), samples (andesite T5-2 and rhyolite C11) and CLAM hydrothermal field. Abbreviations: NOT—northern Okinawa Trough; MOT—middle Okinawa Trough; SOT—southern Okinawa Trough. IG—Iheya Graben; AG—Aguni Graben, KG—Kerama Graben, and SG—Sakishima Graben.

The Iheya Ridge is a volcanic ridge located in the central axis of the Iheya Graben, which also represents the center of an anomalous volcanic zone (known as the VAMP area) (Figure 1). The volcanic rocks in the Iheya Ridge consist of basalts, andesites, and rhyolites. Multi-level magma storage and pervasive magma mixing are the key features of the magma plumbing system beneath this area [26]. An active hydrothermal field (CLAM) has been reported on the Iheya Ridge.

The samples used in this study included andesite (T5-2) and rhyolite (C11) collected from the Iheya Ridge. The detailed whole-rock geochemistry and mineralogical compositions of the Iheya Ridge have been reported previously [26]. Andesite T5-2 contains three groups of minerals that are in equilibrium with (1) rhyolitic melt, including low-Ca plagioclase, low-Mg orthopyroxene, and clinopyroxene with reverse zonations, (2) basaltic melt, including high-Ca plagioclase and high-Mg clinopyroxene with normal zonations, and (3) andesitic melt, including unzoned clinopyroxene and weakly zoned orthopyroxene (Figure S1). This indicates that the andesite used in this study was a mixture of basaltic and rhyolitic melts. Rhyolite C11 contains two groups of minerals that are in equilibrium with (1) host melt, including reversely zoned plagioclase and orthopyroxene, and (2) a less-evolved rhyolitic melt, including normally zoned plagioclase, unzoned orthopyroxene, and clinopyroxene (Figure S2). This suggests that the rhyolite used in this study was formed from a hybridization of rhyolitic melts with different evolution degrees.

3. Methods

We estimated the magma residence time from mixing to eruption beneath the Iheya Ridge using Mg diffusion in olivine, orthopyroxene, and plagioclase crystals. We chose olivine and reversely zoned orthopyroxene for andesite T5-2 and reversely zoned orthopyroxene and plagioclase and normally zoned plagioclase for rhyolite C11. These minerals all have euhedral crystal shapes, except for olivine (Figures 2 and 3). The olivine was corroded due to it mixing with rhyolitic melt (Figure 2a). No recrystallization of the olivine occurred because the hybrid andesitic melt destabilized the olivine. Instead, some tiny pigeonites were found to surround the olivine due to the reaction between olivine and rhyolitic melt (Figure 2a) [26].

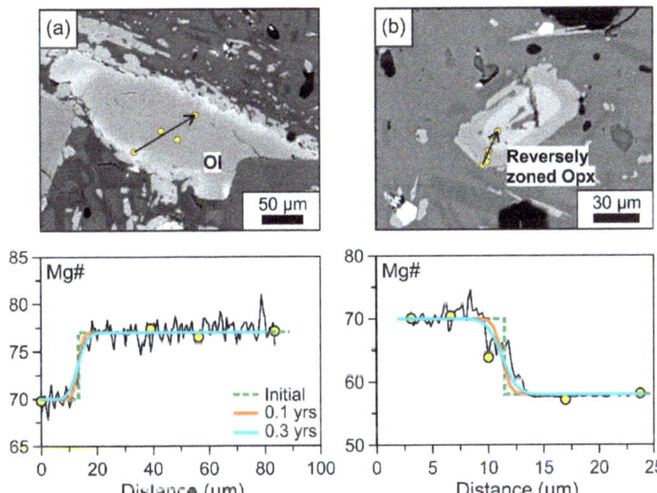

Figure 2. Modeling results of Mg diffusion in the olivine (**a**) and orthopyroxene (**b**) in andesite T5-2. The black curves were obtained from the gray values of BSE images. The yellow spots denote the data obtained by EMPA.

Because the Mg concentrations in olivine and orthopyroxene are proportional to the greyscale of the back-scattered electron (BSE) images of these minerals, the Mg profiles of the olivine and orthopyroxene in this study were calibrated from the greyscale values of high-resolution BSE images [29]. This method can obtain high-resolution Mg profiles. The BSE images were captured using an electron probe (EMPA) at the Ocean University of China under a beam current of 20 nA and at an accelerating voltage of 15 kV. The greyscale values of the BSE images were extracted using the ImageJ software. The greyscale values were then converted into Mg concentrations based on several (~4–6 points) EMPA analyses along the profiles. The Mg concentrations in the plagioclase were analyzed using laser ablation inductively coupled plasma mass spectrometry (LA-ICP-MS) at the State Key Laboratory of Isotope Geochemistry, Guangzhou Institute of Geochemistry, Chinese Academy of Sciences. The detailed analytical methods for EMPA and LA-ICP-MS are given in [26]. The analytical results are given in Data S1.

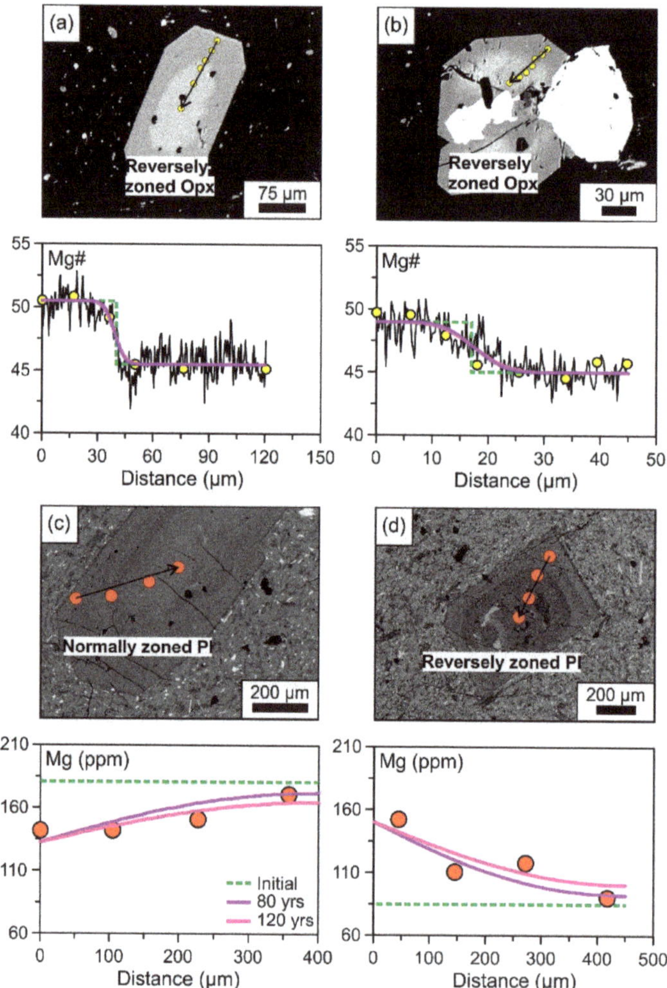

Figure 3. Modeling results of Mg diffusion in the orthopyroxene (**a**,**b**) and plagioclase (**c**,**d**) in rhyolite C11. The black curves in (**a**,**b**) were obtained from the gray values of BSE images. The yellow and red spots denote the data obtained by EMPA and LA-ICP-MS, respectively.

The Mg profiles of the crystals and their evolution with time can be modelled using the equation:

$$\partial C_i / \partial t = D_i \times \partial^2 C_i / \partial x^2 \tag{1}$$

where C_i and D_i represent the concentration and diffusion coefficient of the element i, respectively. The diffusion coefficients and P–T conditions used in the modelling are listed in Table 1. The crystal orientation was estimated according to the crystal shapes. Note that Mg diffusion in plagioclase is isotropic [30]. We assumed that the plagioclases had a homogeneous initial MgO concentration distribution. For the orthopyroxenes and olivines, we assumed that they had a homogeneous core with a higher/lower Mg# value surrounded by a homogeneous mantle with lower/higher Mg# value at the beginning of diffusion. The outermost rim compositions were used as the boundary conditions.

Table 1. Mg diffusion coefficients in olivine, orthopyroxene, and plagioclase.

	DMg ($\mu m^2/s$)	Orientation	P (kbar) [1]	T (°C) [1]	fO_2 ($\triangle NNO$) [1]	X_{Mg}/X_{An}	Reference
Ol_T5-2	4.08×10^{-7}	[010]	2	970	−0.3	0.8	[31]
Opx_T5-2	6.22×10^{-8}	[001]	2	970	−0.3	\	[32]
Opx_C11	3.71×10^{-9}	[001]	2	870	−0.3	\	[32]
Pl_C11	1.11×10^{-5}	\	2	870	−0.3	0.35	[30]

[1] According to [26].

4. Results and Discussions

4.1. Timescales of Magma Mixing Beneath the Iheya Ridge

The diffusion modelling results are shown in Figures 2 and 3. In the andesite T5-2, the Mg diffusion in the olivine yielded a time of 0.1–0.3 years (Figure 2a). While the olivine crystal was slightly corroded, the Fe-rich rim had constant Mg# values, suggesting that the corrosion had little effect on the Mg diffusion profile across the Fe-rich and Mg-rich boundary. This time scale was quite consistent with that obtained from the Mg diffusion in orthopyroxene (0.1–0.3 years), which had an euhedral crystal shape (Figure 2b). In addition, the textures of the olivines in the T5-2 andesite (Figure 2a) were similar to those in the experiment of [33], where the olivines contacted with rhyodacitic and rhyolitic melt for up to 622 h at 800–885 °C and 50–150 MPa. This further suggested that the olivines in the andesite T5-2 contacted with the mixed magma for a very short time. In the rhyolite C11, the orthopyroxene crystals yielded a much longer magma residence time (80–120 years) (Figure 3a,b), which was consistent with the results calculated from the normally zoned plagioclase and the reversely zoned plagioclase (Figure 3c,d), although the latter were not well constrained due to the limited data.

It is noteworthy that the timescale calculation here was not a precise dating, as the diffusion coefficients were not constant but were dependent on the crystal compositions, crystal orientation, temperature, and so on [15,17,30–32]. Other uncertainties came from the choice of the initial and boundary conditions [17]. For example, the initial Mg concentrations of the plagioclases in the rhyolite C11 may not have been homogeneous. However, similar timescales yielded by multiple crystals indicated that our results were a plausible estimation.

Despite this uncertainty, our results suggested different magma residence times (from <1 year to ~100 years) beneath the Iheya Graben. The controlling factors on the timescales of magma mixing might be diverse, e.g., tectonic settings and the physical and chemical conditions of the magmas. Because the andesite T5-2 and rhyolite C11 used in this study were from the same location, their different magma residence times might be in part related to the different rheology and temperature of the mixed magmas. For example, rhyolite C11 is a mixture of rhyolitic melts [26], which have higher viscosities due to their higher SiO_2 concentrations and relatively lower temperatures. A high viscosity and a lower temperature would retard the magma flow and reduce the efficiency of magma mixing [34]. A long magma residence time (~600 year) was also reported in the southernmost part of the SOT, where the rhyolitic magma was intruded by small amount of basaltic magma (present as a mafic magma enclave) [25], suggesting inefficient magma mixing and a relatively high viscosity. In comparison, the andesite T5-2 in this study was generated by the mixing of a basaltic magma with a rhyolitic magma at a ratio of 4:6 [26]. The injection of a high proportion (40%) of high-temperature and low-SiO_2 basaltic magma would form a mixed magma with relatively low viscosity.

4.2. Rapid Magma Mixing and Eruption of Andesitic Magma in Bimodal Volcanic Series

The volcanic rocks in the OT are characterized by bimodal distributions, that is, basalts and rhyolites are dominant in the OT with scarce andesites [35–37]. The andesites have been suggested to have mainly been sourced from the mixing of basaltic magma with rhyolitic magma [26]. Therefore, the amount of andesites in the OT is probably dependent on the efficiency of magma mixing and eruption.

Magma mixing is considered to be an important trigger for eruptions [9,10]. An extremely short time between mixing and eruption has been reported (e.g., days to months) [21,38]. The short residence time (~0.1–0.3 years) of T5-2 magma after the magma mixing implied that the injection of basaltic magma into the rhyolitic magma may have led to the instability of the magma reservoir and the subsequent eruption. The efficient magma mixing suggested that the Iheya Ridge is a place that is favorable for the eruption of andesitic magma, which was consistent with the observation that the andesites recovered in the OT (excluding the modern arc fronts) so far were mostly from the Iheya Ridge and adjacent regions [26,36].

4.3. Implications for the Development and Preservation of Seafloor Hydrothermal Activities in the OT

The formation of seafloor hydrothermal activities involves the penetration of seawater into the crust, fluid–rock interactions heated by shallow magma, the rise of hydrothermal fluids, and the precipitation of hydrothermal products (e.g., sulfides), which may take from years to decades to occur [39]. Therefore, a long-term stable sub-seafloor magma system is essential for the development of a seafloor hydrothermal system with potential mineral resources. For example, the formation of a CLAM hydrothermal field on the Iheya Ridge might be attributed to a rhyolitic magma reservoir similar to that of C11 rhyolite. Similarly, long-term magma storage (~600 years) has also been reported near the Tangyin hydrothermal field [25]. However, the coexistence of a short magma storage suggests that the CLAM hydrothermal field may not have good prospects for resource exploitation because the rapid mixing and eruption could ruin the previously formed hydrothermal deposits and also present a great threat for mining activities. Therefore, evaluating the stability of sub-seafloor magmatic systems is a critical step in the assessment of seafloor hydrothermal resources.

5. Conclusions

The timescale of magma mixing beneath the Iheya Ridge was estimated by using Mg diffusion chronometry. The andesitic magma generated by the mixing between the basaltic and rhyolitic magmas experienced a very-short-time (~0.1–0.3 years) residence prior to eruption. In comparison, the rhyolitic magma generated by rhyolite–rhyolite mixing resided in the magma chamber for a much longer time (~80–120 years), which might have been in part related to the higher viscosity of the mixed magma. The rapid basalt–rhyolite mixing and eruption was favorable to the appearance of andesites in a bimodal volcanic series, but this is a disadvantage for the development and preservation of seafloor hydrothermal resources. This study thus emphasized the importance of evaluating the stability of sub-seafloor magma systems during the assessment of seafloor sulfide resources and mining prospects.

Supplementary Materials: The following supporting information can be downloaded at: https://www.mdpi.com/article/10.3390/jmse11020375/s1, Figure S1: Photomicrographs of three groups of minerals in andesite T5-2; Figure S2. Photomicrographs of two groups of minerals in rhyolite C11; Data S1: Mg concentrations of olivine, orthopyroxene, and plagioclase.

Author Contributions: Conceptualization, Y.Z.; methodology, Y.Z.; investigation, Y.Z.; resources, Z.Z.; data curation, Y.Z.; writing—original draft preparation, Y.Z.; writing—review and editing, Z.C. and Z.Z.; funding acquisition, Y.Z. and Z.Z. All authors have read and agreed to the published version of the manuscript.

Funding: This research was funded by the National Natural Science Foundation of China (91958213 and 42206058), the Strategic Priority Research Program (B) of Chinese Academy of Sciences (XDB42020402), the Shandong Provincial Natural Science Foundation, China (ZR2020QD068), and the Taishan Scholars Program.

Institutional Review Board Statement: Not applicable.

Informed Consent Statement: Not applicable.

Data Availability Statement: All the data that support the findings in this study are given in the Supporting Information or by contacting the corresponding authors.

Acknowledgments: We are grateful for the crew on the HOBAB4 cruise in 2016 for their help in collecting the samples.

Conflicts of Interest: The authors declare no conflict of interest.

References

1. Yang, K.; Scott, S.D. Possible contribution of a metal-rich magmatic fluid to a sea-floor hydrothermal system. *Nature* **1996**, *383*, 420–423. [CrossRef]
2. Ishibashi, J I.; Urabe, T. Hydrothermal activity related to arc-backarc magmatism in the Western Pacific. In *Back Arc Basins*; Taylor, B., Ed.; Plenum Press: New York, NY, USA, 1995; pp. 451–495.
3. Butterfield, D.A.; Nakamura, K.-I.; Takano, B.; Lilley, M.D.; Lupton, J.E.; Resing, J.A., Roe, K.K. High SO_2 flux, sulfur accumulation, and gas fractionation at an erupting submarine volcano. *Geology* **2011**, *39*, 803–806. [CrossRef]
4. Kienle, J.; Kowalik, Z.; Murty, T.S. Tsunamis generated by eruptions from Mount St. Augustine volcano, Alaska. *Science* **1987**, *236*, 1442–1447. [CrossRef] [PubMed]
5. McCoy, F.W.; Heiken, G. Tsunami Generated by the Late Bronze Age Eruption of Thera (Santorini), Greece. *Pure Appl. Geophys.* **2000**, *157*, 1227–1256. [CrossRef]
6. Esposito, A.; Giordano, G.; Anzidei, M. The 2002–2003 submarine gas eruption at Panarea volcano (Aeolian Islands, Italy): Volcanology of the seafloor and implications for the hazard scenario. *Mar. Geol.* **2006**, *227*, 119–134. [CrossRef]
7. Cashman, K.V.; Sparks, R.S.J.; Blundy, J.D. Vertically extensive and unstable magmatic systems: A unified view of igneous processes. *Science* **2017**, *355*, eaag3055. [CrossRef]
8. Geiger, H.; Troll, V.R.; Jolis, E.M.; Deegan, F.M.; Harris, C.; Hilton, D.R.; Freda, C. Multi-level magma plumbing at Agung and Batur volcanoes increases risk of hazardous eruptions. *Sci. Rep.* **2018**, *8*, 10547. [CrossRef]
9. Murphy, M.D.; Sparks, R.S.J.; Barclay, J.; Carroll, M.R.; Lejeune, A.M.; Brewer, T.S.; Macdonald, R.; Black, S.; Young, S. The role of magma mixing in triggering the current eruption at the Soufriere Hills Volcano, Montserrat, West Indies. *Geophys. Res. Lett.* **1998**, *25*, 3433–3436. [CrossRef]
10. Sparks, S.R.J.; Sigurdsson, H.; Wilson, L. Magma mixing: A mechanism for triggering acid explosive eruptions. *Nature* **1977**, *267*, 315–318. [CrossRef]
11. Kahl, M., Chakraborty, S.; Costa, F.; Pompilio, M. Dynamic plumbing system beneath volcanoes revealed by kinetic modeling, and the connection to monitoring data; An example from Mt. Etna. *Earth Planet Sci. Lett.* **2011**, *308*, 11–22. [CrossRef]
12. Kahl, M.; Chakraborty, S.; Pompilio, M.; Costa, F. Constraints on the Nature and Evolution of the Magma Plumbing System of Mt. Etna Volcano (1991–2008) from a Combined Thermodynamic and Kinetic Modelling of the Compositional Record of Minerals. *J. Petrol.* **2015**, *56*, 2025–2068. [CrossRef]
13. Albert, H.; Costa, F.; Martí, J. Timing of magmatic processes and unrest associated with mafic historical monogenetic eruptions in Tenerife Island. *J. Petrol.* **2015**, *56*, 1945–1966. [CrossRef]
14. Cooper, G.F.; Morgan, D.J.; Wilson, C.J.N. Rapid assembly and rejuvenation of a large silicic magmatic system: Insights from mineral diffusive profiles in the Kidnappers and Rocky Hill deposits, New Zealand. *Earth Planet Sci. Lett.* **2017**, *473*, 1–13. [CrossRef]
15. Costa, F., Chakraborty, S.; Dohmen, R. Diffusion coupling between trace and major elements and a model for calculation of magma residence times using plagioclase. *Geochim. Cosmochim. Acta* **2003**, *67*, 2189–2200. [CrossRef]
16. Costa, F.; Coogan, L.A.; Chakraborty, S. The time scales of magma mixing and mingling involving primitive melts and melt–mush interaction at mid-ocean ridges. *Contrib. Mineral. Petrol.* **2010**, *159*, 371–387. [CrossRef]
17. Costa, F.; Dohmen, R.; Chakraborty, S. Time scales of magmatic processes from modeling the zoning patterns of crystals. *Rev. Mineral. Geochem.* **2008**, *69*, 545–594. [CrossRef]
18. Lynn, K.J.; Garcia, M.O.; Shea, T.; Costa, F.; Swanson, D.A. Timescales of mixing and storage for Keanakāko'i Tephra magmas (1500–1820 C.E.), Kīlauea Volcano, Hawai'i. *Contrib. Mineral. Petrol.* **2017**, *172*, 76. [CrossRef]
19. Nishi, Y.; Ban, M.; Takebe, M.; Álvarez-Valero, A.M.; Oikawa, T.; Yamasaki, S. Structure of the shallow magma chamber of the active volcano Mt. Zao, NE Japan: Implications for its eruptive time scales. *J. Volcanol. Geotherm. Res.* **2019**, *371*, 137–161. [CrossRef]

20. Petrone, C.M.; Braschi, E.; Francalanci, L.; Casalini, M.; Tommasini, S. Rapid mixing and short storage timescale in the magma dynamics of a steady-state volcano. *Earth Planet Sci. Lett.* **2018**, *492*, 206–221. [CrossRef]
21. Tomiya, A.; Miyagi, I.; Saito, G.; Geshi, N. Short time scales of magma-mixing processes prior to the 2011 eruption of Shinmoedake volcano, Kirishima volcanic group, Japan. *Bull. Volcanol.* **2013**, *75*, 750. [CrossRef]
22. Glasby, G.P.; Notsu, K. Submarine hydrothermal mineralization in the Okinawa Trough, SW of Japan: An overview. *Ore Geol. Rev.* **2003**, *23*, 299–339. [CrossRef]
23. Sibuet, J.C.; Deffontaines, B.; Hsu, S.K.; Thareau, N.; Formal, L.; Liu, C.S. Okinawa Trough backarc basin: Early tectonic and magmatic evolution. *J. Geophys. Res. Solid Earth* **1998**, *103*, 30245–30267. [CrossRef]
24. Zhang, L.; Luan, X.; Zeng, Z.; Liu, H. Key parameters of the structure and evolution of the Okinawa Trough: Modelling results constrained by heat flow observations. *Geol. J.* **2019**, *54*, 3542–3555. [CrossRef]
25. Chen, Z.; Zeng, Z.; Wang, X.; Peng, X.; Zhang, Y.; Yin, X.; Chen, S.; Zhang, L.; Qi, H. Element and Sr isotope zoning in plagioclase in the dacites from the southwestern Okinawa Trough: Insights into magma mixing processes and time scales. *Lithos* **2020**, *376–377*, 105776. [CrossRef]
26. Zhang, Y.; Zeng, Z.; Gaetani, G.; Zhang, L.; Lai, Z. Mineralogical constraints on the magma mixing beneath the iheya graben, an active back-arc spreading center of the Okinawa trough. *J. Petrol.* **2020**, *61*, egaa098. [CrossRef]
27. Sibuet, J.C.; Letouzey, J.; Barbier, F.; Charvet, J.; Foucher, J.P.; Hilde, T.W.; Kimura, M.; Chiao, L.Y.; Marsset, B.; Muller, C. Back arc extension in the Okinawa Trough. *J. Geophys. Res. Solid Earth* **1987**, *92*, 14041–14063. [CrossRef]
28. Zhang, Y.; Zeng, Z.; Wang, X.; Chen, S.; Yin, X.; Chen, Z. Geologic control on hydrothermal activities in the Okinawa Trough. *Adv. Earth Sci.* **2020**, *35*, 678–690.
29. Fabbro, G.N.; Druitt, T.H.; Costa, F. Storage and Eruption of Silicic Magma across the Transition from Dominantly Effusive to Caldera-forming States at an Arc Volcano (Santorini, Greece). *J. Petrol.* **2017**, *58*, 2429–2464. [CrossRef]
30. Van Orman, J.A.; Cherniak, D.J.; Kita, N.T. Magnesium diffusion in plagioclase: Dependence on composition, and implications for thermal resetting of the ^{26}Al–^{26}Mg early solar system chronometer. *Earth Planet Sci. Lett.* **2014**, *385*, 79–88. [CrossRef]
31. Dohmen, R.; Chakraborty, S. Fe–Mg diffusion in olivine II: Point defect chemistry, change of diffusion mechanisms and a model for calculation of diffusion coefficients in natural olivine. *Phys. Chem. Miner.* **2007**, *34*, 409–430. [CrossRef]
32. Dohmen, R.; Ter Heege, J.H.; Chakraborty, S.; Becker, H.-W. Fe-Mg interdiffusion in orthopyroxene. *Am. Mineral.* **2016**, *101*, 2210–2221. [CrossRef]
33. Coombs, M.L.; Gardner, J.E. Reaction rim growth on olivine in silicic melts: Implications for magma mixing. *Am. Mineral.* **2004**, *89*, 748–758. [CrossRef]
34. Laumonier, M.; Scaillet, B.; Pichavant, M.; Champallier, R.; Andujar, J.; Arbaret, L. On the conditions of magma mixing and its bearing on andesite production in the crust. *Nat. Commun.* **2014**, *5*, 5607. [CrossRef]
35. Shinjo, R.; Chung, S.L.; Kato, Y.; Kimura, M. Geochemical and Sr-Nd isotopic characteristics of volcanic rocks from the Okinawa Trough and Ryukyu Arc: Implications for the evolution of a young, intracontinental back arc basin. *J. Geophys. Res. Solid Earth* **1999**, *104*, 10591–10608. [CrossRef]
36. Shinjo, R.; Kato, Y. Geochemical constraints on the origin of bimodal magmatism at the Okinawa Trough, an incipient back-arc basin. *Lithos* **2000**, *54*, 117–137. [CrossRef]
37. Zhang, Y.; Zeng, Z.; Chen, S.; Wang, X.; Yin, X. New insights into the origin of the bimodal volcanism in the middle Okinawa Trough: Not a basalt-rhyolite differentiation process. *Front. Earth Sci.* **2018**, *12*, 325–338. [CrossRef]
38. Siegburg, M.; Klügel, A.; Rocholl, A.; Bach, W. Magma plumbing and hybrid magma formation at an active back-arc basin volcano: North Su, eastern Manus basin. *J. Volcanol. Geotherm. Res.* **2018**, *362*, 1–16. [CrossRef]
39. Alt, J.C. Subseafloor processes in mid-ocean ridge hydrothennal systems. In *Seafloor Hydrothermal Systems: Physical, Chemical, Biological, and Geological Interactions*; Humphris, S.E., Zierenberg, R.A., Mullineaux, L.S., Thomson, R.E., Eds.; American Geophysical Union: Washington, WA, USA, 1995; pp. 85–114.

Journal of
*Marine Science
and Engineering*

Article

Using Apatite to Track Volatile Evolution in the Shallow Magma Chamber below the Yonaguni Knoll IV Hydrothermal Field in the Southwestern Okinawa Trough

Zuxing Chen [1,2,3,*], Landry Soh Tamehe [4], Haiyan Qi [1,3], Yuxiang Zhang [1,2,3], Zhigang Zeng [1,2,3,5,*] and Mingjiang Cai [6]

[1] CAS Key Laboratory of Marine Geology and Environment, Institute of Oceanology, Chinese Academy of Sciences, Qingdao 266071, China
[2] Laboratory for Marine Mineral Resources, Qingdao National Laboratory for Marine Science and Technology, Qingdao 266071, China
[3] Center for Ocean Mega-Science, Chinese Academy of Sciences, 7 Nanhai Road, Qingdao 266071, China
[4] China Nonferrous Metals (Guilin) Geology and Mining Co. Ltd., Guilin 541004, China
[5] College of Marine Sciences, University of Chinese Academy of Sciences, Beijing 100049, China
[6] School of Resources and Environmental Science, Quanzhou Normal University, Quanzhou 362000, China
[*] Correspondence: chenzuxing@qdio.ac.cn (Z.C.); zgzeng@qdio.ac.cn (Z.Z.)

Abstract: The Yonaguni Knoll IV is an active seafloor hydrothermal system associated with submarine silicic volcanism located in the "cross back-arc volcanic trail" (CBVT) in the southwestern Okinawa Trough. However, the behavior of volatiles during magmatic differentiation in the shallow silicic magma chamber is unclear. Here, the volatile contents of apatite inclusions trapped in different phenocrysts (orthopyroxene and amphibole) and microphenocrysts in the rhyolite from the Yonaguni Knoll IV hydrothermal field were analyzed by using electron microprobe analysis, which aims to track the behavior of volatiles in the shallow magma chamber. Notably, the 'texturally constrained' apatites showed a decreasing trend of X_{Cl}/X_{OH} and X_F/X_{Cl} ratios. Based on the geochemical analyses in combination with thermodynamic modeling, we found that the studied apatites were consistent with the mode of volatile-undersaturated crystallization. Therefore, volatiles were not saturated in the early stage of magmatic differentiation in the shallow rhyolitic magma chamber, and consequently, the metal elements were retained in the rhyolitic melt and partitioned into crystalline magmatic sulfides. Additionally, previous studies suggested that the shallow rhyolitic magma chamber was long-lived and periodically replenished by mafic magma. The injection of volatile-rich and oxidized subduction-related mafic magmas can supply abundant volatiles and dissolve magmatic sulfide in the shallow magma chamber. These processes are important for the later-stage of volatile exsolution, while the forming metal-rich magmatic fluids contribute to the overlying Yonaguni Knoll IV hydrothermal system.

Keywords: apatite; volatiles; thermodynamic modeling; Yonaguni Knoll IV hydrothermal field; Okinawa Trough

Citation: Chen, Z.; Soh Tamehe, L.; Qi, H.; Zhang, Y.; Zeng, Z.; Cai, M. Using Apatite to Track Volatile Evolution in the Shallow Magma Chamber below the Yonaguni Knoll IV Hydrothermal Field in the Southwestern Okinawa Trough. *J. Mar. Sci. Eng.* **2023**, *11*, 583. https://doi.org/10.3390/jmse11030583

Academic Editor: Dimitris Sakellariou

Received: 27 January 2023
Revised: 17 February 2023
Accepted: 6 March 2023
Published: 9 March 2023

1. Introduction

Volatiles (halogens, H_2O, CO_2, and S) are minor elements in magmas but important in volcanic processes [1–5], which influence the physical properties of silicate melts [6–8] and control the eruptive styles [9–12]. Furthermore, Cl and S are fundamental for transporting metals (e.g., Cu and Zn) from silicic magmas to exsolved hydrothermal fluids for formatting active hydrothermal systems and related ore deposits within active volcanoes [13,14]. Nevertheless, it is not easily demonstrated that any particular magmatic system has evolved an aqueous phase [15].

Apatite, $Ca_5(PO_4)_3(OH, F, Cl)$, is a ubiquitous accessory mineral in volcanic rocks [5], and its crystal structure can contain common magmatic volatile species such as H_2O, F, Cl, and S [16]. Thus, apatite is a useful magmatic volatile 'witness' and can reliably record the melt volatile compositions [16–18]. Accordingly, the halogen contents in apatite can be used to estimate the F, Cl, and H_2O compositions of the host melt [19–23]. Moreover, the variation trend in the halogen ratios in apatite can be evaluated whether the host melt is saturated in volatiles or not [15–17,24–28]. Specifically, the F:Cl:OH ratio of apatite remains approximately constant in vapor-undersaturated magma. In contrast, the Cl/F ratio of apatite decreases sharply when a vapor is exsolved [15–17,29]. Combined with the textural features of apatite, these trends in apatite are useful for tracking the vapor evolution in magmatic systems [16,17].

In this study, the volatile contents of apatites that are hosted in the groundmass and phenocryst phases (orthopyroxene and amphibole) of the rhyolite from the Yonaguni Knoll IV hydrothermal field of the southern Okinawa Trough were analyzed. According to the thermodynamic models of Candela [15], Stock et al. [16,17], and Piccoli and Candela [30], the 'texturally constrained' apatites reveal the magmatic volatile behavior in the rhyolitic magma chamber below the Yonaguni Knoll IV hydrothermal field. Furthermore, the implications for magmatic contribution to the overlying hydrothermal system are discussed.

2. Geological Setting

The Okinawa Trough (OT) is a present-day active back-arc basin of the Ryukyu subduction zone with affinities to the subduction of the Philippine Sea Plate beneath the Eurasian Plate at a velocity of ~5 to ~7 cm/year [31,32]. In the southwestern Okinawa Trough, an area of anomalous volcanism has been identified at the cross back-arc volcanic trail (CBVT) [32]. A cluster of more than 70 submarine volcanoes is distributed along the CBVT [33,34]. The abnormally voluminous volcanoes of CBVT might be related to the subduction of the Gagua Ridge on the Philippine seafloor since the early Pleistocene (Figure 1) [32,34]. Volcanic products of this region are dominated by rhyolites and dacites [35–38]. These silicic magmas are influenced by crustal assimilation and magma mixing [38–41].

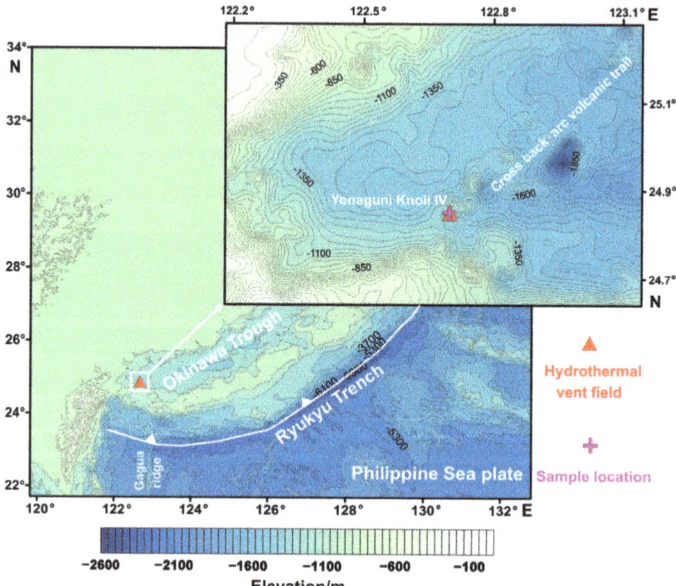

Figure 1. Regional and local bathymetric map showing the location of the studied rhyolite collected in the Yonaguni Knoll IV hydrothermal vent field of the southwestern Okinawa Trough.

The Yonaguni Knoll IV hydrothermal field (24°51′N, 122°42′E) is situated in an elongated valley within the CBVT zone [36,42]. Hydrothermal mineralization at the Yonaguni

Knoll IV includes Zn–Pb–Cu sulfides, anhydrite-rich chimneys, Ba–As chimneys, Mn-rich chimneys, and native sulfur or barite [42,43]. The diverse range of mineralization reflects the contributions of organic-rich continent-derived sediment and the volatile-rich silicic magma [42,43].

3. Sampling and Methods

3.1. Sample and Petrography

The studied rhyolite was collected at the station HOBAB3-T9′ (122°41′55.877″E, 24°50′57.774″N) in the Yonaguni Knoll IV hydrothermal field (Figure 1) by using a TV grab during the R/V Kexue Hao cruise in 2014. The rhyolite sample has been well-studied and has a porphyritic texture with a mineral assemblage consisting of plagioclase, quartz, amphibole, Fe–Ti oxides, and orthopyroxene. Apatite and zircon are common accessory minerals in the rhyolite [36,38,39,44]. Chen et al. [38] reported the bulk-rock major elements, trace elements, and isotopic compositions of the rhyolite. The mineral chemistry of phenocrysts (e.g., orthopyroxene, amphibole, and plagioclase) and accessory minerals (i.e., zircon) were also analyzed for this rock [36,39,44]. Here, we made thin sections of the rhyolite for petrologic textural analysis and in situ chemical analyses of the apatite grains. In the studied rhyolite, apatite occurred as microphenocrysts within the groundmass and as inclusions within the phenocrysts composed of orthopyroxene and amphibole (Figure 2).

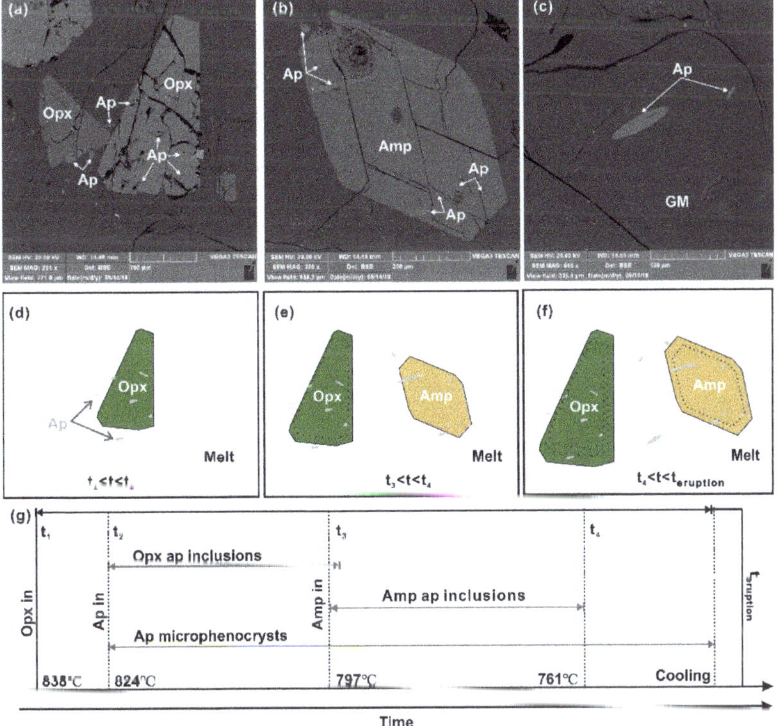

Figure 2. Backscattered-electron (BSE) images displaying orthopyroxene (**a**) and amphibole (**b**) phenocrysts, and apatite microphenocrysts (**c**). (**d**–**f**) Sketch map showing the progressive growth of orthopyroxene and amphibole with the entrapment of apatite inclusions through time during cooling; (**g**) t1–t3 refer to the onset of crystallization of minerals (opx-in, ap-in, and amp-in), while t4 refers to the end of the crystallization of amphibole. The crystallization temperatures of minerals are according to previous studies [36,39]. Ap, apatite; Opx, orthopyroxene; Amp, amphibole; GM, groundmass.

3.2. Analytical Methods

Polished thin sections of the rhyolite were first observed using a VEGA3 scanning electron microscope at the Institute of Oceanology, Chinese Academy of Sciences. Carbon coated polished sections were prepared to identify different texturally-constrained apatites. BSE images were obtained by using a scanning electron microscope operating with an acceleration voltage of 15 keV.

After detailed observation, apatites were analyzed for major elements in the same thin sections. All apatite analyses were made close to the center of the mineral grains. The major element compositions of apatites were determined using a JXA-8230 electron microprobe (EMP) at the State Key Laboratory of Continental Dynamics, Northwest University, Xian, China. The instrument was operated at an acceleration voltage of 15 kV, a beam current of 10 nA, and a beam diameter of 2 μm. The following standards were used: apatite (Ca, P), diopside (Si), almandine (Fe), fluorite (F), and NaCl (Cl). Analytical precision was $\pm$0.01 wt.% for Cl and F, and 0.01–0.2 wt.% for the other elements.

4. Results

A total of 65 analyses were performed on 65 grains to identify the chemical composition of apatites in the studied rhyolite. The compositional data for all apatites are reported in Tables 1–3. Apatites have high concentrations of CaO (50.82 to 54.98 wt.%) and P_2O_5 (38.62 to 42.59 wt.%). They are fluorapatites exhibiting F and Cl contents ranging from 1.76 to 3.08 wt.% and from 0.50 to 0.73 wt.%, respectively (Tables 1–3).

Table 1. Major and volatile element analyses of the orthopyroxene-hosted apatite inclusions (wt.%).

	AP1	AP2	AP3	AP4	AP5	AP6	AP7	AP8	AP9	AP10	AP11	AP12	AP13
P_2O_5	42.47	41.77	40.19	39.94	42.16	38.66	41.18	40.99	42.08	41.87	41.73	41.68	42.07
CaO	53.93	54.21	53.88	53.06	53.83	50.82	53.66	53.28	52.93	53.26	54.19	54.11	54.28
SiO_2	0.25	0.60	0.55	0.32	0.47	3.90	0.16	0.47	0.75	0.22	0.25	0.27	0.35
FeO	1.03	1.17	1.48	1.25	1.23	1.33	1.14	1.15	1.42	0.96	1.17	0.88	1.37
F	2.89	3.01	2.63	2.35	2.70	2.08	2.13	2.07	2.51	2.08	2.36	3.08	2.32
Cl	0.64	0.64	0.65	0.58	0.71	0.67	0.70	0.67	0.69	0.58	0.61	0.63	0.68
Total	101.46	101.58	99.51	97.66	101.30	97.71	99.16	98.75	100.58	99.10	100.49	100.84	101.15
$^1\,X_F$	0.77	0.80	0.70	0.62	0.72	0.55	0.56	0.55	0.67	0.55	0.63	0.82	0.62
$^2\,X_{Cl}$	0.09	0.09	0.10	0.09	0.10	0.10	0.10	0.10	0.10	0.09	0.09	0.09	0.10
$^3\,X_{OH}$	0.14	0.11	0.21	0.29	0.18	0.35	0.33	0.35	0.23	0.36	0.28	0.09	0.28
$X_{F/OH}$	5.51	7.60	3.40	2.15	4.03	1.59	1.70	1.57	2.87	1.53	2.21	9.25	2.17
$X_{Cl/OH}$	0.67	0.90	0.47	0.30	0.59	0.28	0.31	0.28	0.44	0.24	0.31	1.05	0.35
$X_{F/Cl}$	8.23	8.46	7.27	7.28	6.84	5.62	5.50	5.56	6.53	6.44	7.02	8.79	6.13

[1–3] X_F, X_{Cl}, and X_{OH} are the mole fractions of the F-, Cl-, and OH-apatites, respectively ($X_F = C_F/3.767$, $X_{Cl} = C_{Cl}/6.809$, $X_{OH} = 1- X_F - X_{Cl}$; where C_F and C_{Cl} are the contents of F and Cl in apatite in wt %). The calculations of the apatite mole fractions of the end-members are according to Piccoli and Candela [5].

Orthopyroxene-hosted apatite inclusions exhibited F and Cl concentrations ranging from 2.07 to 3.08 wt.% (X_F = 0.55 to 0.82) and from 0.58 to 0.71 wt.% (X_{Cl} = 0.09 to 0.10), respectively (Table 1). X_{OH} ranged from 0.09 to 0.36. Correspondingly, the X_{Cl}/X_{OH}, X_F/X_{OH}, and X_F/X_{Cl} ratios varied from 0.24 to 1.05, 1.53 to 9.25, and 5.50 to 8.79 (where X_F, X_{Cl}, and X_{OH} are the mole fractions of F, Cl, and OH, respectively) (Figure 3; Table 1).

The amphibole-hosted apatite inclusions had F and Cl contents ranging from 2.14 to 3.02 wt.% (X_F = 0.57 to 0.80) and 0.50 to 0.72 wt.% (X_{Cl} = 0.07 to 0.11), respectively (Table 2). X_{OH} ranged from 0.10 to 0.33. Correspondingly, the X_{Cl}/X_{OH}, X_F/X_{OH}, and X_F/X_{Cl} ratios varied from 0.29 to 0.91, 1.71 to 7.70, and 5.58 to 9.33, respectively (Table 2; Figure 3).

Table 2. Major and volatile element analyses of the amphibole-hosted apatite inclusions (wt.%).

	AP1	AP2	AP3	AP4	AP5	AP6	AP7	AP8	AP9	AP10	AP11	AP12	AP13
P_2O_5	41.59	40.67	40.69	40.95	41.30	40.39	39.14	41.29	40.87	41.12	42.59	41.72	41.46
CaO	53.88	54.28	54.56	54.06	53.21	53.57	51.67	54.49	52.87	53.46	54.34	53.83	53.60
SiO_2	0.43	0.61	0.64	0.51	0.50	0.46	2.06	0.44	0.91	0.47	0.37	0.41	0.42
FeO	0.69	0.93	0.92	0.76	0.98	0.96	1.38	0.90	1.16	0.98	0.88	1.02	0.93
F	2.21	2.98	2.59	2.45	2.33	2.33	2.14	2.84	2.28	2.59	2.67	2.34	2.16
Cl	0.72	0.61	0.50	0.66	0.72	0.64	0.67	0.68	0.66	0.67	0.57	0.67	0.66
Total	99.52	100.07	99.90	99.39	99.04	98.34	97.08	100.65	98.76	99.28	101.42	99.99	99.23
X_F	0.59	0.79	0.69	0.65	0.62	0.62	0.57	0.75	0.60	0.69	0.71	0.62	0.57
X_{Cl}	0.11	0.09	0.07	0.10	0.11	0.09	0.10	0.10	0.10	0.10	0.08	0.10	0.10
X_{OH}	0.31	0.12	0.24	0.25	0.28	0.29	0.33	0.15	0.30	0.22	0.21	0.28	0.33
$X_{F/OH}$	1.91	6.60	2.88	2.59	2.24	2.16	1.71	5.19	2.02	3.20	3.41	2.21	1.73
$X_{Cl/OH}$	0.34	0.75	0.31	0.39	0.38	0.32	0.30	0.69	0.32	0.46	0.40	0.35	0.29
$X_{F/Cl}$	5.58	8.78	9.33	6.70	5.87	6.64	5.76	7.51	6.24	7.01	8.49	6.35	5.89

	AP14	AP15	AP16	AP17	AP18	AP19	AP20	AP21	AP22	AP23	AP24	AP25	AP26
P_2O_5	40.54	40.80	41.47	40.82	41.08	42.14	42.05	41.33	41.38	40.93	41.27	42.15	41.15
CaO	53.43	54.05	54.06	53.55	53.55	54.98	54.83	53.79	54.02	53.91	54.73	54.73	54.65
SiO_2	0.56	0.44	0.53	0.32	0.77	0.44	0.38	0.31	0.31	0.37	0.23	0.24	0.78
FeO	0.95	0.66	0.72	0.69	1.16	1.00	0.78	0.79	0.62	0.67	0.90	0.93	0.87
F	2.28	2.74	2.53	2.38	2.63	2.53	2.53	2.49	2.24	2.40	3.02	2.95	2.86
Cl	0.66	0.66	0.70	0.66	0.67	0.62	0.59	0.68	0.68	0.69	0.64	0.70	0.64
Total	98.42	99.35	100.01	98.42	99.86	101.70	101.15	99.38	99.25	98.96	100.78	101.70	100.95
X_F	0.61	0.73	0.67	0.63	0.70	0.67	0.67	0.66	0.60	0.64	0.80	0.78	0.76
X_{Cl}	0.10	0.10	0.10	0.10	0.10	0.09	0.09	0.10	0.10	0.10	0.09	0.10	0.09
X_{OH}	0.30	0.18	0.23	0.27	0.20	0.24	0.24	0.24	0.31	0.26	0.10	0.11	0.15
$X_{F/OH}$	2.04	4.15	2.97	2.33	3.42	2.80	2.77	2.76	1.95	2.42	7.70	6.86	5.17
$X_{Cl/OH}$	0.33	0.56	0.45	0.36	0.48	0.38	0.36	0.42	0.33	0.38	0.89	0.91	0.64
$X_{F/Cl}$	6.24	7.47	6.56	6.54	7.14	7.42	7.76	6.65	5.99	6.31	8.61	7.57	8.11

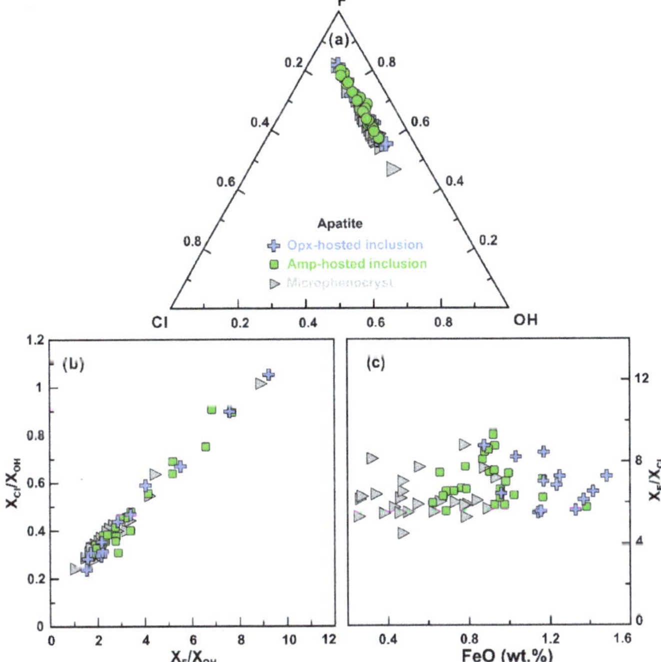

Figure 3. Volatile compositions of orthopyroxene- and amphibole-hosted apatite inclusions and apatite microphenocrysts in the rhyolite from the Yonaguni Knoll IV hydrothermal field. Data are depicted in the ternary diagram of Cl–OH–F (**a**) and binary plots of X_{Cl}/X_{OH} vs. X_F/X_{OH} (**b**), and X_F/X_{Cl} vs FeO (**c**).

Table 3. The major and volatile element analyses of the apatite microphenocrysts (wt.%).

	AP1	AP2	AP3	AP4	AP5	AP6	AP7	AP8	AP9	AP10	AP11	AP12	AP13
P_2O_5	41.21	42.31	42.22	41.02	41.51	41.76	41.36	41.25	41.55	41.79	41.02	42.32	42.27
CaO	53.38	54.08	53.91	53.67	53.19	53.88	52.82	53.33	53.33	52.79	53.32	53.50	53.54
SiO_2	0.48	0.51	0.61	0.24	0.35	0.20	0.32	0.27	0.32	0.54	0.22	0.37	0.35
FeO	0.79	0.63	0.80	0.50	0.39	0.66	0.44	0.55	0.35	0.90	0.47	0.74	0.46
F	2.04	2.21	2.19	2.18	2.14	2.19	2.28	2.22	2.47	2.23	2.42	2.32	2.12
Cl	0.70	0.72	0.67	0.71	0.70	0.66	0.71	0.68	0.70	0.71	0.71	0.70	0.70
Total	98.70	100.57	100.50	98.50	98.43	99.50	98.01	98.49	98.96	99.14	98.37	100.06	99.61
X_F	0.54	0.59	0.58	0.58	0.57	0.58	0.60	0.59	0.66	0.59	0.64	0.62	0.56
X_{Cl}	0.10	0.11	0.10	0.10	0.10	0.10	0.10	0.10	0.10	0.10	0.10	0.10	0.10
X_{OH}	0.36	0.31	0.32	0.32	0.33	0.32	0.29	0.31	0.24	0.30	0.25	0.28	0.33
$X_{F/OH}$	1.52	1.91	1.81	1.83	1.72	1.81	2.07	1.89	2.72	1.95	2.54	2.19	1.69
$X_{Cl/OH}$	0.29	0.34	0.31	0.33	0.31	0.30	0.36	0.32	0.42	0.34	0.41	0.36	0.31
$X_{F/Cl}$	5.29	5.55	5.91	5.54	5.48	6.05	5.78	5.92	6.41	5.70	6.21	6.02	5.50

	AP14	AP15	AP16	AP17	AP18	AP19	AP20	AP21	AP22	AP23	AP24	AP25	AP26
P_2O_5	40.02	41.69	39.90	40.20	41.25	41.29	40.32	40.94	40.48	40.45	39.76	41.44	38.62
CaO	52.80	52.90	53.49	53.08	53.02	54.02	52.86	53.85	52.82	53.31	52.75	53.91	52.07
SiO_2	0.34	0.42	0.63	0.89	0.43	0.41	0.63	0.38	0.62	0.48	1.63	0.33	2.13
FeO	0.48	0.78	0.56	0.33	0.48	0.48	0.85	0.78	0.26	0.27	0.28	0.94	0.88
F	1.76	2.31	2.76	2.64	2.21	2.76	2.39	3.08	2.14	2.22	2.33	2.54	2.65
Cl	0.71	0.71	0.64	0.58	0.61	0.71	0.71	0.63	0.73	0.65	0.67	0.64	0.62
Total	96.29	98.97	98.10	97.92	98.13	99.82	97.94	99.76	97.26	97.42	97.60	100.11	97.21
X_F	0.47	0.61	0.73	0.70	0.59	0.73	0.63	0.82	0.57	0.59	0.62	0.67	0.70
X_{Cl}	0.10	0.10	0.09	0.09	0.09	0.10	0.10	0.09	0.11	0.10	0.10	0.09	0.09
X_{OH}	0.43	0.28	0.17	0.21	0.32	0.16	0.26	0.09	0.32	0.32	0.28	0.23	0.21
$X_{F/OH}$	1.09	2.18	4.23	3.28	1.81	4.48	2.42	8.95	1.75	1.87	2.19	2.88	3.41
$X_{Cl/OH}$	0.24	0.37	0.55	0.40	0.28	0.64	0.40	1.01	0.33	0.30	0.35	0.40	0.44
$X_{F/Cl}$	4.50	5.87	7.74	8.17	6.55	7.04	6.10	8.82	5.32	6.17	6.31	7.20	7.69

The apatite microphenocrysts displayed a large compositional range of F (1.76 to 3.08 wt.%; X_F = 0.47 to 0.82), Cl (0.58 to 0.74 wt.%; X_{Cl} = 0.09 to 0.11), and X_{OH} (0.09 to 0.43) compared with the apatite inclusions (Figure 3; Table 3). Correspondingly, the X_{Cl}/X_{OH}, X_F/X_{OH}, and X_F/X_{Cl} ratios ranged from 0.24 to 1.01, 1.09 to 8.95, and 4.50 to 8.82, respectively (Table 3). Apatite microphenocrysts were plotted on the same compositional trends with the apatite inclusions, but typically extended to more FeO-depleted compositions (Figure 3).

5. Discussion

5.1. Crystallization Sequence of the Texturally Constrained Apatites

Experiments have demonstrated that apatite crystallization is primarily a function of the SiO_2 and P_2O_5 concentrations and temperature in silicate melts [5,45]. The saturation temperature of apatite calculated from the bulk rock compositions provides the minimum estimates of the initial temperature when apatite crystallized from the magma [5]. Based on the calculation formula of Piccoli and Candela [5], the apatite-saturation temperature of the rhyolite (SiO_2 = 73.06 wt.% and P_2O_5 = 0.03 wt.%; [38]) was 824 °C, which was higher than the crystallization temperature of amphiboles in the rhyolite (761–797 °C; [36]). Moreover, quartz–amphibole intergrowth in the rhyolite [39] suggests that amphibole crystallized at a relatively late stage during magmatic differentiation. The orthopyroxene phenocrysts were in equilibrium with the host rhyolitic magma, having a crystallization temperature ranging between 792 and 838 °C [39]. Therefore, orthopyroxene began crystallizing before the amphibole. As orthopyroxene is enriched in FeO content (~34 wt.%, [39]), orthopyroxene fractional crystallization results in the residual melt toward decreasing FeO. On the whole, apatite microphenocrysts have lower FeO contents than those of apatite inclusions in orthopyroxene and amphibole phenocrysts (Figure 3), which is consistent with a comparative decrease in the melt FeO contents during magmatic differentiation. Thus, apatite hosted in phenocrysts generally crystallized earlier than their equivalents in the groundmass.

5.2. Behavior of the Volatiles during the Rhyolitic Magmatic Differentiation

The volatile contents of 'texturally constrained' apatites are a useful tracer for volatile evolution in the shallow magma chamber [16,17]. The factors controlling the variation trend of the F/Cl ratio in magmatic apatites during magma evolution include: (1) temperature and pressure conditions in open systems; (2) fractional crystallization; and (3) variable degrees of degassing or fluid exsolution [15–17,30]. We discuss these factors in the following sections.

5.2.1. Influence of Cooling and Decompression on Compositional Variability of Apatite Volatiles

Since temperature and pressure affect the halogen–OH exchange coefficients between apatite and melt [30,46], the volatile compositions of apatite are likely variable in the constant melt composition [17]. To model the apatite volatile compositional evolution during cooling or decompression at a given melt composition (Figure 4), we used the method according to Piccoli and Candela [5] and Stock et al. [16].

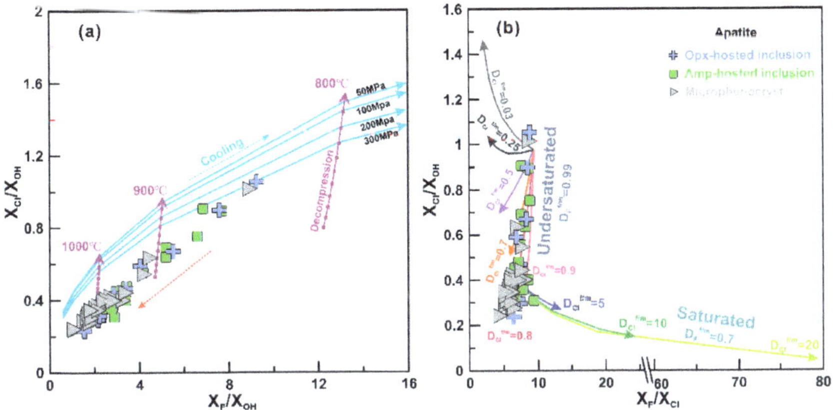

Figure 4. Volatile compositions of apatites in the studied rhyolite. (**a**) X_F/X_{OH} versus X_{Cl}/X_{OH}. Purple and blue arrows are the modeled trends of apatite compositional evolution at a constant melt composition under isothermal decompression and isobaric cooling conditions, respectively. The approach outlined by Stock et al. [16] was used to model the effects of physical processes such as the cooling and decompression on apatite compositions. (**b**) X_F/X_{Cl} versus X_{Cl}/X_{OH}. Arrows show the modeled trends of apatite composition during 80% volatile undersaturated crystallization, followed by 20% H_2O-saturated crystallization. The various fluid–melt partition coefficients ($D_{Cl}^{f/m}$) in the thermodynamic modeling during H_2O-undersaturated and saturated crystallization were according to Stock et al. [17]. Although the modeled results may not be quantitatively robust, these calculations provide a reference frame to outline the different processes on the apatite compositions.

As shown in Figure 4a, the X_{Cl}/X_{OH} and X_F/X_{OH} ratios of apatite strongly increased with decreasing temperature, which is consistent with the experimental results of a preference for the OH end-member in apatite at higher temperatures [16]. In contrast, pressure had little effect on the apatite volatile compositions at constant melt composition [30,46], with depressurization causing a slight increase in the Cl component, corresponding to an almost exclusive increase in the X_{Cl}/X_{OH} ratios [16,30]. As apatite inclusions generally crystallized earlier than their equivalents in groundmass and thus displayed a decreased trend of X_{Cl}/X_{OH} and X_F/X_{OH} ratios during magmatic evolution (Figure 4a), this suggests that cooling or depressurization is not the major control in driving the volatile composition variations of apatites. Although the injection of a mantle-derived basaltic magma into the shallow rhyolitic magma chamber and subsequent magma mixing is likely to lead to temporary heating and pressurization [22,35,38], the unzoned composition of apatite-hosted phenocrysts (amphibole and orthopyroxene) [36,39] suggests the crystallization of apatite under decreasing temperature. Collectively, the volatile composition trend in the studied apatites was not attributed to temperature and pressure variations. Alternatively, the progressive variation in the melt volatile composition during magmatic differentiation might explain the volatile composition trend observed in the studied apatites.

5.2.2. Compositional Variability of Apatite Volatiles during Magmatic Differentiation

As magmatic differentiation progressively increases the volatile contents of the residual melt, the volatile-rich subduction-related magmas may exsolve an aqueous fluid phase in the shallow magma chamber [15]. Before volatile unsaturation, the halogens (e.g., F and Cl) and H_2O in a silicate melt were incompatible [15,17]. In contrast, when H_2O was saturated, Cl will preferably partition into the fluid while F will be retained in the coexisting silicate melt [10,15,27,30]. For example, Cl partition coefficients between the fluid and silicate melt varied from 10 to over 200 in contrast to 0.7 for F [17,27] (Figure 4b). Thus, fluid-present magmatic differentiation can cause a strong fractionation of F from Cl in the residual melt [47]. Correspondingly, the volatile ratios of apatite (e.g., X_F/X_{Cl}) are sensitive to aqueous fluid exsolution during magmatic evolution [15–17,24,26,27]. We then used the two-step thermodynamic model of Stock et al. [17] to constrain the volatile behaviors of the studied apatites during magmatic differentiation.

As shown in Figure 4, the X_F/X_{OH} and X_F/X_{Cl} ratios of apatite typically displayed a decreasing trend during H_2O-undersaturated crystallization. However, the X_{Cl}/X_{OH} ratios of apatite either decreased or increased, which is controlled by the exact values of the F, Cl, and OH partition coefficients between apatite and the melt (Figure 4b). In contrast, the X_F/X_{Cl} ratios of apatite strongly increased during H_2O-saturated crystallization (Figure 4b). The decreased trend in the X_{Cl}/X_{OH} and X_F/X_{OH} ratios of the studied apatites was consistent with the mode of H_2O-undersaturated crystallization [48] (Figure 4b). Therefore, the rhyolitic magma in a shallow magma chamber was H_2O-unsaturated during the period of apatite crystallization.

5.3. Implications for Magmatic Contribution to Hydrothermal System

Generally, there are at least two hypotheses for the origin of the metals in submarine massive sulfide deposits. One hypothesis is that the magmatic intrusion in the hydrothermal zone was leached by heated seawater, thereby indirectly contributing metal elements to the hydrothermal system [49–51]. The other view suggests that an evolved magma system can exsolve metal-rich magmatic fluids, thereby directly supplying metallic elements to the overlying hydrothermal system [52–54]. Moreover, the model of contribution of metal-rich magmatic fluids to the OT hydrothermal systems has been proposed by several scholars [41,55,56]. Nevertheless, it remains unclear whether and how the silicic magma exsolve metal-rich magmatic fluids.

As volatiles are ligands of metals transported in a magmatic–hydrothermal system [57], volatile exsolution is an essential precondition for the formation of metal-rich magmatic fluids [57,58]. The volatile exsolution from a silicate melt is expected to be recorded in the apatite with a drastic increase in the X_F/X_{Cl} ratios (Figure 4b). Apatite occurs as inclusions in orthopyroxene and amphibole phenocrysts in the studied rhyolite (Figure 2), thus suggesting an early apatite saturation in the melt during magmatic differentiation [22]. However, the decreasing trend of X_{Cl}/X_{OH} and X_F/X_{Cl} ratios of the studied apatites in the rhyolite argues against H_2O-saturated crystallization (Figure 4b). Therefore, the volatiles were not saturated and thus lacked exsolution of the metal-rich magmatic fluids in the early stage of magmatic differentiation in the shallow rhyolitic magma chamber. Metallic elements (e.g., Cu) are generally incompatible with silicate minerals but highly compatible with magmatic sulfides [57,58]. Consequently, the metal elements were partitioned into crystalline magmatic sulfides and/or dissolved in the rhyolitic melt [36,41]. Correspondingly, magmatic sulfide is common in silicic rocks from the CBVT region [41]. Additionally, the low Cu content of the rhyolite (13 μg/g; [55]) also indicates that magmatic sulfide was saturated during the rhyolitic magma evolution.

The prolonged zircon U–Th ages in the rhyolite reveal a long-lived (at least 100 ka) silicic magma chamber beneath the Yonaguni Knoll IV hydrothermal field [44]. The occurrence of mafic magmatic enclaves and compositional zoned plagioclase in the rhyolite [38,39] suggest multi-stages of mafic magma injection into the silicic magma chamber [22,35,40]. As subduction-related mafic magmas are enriched in volatiles [59], periodic injection of mafic magma can supply abundant volatiles into the shallow magma chamber, which is essential for the later-stage of volatile exsolution. Since subduction-related mafic magmas are oxidized [60,61], the injection of oxidized mafic magmas into the shallow rhyolitic magma chamber may lead to the oxidative dissolution of the earlier magmatic sulfides [41,55]. Consistently, the rhyolite magma has high oxygen fugacity at around QFM + 1 (where QFM is a quartz–fayalite–magnetite buffer) [36]. The phenomenon of magmatic sulfide dissolution is common in silicic rocks from the CBVT region [41]. Because magmatic sulfide is a significant sink of ore metals, the oxidative dissolution process facilitates the formation of metal-rich magmatic fluids [55,62,63]. Collectively, there is a lack of metal-rich magmatic fluid exsolved from the rhyolitic melt during the early stage of magmatic differentiation, but the subsequent periodic injection of volatile-rich and oxidized mafic magmas into the rhyolitic magma chamber, accompanied by the exsolution of metal-rich magmatic fluids, contribute to the submarine hydrothermal systems.

6. Conclusions

Apatites trapped in orthopyroxene and amphibole phenocrysts suggest early apatite saturation during the rhyolitic magmatic differentiation. The decreasing trend in the X_{Cl}/X_{OH} and X_F/X_{Cl} ratios of the apatite inclusions and microphenocrysts revealed that the magma chamber beneath the Yonaguni Knoll IV hydrothermal systems remained water-undersaturated during the early stage of magmatic differentiation. The lack of magmatic fluid exsolution resulted in the metal elements remaining dissolved in the rhyolitic melt and partitioning into magmatic sulfides. Subsequently, the injection of volatile-rich and oxidized basaltic magmas into the shallow rhyolitic magma chamber promoted the later-stage of volatile exsolution and forming metal-rich magmatic fluids to contribute to the overlying Yonaguni Knoll IV submarine hydrothermal system.

Author Contributions: Z.C.—Conceptualization, data curation, formal analysis, validation, writing and editing. L.S.T.—review and editing. H.Q.—Data curation, methodology. Y.Z.—Review and editing. Z.Z.—project administration and funding acquisition. M.C.—Review and editing. All authors have read and agreed to the published version of the manuscript.

Funding: This research was supported by the National Natural Science Foundation of China (Grant Nos. 42106083 and 91958213), the Open Fund of the Key Laboratory of Marine Geology and Environment, the Chinese Academy of Sciences (Grant Nos. MGE2022KG10 and MGE2020KG11), the Strategic Priority Research Program (B) of the Chinese Academy of Sciences (XDB42020402), the Shandong Provincial Natural Science Foundation, China (Grant Nos. ZR2020QD069 and ZR2020MD068), the National Basic Research Program of China (Grant No. 2013CB429700), and the Special Fund for the Taishan Scholar Program of Shandong Province (Grant No. ts201511061).

Institutional Review Board Statement: Not applicable.

Informed Consent Statement: Not applicable.

Data Availability Statement: Data can be obtained from the corresponding author and are available in the main text.

Acknowledgments: We would like to thank the crews of the R/V Kexue during the HOBAB 3 cruise for their help with the sample collection. The authors thank Wenqiang Yang for his assistance in the EPMA analysis. We are grateful to three anonymous reviewers for their detailed and constructive comments and suggestions, which greatly improved an earlier version of the manuscript. We thank the MDPI editors for their efficient handling of the manuscript.

Conflicts of Interest: The authors declare no conflict of interest.

References

1. Aiuppa, A.; Baker, D.; Webster, J. Halogens in volcanic systems. *Chem. Geol.* **2009**, *263*, 1–18. [CrossRef]
2. De Vivo, B.; Lima, A.; Webster, J.D. Volatiles in magmatic-volcanic systems. *Elements* **2005**, *1*, 19–24. [CrossRef]
3. Edmonds, M.; Wallace, P.J. Volatiles and exsolved vapor in volcanic systems. *Elements* **2017**, *13*, 29–34. [CrossRef]
4. Edmonds, M.; Woods, A.W. Exsolved volatiles in magma reservoirs. *J. Volcanol. Geotherm. Res.* **2018**, *368*, 13–30. [CrossRef]
5. Piccoli, P.M.; Candela, P.A. Apatite in Igneous Systems. *Rev. Mineral. Geochem.* **2002**, *48*, 255–292. [CrossRef]
6. Scott, J.A.; Humphreys, M.C.; Mather, T.A.; Pyle, D.M.; Stock, M.J. Insights into the behaviour of S, F, and Cl at Santiaguito Volcano, Guatemala, from apatite and glass. *Lithos* **2015**, *232*, 375–394. [CrossRef]
7. Zimova, M.; Webb, S.L. The combined effects of chlorine and fluorine on the viscosity of aluminosilicate melts. *Geochim. Cosmochim. Acta* **2007**, *71*, 1553–1562. [CrossRef]
8. Giordano, D.; Romano, C.; Dingwell, D.; Poe, B.; Behrens, H. The combined effects of water and fluorine on the viscosity of silicic magmas. *Geochim. Cosmochim. Acta* **2004**, *68*, 5159–5168. [CrossRef]
9. Moretti, R.; Arienzo, I.; Di Renzo, V.; Orsi, G.; Arzilli, F.; Brun, F.; D'Antonio, M.; Mancini, L.; Deloule, E. Volatile segregation and generation of highly vesiculated explosive magmas by volatile-melt fining processes: The case of the Campanian Ignimbrite eruption. *Chem. Geol.* **2019**, *503*, 1–14. [CrossRef]
10. Popa, R.G.; Tollan, P.; Bachmann, O.; Schenker, V.; Ellis, B.; Allaz, J.M. Water exsolution in the magma chamber favors effusive eruptions: Application of Cl-F partitioning behavior at the Nisyros-Yali volcanic area. *Chem. Geol.* **2021**, *570*, 120170. [CrossRef]
11. Roggensack, K. Explosive Basaltic Volcanism from Cerro Negro Volcano: Influence of Volatiles on Eruptive Style. *Science* **1997**, *277*, 1639–1642. [CrossRef]
12. Shinohara, H. Excess degassing from volcanoes and its role on eruptive and intrusive activity. *Rev. Geophys.* **2008**, *46*, RG4005. [CrossRef]
13. Mason, E.; Wieser, P.E.; Liu, E.J.; Edmonds, M.; Ilyinskaya, E.; Whitty, R.C.; Mather, T.A.; Elias, T.; Nadeau, P.A.; Wilkes, T.C. Volatile metal emissions from volcanic degassing and lava–seawater interactions at Kīlauea Volcano, Hawai'i. *Commun. Earth Environ.* **2021**, *2*, 79. [CrossRef]
14. Mason, E.; Hogg, O. Volcanic Outgassing of Volatile Trace Metals. *Annu. Rev. Earth Planet. Sci.* **2022**, *50*, 79–98.
15. Candela, P.A. Toward a thermodynamic model for the halogens in magmatic systems: An application to melt-vapor-apatite equilibria. *Chem. Geol.* **1986**, *57*, 289–301. [CrossRef]
16. Stock, M.J.; Humphreys, M.C.S.; Smith, V.C.; Isaia, R.; Pyle, D.M. Late-stage volatile saturation as a potential trigger for explosive volcanic eruptions. *Nat. Geosci.* **2016**, *9*, 249–254. [CrossRef]
17. Stock, M.J.; Humphreys, M.C.S.; Smith, V.C.; Isaia, R.; A Brooker, R.; Pyle, D.M. Tracking Volatile Behaviour in Sub-volcanic Plumbing Systems Using Apatite and Glass: Insights into Pre-eruptive Processes at Campi Flegrei, Italy. *J. Petrol.* **2018**, *59*, 2463–2492. [CrossRef]
18. Zafar, T.; Leng, C.-B.; Zhang, X.-C.; Rehman, H.U. Geochemical attributes of magmatic apatite in the Kukaazi granite from western Kunlun orogenic belt, NW China: Implications for granite petrogenesis and Pb-Zn (-Cu-W) mineralization. *J. Geochem. Explor.* **2019**, *204*, 256–269. [CrossRef]
19. Li, W.; Costa, F. A thermodynamic model for F-Cl-OH partitioning between silicate melts and apatite including non-ideal mixing with application to constraining melt volatile budgets. *Geochim. Cosmochim. Acta* **2020**, *269*, 203–222. [CrossRef]
20. Li, H.; Hermann, J. Chlorine and fluorine partitioning between apatite and sediment melt at 2.5 GPa, 800 °C: A new experimentally derived thermodynamic model. *Am. Mineral.* **2017**, *102*, 580–594. [CrossRef]
21. Li, H.; Hermann, J. Apatite as an indicator of fluid salinity: An experimental study of chlorine and fluorine partitioning in subducted sediments. *Geochim. Cosmochim. Acta* **2015**, *166*, 267–297. [CrossRef]
22. Chen, Z.; Zeng, Z.; Wang, X.; Yin, X.; Chen, S.; Guo, K.; Lai, Z.; Zhang, Y.; Ma, Y.; Qi, H.; et al. U-Th/He dating and chemical compositions of apatite in the dacite from the southwestern Okinawa Trough: Implications for petrogenesis. *J. Asian Earth Sci.* **2018**, *161*, 1–13. [CrossRef]
23. Du, X.; Zeng, Z.; Chen, Z. Volatile Characteristics of Apatite in Dacite from the Eastern Manus Basin and Their Geological Implications. *J. Mar. Sci. Eng.* **2022**, *10*, 698. [CrossRef]
24. Boudreau, A.E.; McCallum, I.S. Investigations of the Stillwater Complex: Part V. Apatites as indicators of evolving fluid composition. *Contrib. Mineral. Petrol.* **1989**, *102*, 138–153. [CrossRef]
25. Boudreau, A.E.; Kruger, F.J. Variation in the Composition of Apatite through the Merensky Cyclic Unit in the Western Bushveld Complex. *Econ. Geol.* **1990**, *85*, 737–745. [CrossRef]
26. Cawthorn, R.G. Formation in of chlor- and fluor-apatite layered intrusions. *Mineral. Mag.* **1994**, *38*, 299–306. [CrossRef]

27. Meurer, W.P.; Boudreau, A.E. An evaluation of models of apatite compositional variability using apatite from the Middle Banded series of the Stillwater Complex, Montana. *Contrib. Mineral. Petrol.* **1996**, *125*, 225–236. [CrossRef]

28. Li, W.; Costa, F.; Nagashima, K. Apatite crystals reveal melt volatile budgets and magma storage depths at Merapi volcano, Indonesia. *J. Petrol.* **2021**, *62*, egaa100. [CrossRef]

29. Li, J.-X.; Li, G.-M.; Evans, N.J.; Zhao, J.-X.; Qin, K.-Z.; Xie, J. Primary fluid exsolution in porphyry copper systems: Evidence from magmatic apatite and anhydrite inclusions in zircon. *Miner. Deposita* **2020**, *56*, 407–415. [CrossRef]

30. Piccoli, P.; Candela, P. Apatite in felsic rocks: A model for the estimation of initial halogen concentrations in the Biship Tuff (Long Vally) and Tuolumne Intrusive Suite (Sierra Nevada Batholith) magmas. *Am. J. Sci.* **1994**, *294*, 92–135. [CrossRef]

31. Sibuet, J.-C.; Letouzey, J.; Barbier, F.; Charvet, J.; Foucher, J.-P.; Hilde, T.W.C.; Kimura, M.; Chiao, L.-Y.; Marsset, B.; Muller, C.; et al. Back arc extension in the Okinawa Trough. *J. Geophys. Res. Solid Earth* **1987**, *92*, 14041–14063. [CrossRef]

32. Sibuet, J.-C.; Deffontaines, B.; Hsu, S.-K.; Thareau, N.; Le Formal, J.-P.; Liu, C.-S. Okinawa trough backarc basin- Early tectonic and magmatic evolution. *J. Geophys. Res. Solid Earth* **1998**, *103*, 30245–30267. [CrossRef]

33. Tsai, C.-H.; Hsu, S.-K.; Chen, S.-C.; Wang, S.-Y.; Lin, L.-K.; Huang, P.-C.; Chen, K.-T.; Lin, H.-S.; Liang, C.-W.; Cho, Y.-Y. Active tectonics and volcanism in the southernmost Okinawa Trough back-arc basin derived from deep-towed sonar surveys. *Tectonophysics* **2021**, *817*, 229047. [CrossRef]

34. Lin, J.-Y.; Sibuet, J.-C.; Lee, C.-S.; Hsu, S.-K.; Klingelhoefer, F. Origin of the southern Okinawa Trough volcanism from detailed seismic tomography. *J. Geophys. Res. Solid Earth* **2007**, *112*, 582–596. [CrossRef]

35. Chen, Z.; Zeng, Z.; Wang, X.; Peng, X.; Zhang, Y.; Yin, X.; Chen, S.; Zhang, L.; Qi, H. Element and Sr isotope zoning in plagioclase in the dacites from the southwestern Okinawa Trough: Insights into magma mixing processes and time scales. *Lithos* **2020**, *376–377*, 105776. [CrossRef]

36. Chen, Z.; Zeng, Z.; Qi, H.; Wang, X.; Yin, X.; Chen, S.; Yang, W.; Ma, Y.; Zhang, Y. Amphibole perspective to unravel the rhyolite as a potential fertile source for the Yonaguni Knoll IV hydrothermal system in the southwestern Okinawa Trough. *Geol. J.* **2020**, *55*, 4279–4301. [CrossRef]

37. Chung, S.-L.; Wang, S.-L.; Shinjo, R.; Lee, C.-S.; Chen, C.-H. Initiation of arc magmatism in an embryonic continental rifting zone of the southernmost part of Okinawa Trough. *Terra Nova* **2000**, *12*, 225–230. [CrossRef]

38. Chen, Z.; Zeng, Z.; Yin, X.; Wang, X.; Zhang, Y.; Chen, S.; Shu, Y.; Guo, K.; Li, X. Petrogenesis of highly fractionated rhyolites in the southwestern Okinawa Trough: Constraints from whole-rock geochemistry data and Sr-Nd-Pb-O isotopes. *Geol. J.* **2019**, *54*, 316–332. [CrossRef]

39. Chen, Z.; Zeng, Z.; Wang, X.; Zhang, Y.; Yin, X.; Chen, S.; Ma, Y.; Li, X.; Qi, H. Mineral chemistry indicates the petrogenesis of rhyolite from the southwestern Okinawa Trough. *J. Ocean. Univ. China* **2017**, *16*, 1097–1108. [CrossRef]

40. Guo, K.; Zhai, S.-K.; Wang, X.-Y.; Yu, Z.-H.; Lai, Z.-Q.; Chen, S.; Song, Z.-J.; Ma, Y.; Chen, Z.-X.; Li, X.-H.; et al. The dynamics of the southern Okinawa Trough magmatic system: New insights from the microanalysis of the An contents, trace element concentrations and Sr isotopic compositions of plagioclase hosted in basalts and silicic rocks. *Chem. Geol.* **2018**, *497*, 146–161. [CrossRef]

41. Li, Z.; Chu, F.; Zhu, J.; Ding, Y.; Zhu, Z.; Liu, J.; Wang, H.; Li, X.; Dong, Y.; Zhao, D. Magmatic sulfide formation and oxidative dissolution in the SW Okinawa Trough: A precursor to metal-bearing magmatic fluid. *Geochim. Cosmochim. Acta* **2019**, *258*, 138–155. [CrossRef]

42. Suzuki, R.; Ishibashi, J.-I.; Nakaseama, M.; Konno, U.; Tsunogai, U.; Gena, K.; Chiba, H. Diverse Range of Mineralization Induced by Phase Separation of Hydrothermal Fluid: Case Study of the Yonaguni Knoll IV Hydrothermal Field in the Okinawa Trough Back-Arc Basin. *Resour. Geol.* **2008**, *58*, 267–288. [CrossRef]

43. Zhang, X.; Zhai, S.; Yu, Z.; Yang, Z.; Xu, J. Zinc and lead isotope variation in hydrothermal deposits from the Okinawa Trough. *Ore Geol. Rev.* **2019**, *111*, 102944. [CrossRef]

44. Zeng, Z.-G.; Chen, Z.-X.; Zhang, Y.-X. Zircon record of an Archaean crustal fragment and supercontinent amalgamation in quaternary back-arc volcanic rocks. *Sci. Rep.* **2021**, *11*, 12367. [CrossRef]

45. Harrison, T.M.; Watson, E.B. The behaviour of apatite during crustal anatexis: Equilibrium and kinetic considerations. Geochim Cosmochim Acta. *Geochim. Cosmochim. Acta* **1984**, *48*, 1467–1477. [CrossRef]

46. Riker, J.; Humphreys, M.C.; Brooker, R.A.; De Hoog, J.C. First measurements of OH-C exchange and temperature-dependent partitioning of OH and halogens in the system apatite–silicate melt. *Am. Mineral.* **2018**, *103*, 260–270. [CrossRef]

47. Webster, J.D.; Tappen, C.M.; Mandeville, C.W. Partitioning behavior of chlorine and fluorine in the system apatite–melt–fluid. II: Felsic silicate systems at 200 MPa. *Chem. Geol.* **2009**, *73*, 559–581. [CrossRef]

48. Chen, Z.; Chen, J.; Tamehe, L.S.; Zhang, Y.; Zeng, Z.; Xia, X.; Cui, Z.; Zhang, T.; Guo, K. Heavy Copper Isotopes in Arc-Related Lavas From Cold Subduction Zones Uncover a Sub-Arc Mantle Metasomatized by Serpentinite Derived Sulfate-Rich Fluids. *J. Geophys. Res. Solid Earth* **2022**, *127*, e2022JB024910. [CrossRef]

49. Humphris, S.E.; Klein, F. Progress in Deciphering the Controls on the Geochemistry of Fluids in Seafloor Hydrothermal Systems. *Annu. Rev. Mar. Sci.* **2018**, *10*, 315–343. [CrossRef]

50. Alt, J.C. Subseafloor Processes in Mid-Ocean Ridge Hydrothennal Systems. *Geophys. Monogr. Ser.* **1995**, *91*, 85–114.

51. Zeng, Z.; Chen, Z.; Zhang, Y.; Li, X. Geological, physical, and chemical characteristics of seafloor hydrothermal vent fields. *J. Oceanol. Limnol.* **2020**, *38*, 985–1007. [CrossRef]

52. Yang, K.; Scott, S.D. Possible contribution of a metal-rich magmatic fluid to a sea-floor hydrothermal system. *Nature* **1996**, *383*, 420–423. [CrossRef]

53. Yang, K.; Scott, S.D. Vigorous exsolution of volatiles in the magma chamber beneath a hydrothermal system on the modern sea floor of the eastern Manus back-arc basin, western Pacific: Evidence from melt inclusions. *Econ. Geol.* **2005**, *100*, 1085–1096. [CrossRef]

54. Yang, K.; Scott, S.D. Magmatic degassing of volatiles and ore metals into a hydrothermal system on the modern sea floor of the eastern Manus back-arc basin, western Pacific. *Econ. Geol.* **2002**, *97*, 1079–1100. [CrossRef]

55. Chen, Z.; Zeng, Z.; Tamehe, L.S.; Wang, X.; Chen, K.; Yin, X.; Yang, W.; Qi, H. Magmatic sulfide saturation and dissolution in the basaltic andesitic magma from the Yaeyama Central Graben, southern Okinawa Trough. *Lithos* **2021**, *388–389*, 106082. [CrossRef]

56. Yu, Z.; Zhai, S.; Zhao, G. The Evidence from Inclusions in Pumices for the Direct Degassing of Volatiles from the Magma to the Hydrothermal Fluids in the Okinawa Trough. *J. Ocean. Univ. Qingdao* **2002**, *1*, 171–175.

57. Hedenquist, J.W.; Lowenstern, J.B. The role of magmas in the formation of hydrothermal ore deposits. *Nature* **1994**, *370*, 519–527. [CrossRef]

58. Williams-Jones, A.E.; Heinrich, C.A. 100th Anniversary Special Paper: Vapor Transport of Metals and the Formation of Magmatic-Hydrothermal Ore Deposits. *Econ. Geol.* **2005**, *100*, 1287–1312. [CrossRef]

59. Zellmer, G.F.; Edmonds, M.; Straub, S.M. Volatiles in subduction zone magmatism. *Geol. Soc. Lond. Spec. Publ.* **2015**, *410*, 1–17. [CrossRef]

60. Chen, Z.; Chen, J.; Tamehe, L.S.; Zhang, Y.; Zeng, Z.; Zhang, T.; Shuai, W.; Yin, X. Light Fe isotopes in arc magmas from cold subduction zones: Implications for serpentinite-derived fluids oxidized the sub-arc mantle. *Geochim. Cosmochim. Acta* **2023**, *342*, 1–14. [CrossRef]

61. Kelley, K.A.; Cottrell, E. Water and the oxidation state of subduction zone magmas. *Science* **2009**, *325*, 605–607. [CrossRef] [PubMed]

62. Wilkinson, J.J. Triggers for the formation of porphyry ore deposits in magmatic arcs. *Nat. Geosci.* **2013**, *6*, 917–925. [CrossRef]

63. Nadeau, O.; Williams-Jones, A.E.; Stix, J. Sulphide magma as a source of metals in arc-related magmatic hydrothermal ore fluids. *Nat. Geosci.* **2010**, *3*, 501–505. [CrossRef]

Journal of
Marine Science and Engineering

Article

Volatile Characteristics of Apatite in Dacite from the Eastern Manus Basin and Their Geological Implications

Xiaoning Du [1,2,3], Zhigang Zeng [1,2,4,*] and Zuxing Chen [1,3]

1   CAS Key Laboratory of Marine Geology and Environment, Institute of Oceanology, Chinese Academy of Sciences, Qingdao 266071, China; duxiaoning@qdio.ac.cn (X.D.); chenzuxing@qdio.ac.cn (Z.C.)
2   University of Chinese Academy of Sciences, Beijing 100049, China
3   Center for Ocean Mega-Science, Chinese Academy of Sciences, Qingdao 266071, China
4   Laboratory for Marine Mineral Resources, Pilot National Laboratory for Marine Science and Technology (Qingdao), Qingdao 266237, China
*   Correspondence: zgzeng@ms.qdio.ac.cn

Abstract: As one of the youngest back-arc basins, the evolutionary behavior of magmatic volatiles in the Eastern Manus Basin has been poorly studied. Recently, apatite has received widespread attention for its powerful function in recording information on magmatic volatiles. In this paper, by determining the major element compositions and primary volatile abundances (F, Cl, SO_3) of apatites in dacite, we analyze the compositions of volatiles before magma eruption in the Eastern Manus Basin, as well as their indications for magmatic oxygen fugacity and petrogenesis, so as to improve the study about the evolution of magmatic volatiles in this region. Experimental data indicate that apatite saturation temperatures range from 935 to 952 °C. All the apatites are magmatic apatites with F contents of 0.87−1.39 wt.%, Cl contents of 1.24−1.70 wt.%, and $SO_3 \leq 0.06$ wt.%. Analysis reveals that the apatites in this study crystallized from volatile-undersaturated melts, so their chemical compositions can be used as indicators of magmatic compositions. According to calculations, the minimum S concentrations of the host melts range from 2−65 ppm or 8−11 ppm. The crystallization and separation of magnetite caused the reduction state of melts, and the relatively low oxygen fugacity (ΔFMQ = −0.2 ± 0.9) caused low SO_3 contents in apatites. In addition, F and Cl contents of the host melt were calculated to be 185−448 ppm and 1059−1588 ppm, corresponding to the H_2O contents of 1.4−2.1% and 1.2−1.5% (error ± 30−40%), respectively. The high Cl/F ratio and H_2O contents of samples indicate the addition of slab-derived fluids in the mantle source region of the Eastern Manus Basin. High F contents of the melts may be influenced by F-rich sediments, as well as the release of F from lawsonite and phengite decomposition. High Cl appears to originate from the dual influence of subduction-released fluids and Cl-rich seawater-derived components. Further, it is estimated that 14−21% of the total Cl concentrations in melts were added directly from subduction-released fluids, or higher.

Keywords: Eastern Manus Basin; apatite; dacite; magma evolution; volatile

Citation: Du, X.; Zeng, Z.; Chen, Z. Volatile Characteristics of Apatite in Dacite from the Eastern Manus Basin and Their Geological Implications. *J. Mar. Sci. Eng.* **2022**, *10*, 698. https://doi.org/10.3390/jmse10050698

Academic Editor: Antoni Calafat Frau

Received: 6 April 2022
Accepted: 11 May 2022
Published: 20 May 2022

Publisher's Note: MDPI stays neutral with regard to jurisdictional claims in published maps and institutional affiliations.

1. Introduction

Volatile components of magmas are important for a majority of silicate melts by influencing magma evolution and eruption processes [1]. Despite their comparatively low concentrations, they can influence the thermodynamic stability of minerals, as well as the density and viscosity of melts, thereby playing a primary role in controlling magma storage depth and the crustal-scale structure of sub-volcanic systems [2,3]. Subduction of altered oceanic crust and sediments are involved in global geochemical cycling of volatiles [4], and their migration of H_2O-rich slab fluids to mantle wedge is the primary process of realizing volatile circulation in subduction zones [5–8]. Halogens in volatiles can effectively trace subduction-released fluids, due to their behavior as fluid-mobile elements [9]. The fluid-mobile element Cl can be used as a tracer for slab-derived fluids [10]. Understanding the compositions of volatiles in subduction zone magmas, and the related magmatic processes, is essential for further insight into magma evolution and volatile circulation in subduction zones.

The Manus Basin is a fast spreading back-arc basin between the New Britain and New Ireland arc in the southwestern Pacific, which is bound by the inactive Manus Trench in the north and New Britain Trench in the south (Figure 1) [11]. A series of young volcanoes extend as an en-echelon series in the eastern part of the Manus Basin, which is known as the Eastern Manus Volcanic Zone [11]. Hydrothermal activity is widely developed [11–14]. Rock types in this region include basalt, basaltic andesite, andesite, dacite, rhyolitic dacite, and rhyolite. Research on the volcanic rocks of the Eastern Manus Basin is important for understanding magmatic processes and related mineralization in the back-arc basins, which are at the early stage of sea-floor spreading. Based on the geochemical characteristics of the whole rock and main rock-forming minerals (e.g., olivine and plagioclase), considerable work has been done on the magma source region and evolution of the Eastern Manus Basin ([11,14–20]). However, studies concerning the details of the relevant magmatic processes are still lacking, especially regarding the evolutionary behavior of magmatic volatiles. Only Sun et al. [15] have discussed the geochemical behavior of Cl in arc-type magmas in the Eastern Manus Basin through volcanic glasses.

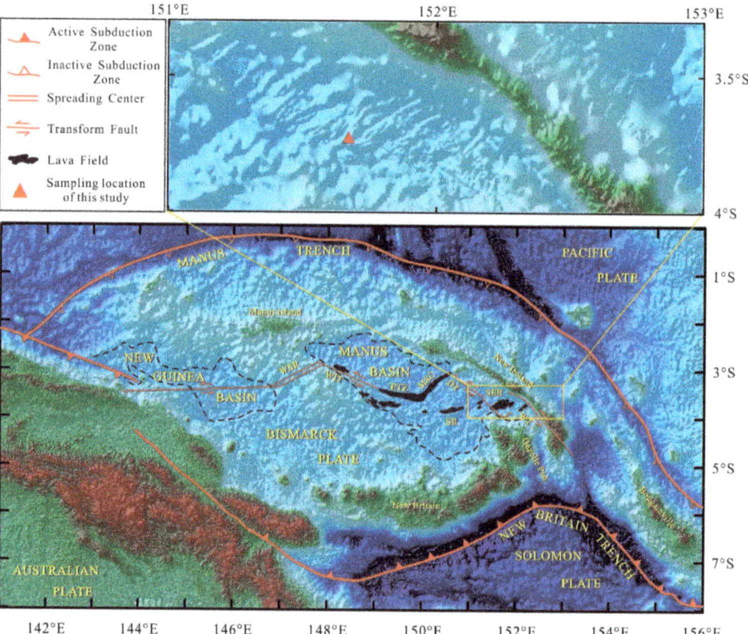

Figure 1. Regional geologic map of the Manus Basin, western Pacific (base map database from https://www.gebco.net/data_and_products/gridded_bathymetry_data/ (accessed on 28 February 2022); modified from ref. [12,21]).

Apatite is widely distributed in igneous rocks as an accessory mineral, and its molecular formula is commonly expressed as $A_5(XO_4)_3Z$ [22–26]. The A-site can accommodate large cations, such as Ca^{2+}, Sr^{2+}, Pb^{2+}, Ba^{2+}, Mg^{2+}, Mn^{2+}, Fe^{2+}, REE^{3+}, Eu^{2+}, Cd^{2+}, and Na^+. The X-site is mainly occupied by PO_4^{3+}, and can also accommodate other small highly charged cations, such as Si^{4+}, S^{6+}, As^{5+}, and V^{5+}. The Z-site is usually occupied by F^- (fluorapatite), Cl^- (chlorapatite), and OH^- (hydroxyapatite) [22]. Zhang et al. [27] have pointed out that submarine volcanic rocks are susceptible to seawater and hydrothermal alteration, and volatiles in volcanic glasses may undergo diffusion during the cooling process [28].

Apatite exhibits high resistance to weathering, alteration, and diffusion processes [29] and, since it can incorporate volatiles, such as F, Cl, and S into its lattice, it has become one of the most useful minerals in magmatic systems, in that it can provide information about the volatile concentrations in melts [28]. Apatite has important implications for assisting in determining the volatile compositions of magmatic systems, as well as the fugacities and concentrations of related compounds [30–33], and it can also provide critical geological information on petrogenesis and mineralization [34,35].

Therefore, we chose apatite, which is highly resistant to alteration and diffusion processes, as the object of this study. By analyzing the F, Cl, OH, and SO_3 contents of apatites, we have determined the behavior of volatiles and the redox state of magmas before magma eruption in the Eastern Manus Basin. Further, we have estimated the concentrations of magmatic volatiles, so as to improve understanding of the evolutionary behavior of magmatic volatiles and to supplement details of magmatic evolution in the Eastern Manus Basin; thus, deepening the understanding of magmatism in this region, as well as in the trench-arc-basin system.

2. Geological Background

The Manus Basin ($3°-4°$ S, $149°-152°$ E) is located in the eastern part of the Bismarck Sea, between the Manus Trench and New Britain Trench [12,16,36,37]. Due to the collision of the Ontong Java Plateau, subduction of the Pacific Plate along the Manus Trench almost stopped, and current volcanic activity in New Britain and northwestern Papua New Guinea is primarily the result of northward subduction of the Solomon Sea Plate along the New Britain Trench [16]. Submarine magnetic anomalies indicate that the Manus Basin was formed by asymmetric sea-floor spreading since 3.5 Ma [21]. GPS defined a rotation vector of $10°12'$ N, $33°30'$ W, $8.75°$ Ma^{-1} for the separation of the North Bismarck Sea plate from the South Bismarck Sea plate, which predicts a spreading rate of $115-145$ mm·a^{-1} for the Manus Basin, indicating that it is one of the fastest spreading basins in the world [16,38]. The spreading of the Manus Basin occurs along several rifts and submarine spreading centers offset by the Willaumez, Djaul, and Weitin transforms, which are known as the Western Spreading Rift (WSR), the Extensional Transform Zone (ETZ), the Manus Spreading Center (MSC), the Southern Rift (SR), and the Southeast Rift (SER), respectively (Figure 1), Their spreading patterns are somewhat different [16,36,37].

Located between the Djaul and Weitin transforms, the SER is the easternmost volcanic zone of the Manus Basin with water depths generally below 2000 m [14], consisting of a series of en-echelon neo-volcanic seafloor ridges and domes [20]. Large areas of exposed submarine lavas, and related high sonar reflectivity, indicate that this area has been the site of recent volcanic activity [39]. Since the SER is located within the shortest distance between the old and new trenches, petrochemistry indicates that it is more intensely influenced by subduction components than several other rifts and spreading centers in the Manus Basin (e.g., [11,39]). GPS shows its total spreading rate to be $137-145$ mm·a^{-1}, indicating that it is the fastest spreading part of the Manus Basin [38], and is in the early phases of sea-floor spreading [40]. Abundant hydrothermal activity has been developed in this region [13,41].

3. Sample and Analytical Techniques

3.1. Sample and Petrography

The samples analyzed in this paper were obtained from five stations (TVG7 ($151°40'21''$ E, $3°43'36.6''$ S, depth 1692 m), TVG 8($151°40'20.747''$ E, $3°43'36.617''$ S, depth 1692 m), TVG 9 ($151°40'19.873''$ E, $3°43'43.341''$ S, depth 1713 m), TVG10 ($151°40'19.712''$ E, $3°43'40.909''$ S, depth 1702 m), TVG 12 ($151°40'18.096''$ E, $3°43'36.946''$ S, depth 1683 m)) by using a TV grab during the cruise of the R/V Kexue Hao in 2015, and all the samples are dacite (Figures 1 and 2).

Figure 2. Hand specimens of dacite from (**a**) TVG 7; (**b**) TVG 8; (**c**) TVG 9; (**d**) TVG 10; (**e**) TVG 12; and transmitted light images of phenocrysts in dacite from (**f**) and (**g**) TVG 9; (**h**) TVG 10 in the Eastern Manus Basin. Abbreviations: Pl—plagioclase, Cpx—clinopyroxene.

The sample hand specimens are black in color and porphyritic, with phenocryst contents <5 vol.%, mainly composed of plagioclase, pyroxene, and minor magnetite. The sample thin sections were observed using a polarized light microscope. The sizes of phenocrysts are 50−600 μm, and they exhibit euhedral to subhedral texture; the matrix is composed of acicular plagioclase micro-crystals, granular pyroxene micro-crystals, and minor granular magnetite micro-crystals with pilotaxitic texture (Figure 2f–h). According to the back-scattering images of the thin sections, apatites exist primarily as micro-crystals in the host dacite, and partly as inclusions in plagioclase, pyroxene, and magnetite phenocrysts, implying the presence of early apatite-saturation in melts (Figure 3a–c) [42]. The apatite micro-crystals in the matrix have a size of ~20 μm, mostly exhibiting equant to sub-equant crystals in euhedral-subhedral texture, and showing uniformly bright surfaces without mineral- or fluid-inclusions (Figure 3d–g), indicating that no hydrothermal replacement has occurred. A very few apatites are highly acicular and large (maximum elongation along the c-axis to more than 300 μm, as in Figure 3h,i), which may precipitate during a rapid quench [24].

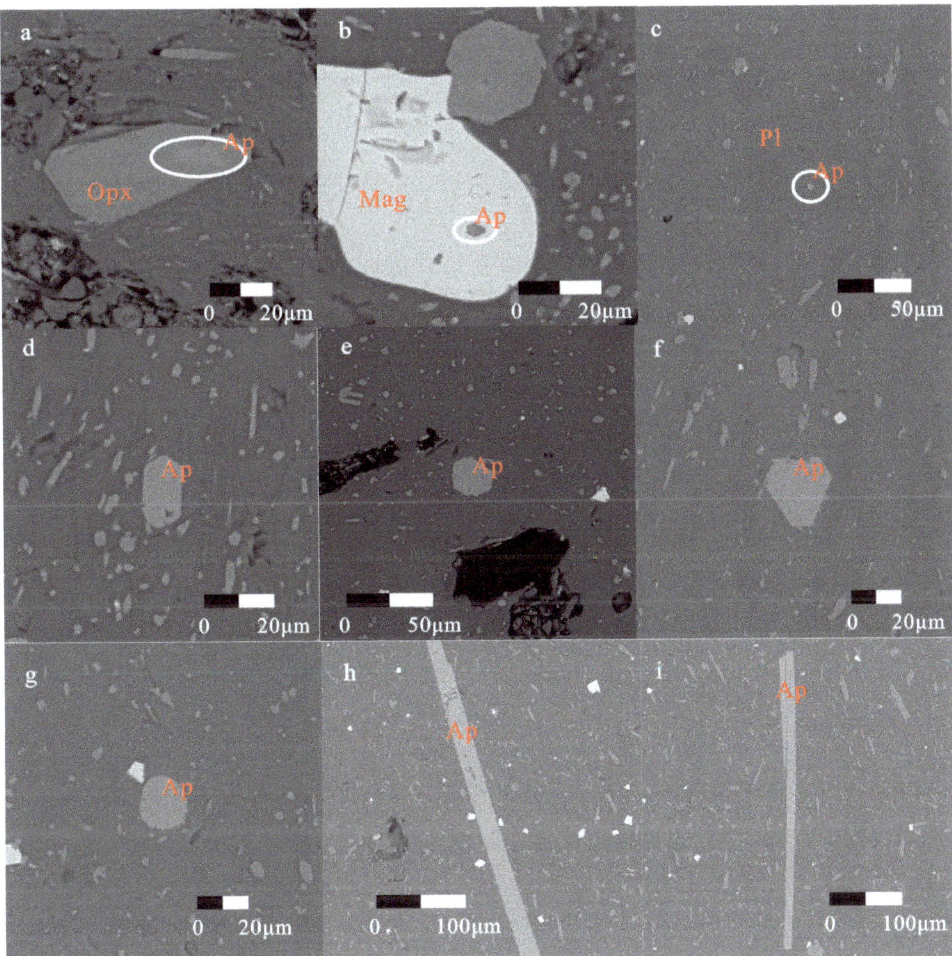

Figure 3. Back-scattering images of apatites in dacite from the Eastern Manus Basin. (**a**) Apatite in orthopyroxene; (**b**) Apatite in magnetite; (**c**) Apatite in plagioclase; (**d**–**g**) Apatite crystals in euhedral-subhedral texture; (**h**,**i**) Apatite crystals in highly acicular texture. Abbreviations: Opx—orthopyroxene, Mag—magnetite, Pl—plagioclase, Ap—apatite.

3.2. Analytical Methods

The samples were made into thin sections that met the requirements for onboard testing, then observed and analyzed using a polarized light microscope, scanning electron microscopy (SEM), equipped with an X-ray energy spectrometer, and an electron probe microanalyzer (EPMA).

Preliminary observations were carried out using Axio Scope A1 polarized light microscopy and VEGA3 TESCAN scanning electron microscopy at CAS Key Laboratory of Marine Geology and Environment, Institute of Oceanology, Chinese Academy of Sciences, Qingdao, China. The operating conditions of the scanning electron microscope included a voltage of 20 kV, an emission current of 1.4–1.9 nA, and a working distance of 15 mm. Qualitative and semi-quantitative analyses of apatites and other minerals in the thin sections were carried out using the Oxford X-Max20 energy spectrometer fitted to the SEM, which employed an excitation voltage of 20 kV. Quantitative analysis of the mineral compositions was performed using quartz, anorthoclase, and pyrite as specimens. A total of 16 apatite microcrystals in 10 thin sections were finally selected for this study.

Microprobe analyses of apatites were performed using a JOEL JXA-8230 electron microprobe at the State Key Laboratory for Mineral Deposits Research, Nanjing University, Nanjing, China. The operating conditions included an accelerating voltage of 15 kV, a beam current of 20 nA, and, depending on the size of the apatites, a circle size with a diameter of 3 or 5 μm was selected. A total of 33 mineral sites were analyzed for matrix apatite micro-crystals. The standard sample used in this work was the apatite from the American National Bureau of Standards, and the precision of the analysis could reach 0.1%. All test data were corrected by the ZAF (Z, atomic number correction factor; A, X-ray absorption correction factor; F, fluorescence correction factor) program.

4. Results

4.1. Compositions of Apatites

The results of the analysis are shown in Table S1, which includes calculations of OH concentrations of apatites using the method of Ketcham [43], and X_F, X_{Cl}, and X_{OH} are molar fractions of F, Cl, and OH. It can be seen that the apatite samples all belong to magmatic apatite with high hydroxyl contents (Table S1 in Supplementary Materials, Figures 4 and 5). The dominant components of apatite samples were CaO and P_2O_5, which exhibited a range of 53.08−54.55% and 40.95−42.55%, respectively. Moreover, the apatite samples exhibited the characteristics of Cl-rich and F-poor with contents of 1.24−1.70% and 0.87–1.39%, respectively, which were between "mantle apatite A" and "mantle apatite B" [44] (Figure 6a). Compared with apatites in dacite from other subduction systems of the western Pacific, our apatites had higher Cl contents, as well as lower F and SO_3 contents [42,45] (Figure 6a,b).

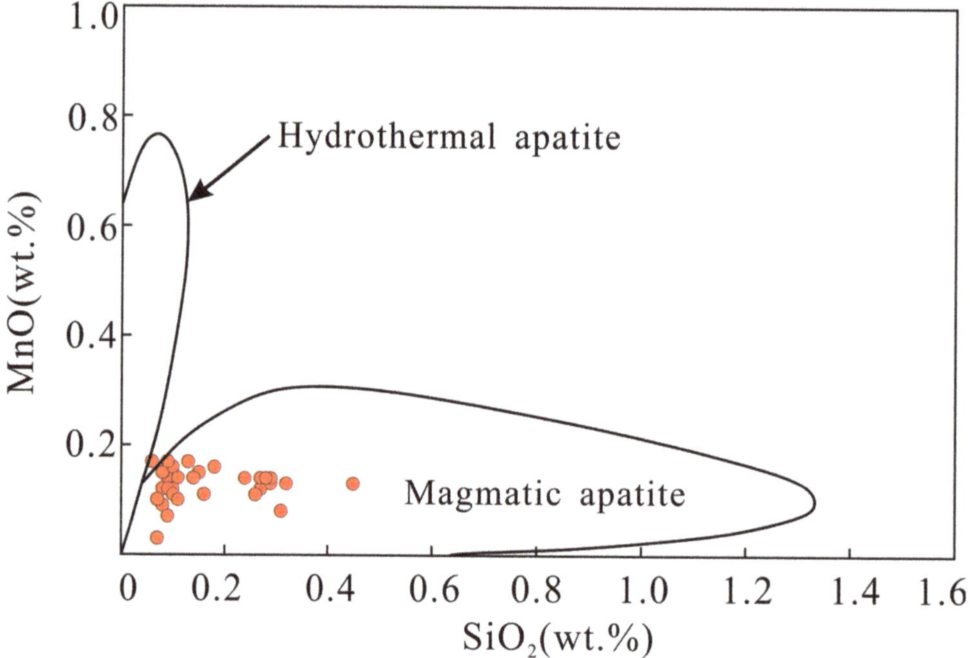

Figure 4. Plots of MnO versus SiO_2 in apatites of dacite (modified from ref. [46]).

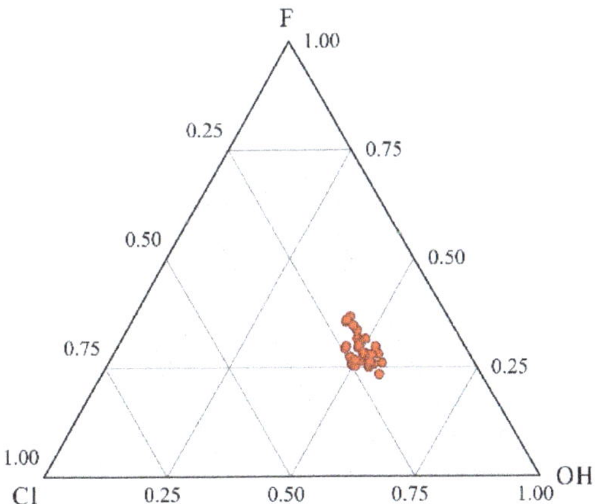

Figure 5. Volatile contents (molar fraction) of apatites from the dacite.

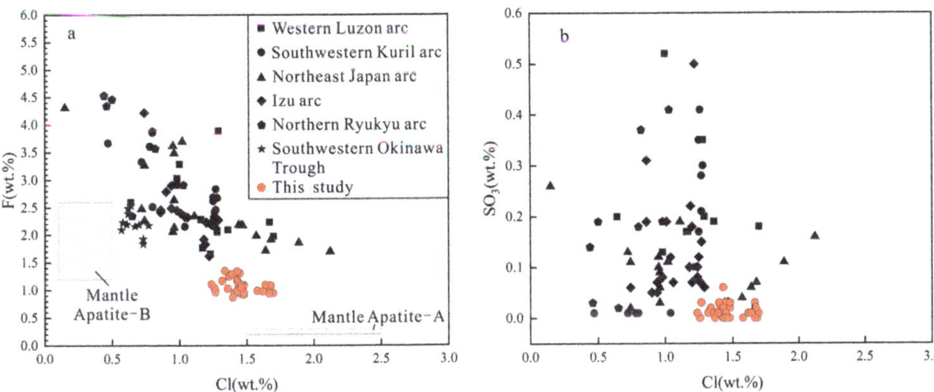

Figure 6. Plots of (**a**) F versus Cl in apatites of dacite; (**b**) SO$_3$ versus Cl in apatites of dacite (data from [42,45] and this study).

4.2. Apatite Saturation Temperatures (AST)

Mostly, the elemental partitioning between apatites and melts varies significantly with temperature [25], so before calculating the individual elemental components of melts, we needed to know the equilibrium temperature of the system apatite-melt. Here we used the apatite saturation temperature $\left(\ln D_P^{apatite/melt} = \dfrac{[8400+(SiO_2-0.5)\times 2.64\times 10^4]}{T} - [3.1 + 12.4\times(SiO_2 - 0.5)]\right)$ proposed by Harrison and Watson [47] to constrain the equilibrium temperature, where the $D_P^{apatite/melt} = P_2O_5^{apatite}/P_2O_5^{melt}$. Remarkably, there is no reliable method to accurately estimate the instantaneous concentration of melt P_2O_5 (i.e., $P_2O_5^{melt}$) at the time of apatite crystallization, but given that the samples in this study were all from felsic melts with low crystallization, the whole rock P_2O_5 concentrations could be approximated here as the P_2O_5 concentrations of melt. Although there will be some errors, the correction should be small and have little effect on the results [22]. As shown in Table 1, the AST of the apatite samples ranged from 935 to 952 °C.

Table 1. Apatite saturation temperatures of apatite samples.

Sample ID	SiO_2 in Whole-Rock (wt.%) *	P_2O_5 in Whole-Rock (wt.%) *	P_2O_5 in Apatites (wt.%)	AST (°C)
7-2-5-1	69.34	0.14	40.96	936
7-2-5-2	69.34	0.14	41.51	935
8-1-6-1	69.54	0.14	41.50	936
8-1-6-2	69.54	0.14	41.61	936
8-1-5-1	69.54	0.14	41.71	935
9-2-1-14-1	67.57	0.19	41.14	951
9-2-1-14-2	67.57	0.19	41.67	949
9-2-1-11-1	67.57	0.19	40.95	951
9-2-1-15-1	67.57	0.19	41.77	949
9-2-1-15-2	67.57	0.19	41.78	949
9-3②-1-7-1	67.60 (2)	0.19 (2)	42.28	949
9-3②-1-7-2	67.60 (2)	0.19 (2)	42.55	948
9-3②-1-8-1	67.60 (2)	0.19 (2)	41.78	950
9-3②-1-8-2	67.60 (2)	0.19 (2)	41.27	952
9-3②-2-2-1	67.60 (2)	0.19 (2)	41.90	950
9-3②-2-2-2	67.60 (2)	0.19 (2)	41.61	951
9-3②-2-2-3	67.60 (2)	0.19 (2)	41.60	951
9-3②-2-2-4	67.60 (2)	0.19 (2)	41.64	951
9-3②-2-3-1	67.60 (2)	0.19 (2)	41.33	951
9-3②-2-3-2	67.60 (2)	0.19 (2)	41.61	951
9-3②-2-3-3	67.60 (2)	0.19 (2)	41.65	951
10-1-1-14-1	67.46	0.19	42.05	948
10-1-1-14-2	67.46	0.19	41.70	949
10-1-1-16-1	67.46	0.19	42.04	948
10-1-1-16-2	67.46	0.19	41.44	950
10-2①-1-13-1	68.57	0.16	42.14	937
10-3-1-7-1	67.51	0.19	42.39	948
10-3-1-7-2	67.51	0.19	42.29	948
12-1-6-1	69.46	0.14	42.10	936
12-1-6-2	69.46	0.14	41.32	938
12-1-6-3	69.46	0.14	41.61	937
12-2-9-1	69.46	0.14	41.48	938
12-2-9-2	69.46	0.14	41.55	938

() is the number of samples used to calculate the average value. * SiO_2 and P_2O_5 contents of whole-rock were determined by X-ray fluorescence (XRF) using a PANalytical AXIOS Minerals at Institute of geology and geophysics, Chinese Academy of Sciences.

5. Discussion

5.1. Behavior of Magmatic Volatiles

When the magmas have reached volatile-saturated conditions, melts will be highly depleted with fluid exsolution, as fluid-mobile elements (e.g., Cl) prefer to enter the fluid phase [48]. Apatites crystallized in such a state could not indicate the compositions of primitive melts well. Hence, before using apatites to extrapolate the compositions of primitive magmas, it was necessary to determine whether volatiles in the melts became saturated before and during the crystallization of apatites. Stock et al. [1] developed a model to determine whether magmatic volatiles have reached saturation by the relative relationship between contents of F, Cl, and OH as well as their ratios in apatites. Combined with this model, as shown in Figures 5 and 7, X_F/X_{Cl} of the apatite samples exhibited a decreasing trend, while X_{Cl}/X_{OH} and X_F/X_{OH} exhibited increasing trends, implying that those apatites crystallized under a volatile-undersaturated condition [1]. Therefore, the compositions of apatites in this study could be useful indicators for their host magmas.

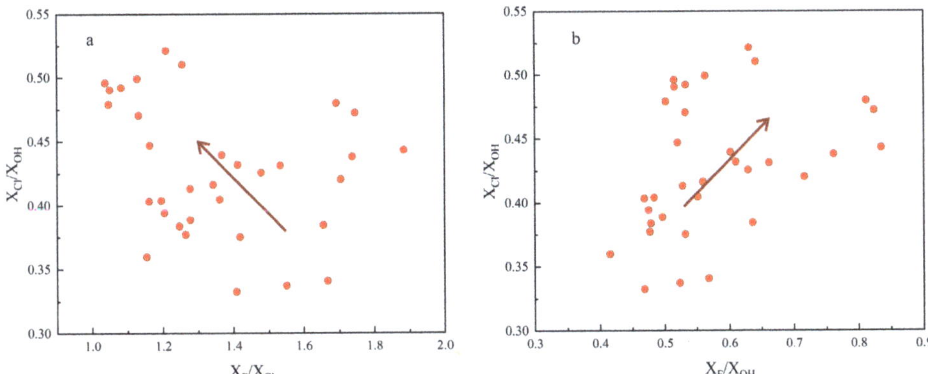

Figure 7. Plots of (**a**) X_{Cl}/X_{OH} versus X_F/X_{Cl} in apatites of dacite; (**b**) X_{Cl}/X_{OH} versus X_F/X_{OH} in apatites of dacite. The arrow shows the trajectory of apatite compositional evolution.

5.2. Magmatic S Contents and Redox State

5.2.1. Magmatic S Concentrations

The S concentration in apatite is related to that in melts during the crystallization of apatite, so it can trace the S contents of melts [49]. Although there is no formula to accurately calculate the S concentrations of magmas from the SO_3 contents of apatites [50], if the chemical exchange between apatites and their host melts can be neglected after their formation, the relative S contents of magmas can be estimated according to two semi-quantitative formulas given by Peng et al. [29] and Parat et al. [28], respectively [49]. The apatite micro-crystals in this study were morphologically intact, small in size, and without zoning, indicating their short contact time with melts after their formation, as well as limited chemical exchange between the two. Thus, we estimated the S contents of melts using the above two semi-quantitative formulae. As shown in Table S2 in Supplementary Materials, the S concentrations of the host melt ranged from 2−65 ppm or 8−11 ppm, which is consistent with the range of S contents in silicate melts produced by partial melting of subduction-related mantle wedge (<4000 ppm [51]). It should be noted that S exists in magmatic systems in five main valence states, including S^{2-}, S^{1-}, S^0, S^{4+}, and S^{6+} [52]. Since S is present mainly as SO_4^{2-} in apatite lattice, the S contents obtained from our analysis were only related to the concentrations of SO_4^{2-} (S^{6+}) in the melts, while there may also be considerably reduced sulfur in the melts (more details will be discussed in Section 5.2.2), so the values we obtained here were probably just the minimum S contents of melts [53].

From previous studies, the SO_3 content of apatite is a complex function of magmatic temperature, sulfur fugacity and redox state [28,29,54]. According to Peng et al. [29], the partitioning of S between apatites and their host melts is highly dependent on the redox state of the latter. Combining Table 1 and Table S2 in Supplementary Materials, the apatites studied crystallized in narrow ranges of temperature and sulfur fugacity, and their SO_3 contents remained essentially stable with magmatic differentiation (Figure 8), which tends to indicate that the magma has a stable redox state during this evolution stage.

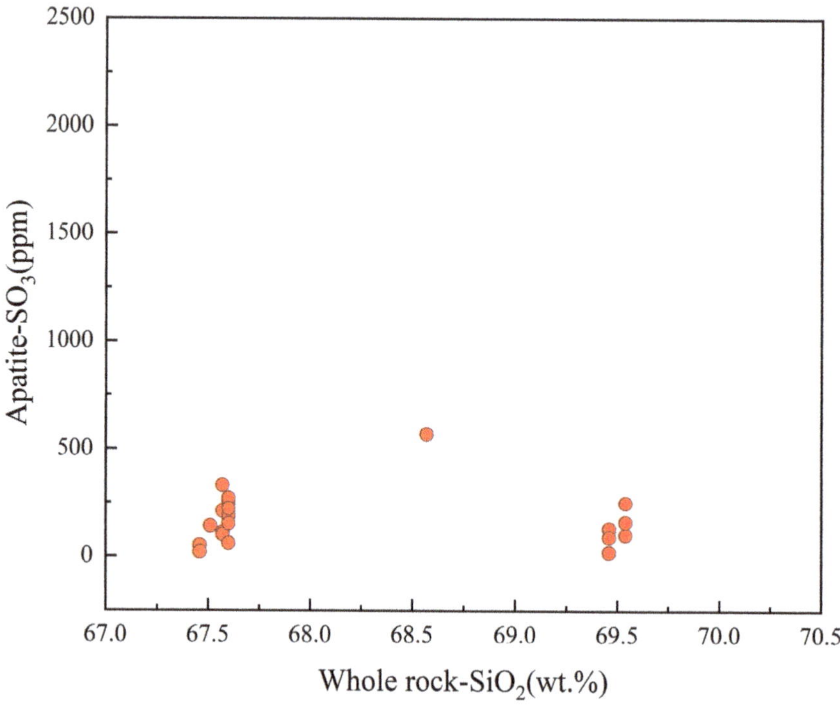

Figure 8. Correlation between SO_3 concentrations in apatites and magmatic differentiation.

5.2.2. Indication for Oxygen Fugacity of Melts

Affected by the magmatic oxygen fugacity, Mn ions may enter the apatite lattice in three valence states: Mn^{2+}, Mn^{3+} and Mn^{5+}. They enter the apatite lattice through the following displacement reactions [55]:

$$Mn^{2+} = Ca^{2+} \tag{1}$$

$$Mn^{3+} + SiO_4^{4-} = Ca^{2+} + PO_4^{3-} \tag{2}$$

$$Mn^{3+} + Na^+ = 2Ca^{2+} \tag{3}$$

$$MnO_4^{3-} = PO_4^{3-} \tag{4}$$

When the above reactions have occurred, there will be an obvious negative correlation between the ions that undergo displacement. In apatite samples, obviously, no negative correlations are shown between Mn and Ca as well as Mn + Na and 2Ca (Figure 9a,b), while a significant negative correlation exists between Mn + Si and Ca + P (Figure 9c), and a moderate negative correlation between Mn and P (Figure 9d). Therefore, it can be concluded that Mn in our samples entered the apatite lattice mainly as Mn^{3+}, and brought SiO_4^{4-} into the lattice simultaneously. In addition, a few Mn replaced PO_4^{3-} as MnO_4^{3-} in the apatite lattice. It indicates that apatites from this study crystallized in a medium oxygen fugacity condition, and the oxygen fugacity only slightly changed during this process.

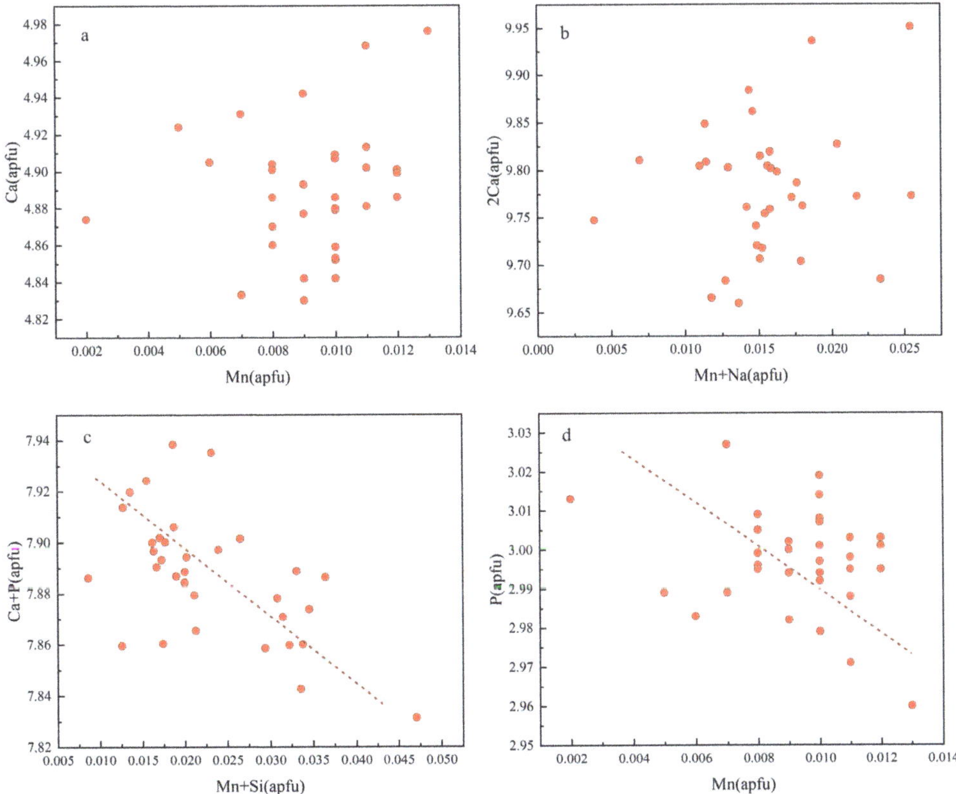

Figure 9. Covariant relations of cations in apatites from the dacite. (**a**) Ca versus Mn; (**b**) 2Ca versus Mn + Na; (**c**) Ca + P versus Mn + Si; (**d**) P versus Mn.

The quantitative estimation of the redox state can be achieved by calculating the oxygen fugacity (f_{O_2}). The oxygen fugacity is a tool to assess the availability and potential of oxygen to participate in reactions between minerals and fluids. Concentrations of redox-sensitive elements, such as Fe, Mn, Ce, and Eu, in accessory minerals are important for obtaining the redox states of magmas [56]. Miles et al. [56] found that the variation of Mn contents of apatites in intermediate-acid silicate melts is relatively independent, so that the redox states of magmas can be reliably estimated based on the Mn contents of apatites. To further clarify the redox state of magmas and their stabilities, we estimated the absolute and relative oxygen fugacities of the host melts according to MnO contents of apatite samples using the equations given by Miles et al. [56], as well as Myers and Eugster [57]. Since the FMQ oxygen buffer is minimally influenced by pressure [58], the relative oxygen fugacities (ΔFMQ) of melts were calculated here with the oxygen fugacities of the FMQ oxygen buffer as the reference (Table S2 in Supplementary Materials).

The mantle wedge above subduction zones is oxidized relative to the mantle from other tectonic settings [59–61], and the water-bearing intermediate silicate magmas in such regions generally have intermediate oxygen fugacity conditions (from general reducing to general oxidizing conditions) [62]. As regards this study, except for one anomalously high value (ΔFMQ = +1.4), all values showed that the host melt was under the reduction condition with a narrow fluctuation range (ΔFMQ = −0.2 ± 0.9, Table S2 in Supplementary Materials). Kleinsasser et al. [62] found experimentally that dacitic melts under redox conditions of ΔFMQ − 0.7~ΔFMQ + 2.08 would remain both sulfide- and sulfate- undersaturated, which is consistent with the observations and calculations of this study. The trace element compositions of dacite in the Eastern Manus Basin distinctly exhibited the characteristics of island arc volcanic rocks, such as enrichment in light rare earth elements (LREE) and large ionic lithophile elements (LILE) [11,14,18]. However, this study indicates that the oxygen fugacity of dacitic magmas in the Eastern Manus Basin is lower than that of island-arc magmas influenced by subduction components (ΔFMQ = +1~+2, [63–65]). Two reasons may account for this phenomenon: (1) As shown in previous studies, processes such as crystallization and separation of closed-system, crustal contamination, and water degassing ($\leq$8 wt.%) have just a slight impact on magmatic oxygen fugacity, while sulfur degassing and addition of subduction components have considerable effects on redox states of magmas (e.g., [66–69]). According to the discussion in 5.1, the host magma has not yet reached fluid saturation, and it is also unlikely to have reached S saturation [70]; the relatively low oxygen fugacities of the dacite samples may be inherited from the mantle source region. However, this is obviously inconsistent with the high oxygen fugacity of the mantle wedge in the subduction zones; (2) Richards [65] found that crystallization and segregation of Fe^{3+}-rich magnetite can result in a reduction of reduced to moderately oxidized magmas, which could be offset by oxidation from degassing of reducing or weakly oxidizing gases (e.g., H_2, H_2S, etc.), as well as dissolution and migration of Fe^{2+} in high-temperature saline hydrothermal fluids. The saturation of apatites in the Eastern Manus Basin was concomitant with the saturation of magnetite, and they crystallized from undegassed melts (see the description in Section 3.1, Figure 3b, and previous studies [71] for details). Furthermore, hydrothermal activity is widely developed in the Eastern Manus Basin, but the oxidation process from high-temperature hydrothermal fluids may be slow [65]. When the apatites in melts reached saturation and started crystallizing, the intensity of oxidation from the hydrothermal fluids may not yet have been sufficient to offset the decrease in oxygen fugacity, due to the crystallization and separation of magnetite. The above factors collectively caused the oxygen fugacity reduced from ΔFMQ $\approx$ +1~+2 in the mantle source region to ΔFMQ = -0.2 ± 0.9 in the host magma recorded by apatites. Apparently, the second explanation is more reasonable. Under such a low oxygen fugacity condition, S existed mainly as low-valence states in melts, which caused the low SO_3 contents of apatites in this study.

5.3. F, Cl and H_2O Contents of Melts

Apatite is one of the rare magmatic minerals containing halogens. For extrusive and hypabyssal rocks, the rapid cooling of magmas allows the preservation of information about magmatic F, Cl, and H_2O contents recorded by apatite during its crystallization [72]. Further, apatites crystallized at temperatures above 500 °C are resistant to hydrothermal alteration [73,74], therefore they can be used to deduce the F, Cl, and H_2O contents of the host magma.

5.3.1. F, Cl Contents of Melts

Before deducing the volatile contents of melts by using that of apatites, the partition behavior of volatiles between the two phases needs to be known [75]. Since F, Cl, and OH occupy the same position in the apatite lattice, their abundances are controlled by stoichiometric numbers, which prevent the use of Nernst partition coefficients for the estimations of F, Cl, and OH contents in melts [76]. To better represent partitioning behavior between apatites and the host melts, previous works have proposed the exchange coefficient K_d involving two volatile elements (e.g., [75,77,78]), which can be expressed as the molar fraction (i.e., X_A^{Ap}, X_B^{Ap}, X_A^{melt}, X_B^{melt}) of two anions (e.g., A and B) in both apatites and the host melt:

$$K_{d\,A-B}^{Ap-melt} = \frac{X_A^{Ap} \cdot X_B^{melt}}{X_A^{melt} \cdot X_B^{Ap}} \tag{5}$$

In this paper, we used the regular ternary solution model for apatite established by Li and Hermann [78], combined with the modified K_d expression by Li and Costa [75], to calculate the F, Cl contents of melts as follows:

$$Cl_{melt}(wt.\%) = \frac{X_{Cl}^{Ap}}{X_{OH}^{Ap}} \times K_{dOH-Cl}^{Ap-melt} \times 10.79(\pm0.52) \tag{6}$$

$$F_{melt}(wt.\%) = \frac{X_{F}^{Ap}}{X_{OH}^{Ap}} \times K_{dOH-F}^{Ap-melt} \times 6.18(\pm0.41) \tag{7}$$

$$K_{dOH-Cl}^{Ap-melt} = e^{-\frac{1}{8.314\times T} \times \left\{ \begin{array}{c} 72900(\pm2900) - 34(\pm0.3) \times T \\ -1000 \times [5(\pm2) \times (X_{Cl}^{Ap} - X_{OH}^{Ap}) - 10(\pm8) \times X_{F}^{Ap}] \end{array} \right\}} \tag{8}$$

$$K_{dOH-F}^{Ap-melt} = e^{-\frac{1}{8.314\times T} \times \left\{ \begin{array}{c} 96400(\pm5600) - 40(\pm0.1) \times T \\ -1000 \times [7(\pm4) \times (X_{F}^{Ap} - X_{OH}^{Ap}) - 11(\pm7) \times X_{Cl}^{Ap}] \end{array} \right\}} \tag{9}$$

where T was substituted by AST calculated in 4.2, in K. With the method described above, the F and Cl contents of the host melt were estimated to be about $185-448$ ppm and $1059-1588$ ppm, respectively (Table S3 in Supplementary Materials), which are consistent with the range of that in subduction-related dacite ($0.01-0.15$ wt.% for F and $0.01-0.3$ wt.% for Cl, [79]).

Magmatic F and Cl contents will differ widely depending on tectonic settings and magma compositions [80]. At concentrations of a few tens to 1000 ppm, they will not affect the fundamental magma properties (e.g., melting behavior and melt rheology) [79]. The Cl content of our dacite samples exhibited relatively high values. Cl at such concentrations may have implications on the composition and timing of exsolved fluids, as well as on the rheology of host magma [81–84]. From Figure 6a, the F and Cl contents of the apatite samples were similar to the "mantle apatite A", which is formed by the account of Cl-, CO_2- and H_2O-rich fluids, indicating that the host melt might have been influenced by exotic Cl-rich components. Moreover, F is highly compatible with melt, while Cl behaves as a fluid-mobile element [79], so the high Cl/F ratio in the samples indicates the addition of slab-derived fluids in the parent magmas [35], which is parallel with the results of previous studies on volcanic rocks in the Eastern Manus Basin (e.g. [16,18,39,85]). Additionally, the non-negative correlation between the Cl/F ratio and SiO_2 contents of samples verifies again that the melt remained in a fluid-undersaturated condition (Figure 10).

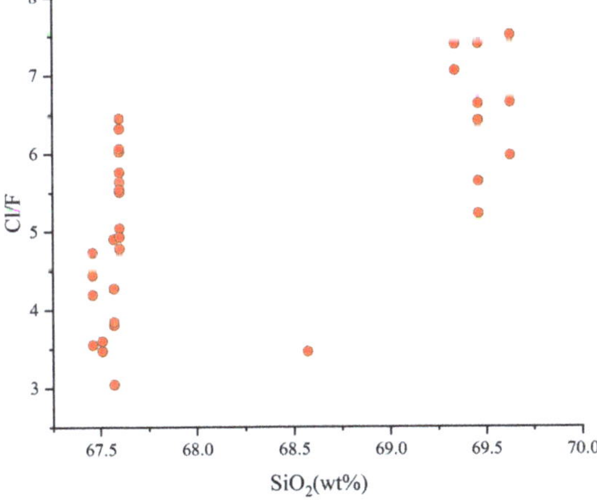

Figure 10. Plots of Cl/F versus SiO_2 in dacitic melts of the Eastern Manus Basin.

5.3.2. Abundance of H_2O in Melts

Boyce and Herving [86] found that the variation of F and OH in apatites can be well-matched with their variation in coexisting melts, indicating that OH is a valid indicator for evaluating the H_2O content of coexisting melts. Since the water in melts exists as both molecular water (H_2O_m) and hydroxyl (OH) [87], the calculation of its molar fraction is complicated. Li and Costa [75] developed an online calculation program based on the thermodynamic model for apatite. This program for the determination of H_2O concentrations in multicomponent silicate melts is based on a series of thermodynamic formulae. As mentioned above, the exchange coefficients K_d between apatite-melt can be obtained by using AST as well as F, Cl, and OH contents of apatites, and from that, $\frac{X_{OH}^{melt}}{X_{Cl}^{melt}}$ and $\frac{X_{OH}^{melt}}{X_{Cl}^{melt}}$ can be calculated. Furthermore, according to Li and Costa [75], X_F^{melt}, X_{Cl}^{melt}, and X_{OH}^{melt} can be expressed by the mass fractions of F, Cl, and H_2O (c_F^{melt}, c_{Cl}^{melt}, and $c_{H_2O}^{melt}$, respectively) in melts as follows:

$$X_F^{melt} = \frac{\frac{c_F^{melt}}{19}}{\frac{c_{H_2O}^{melt}}{18} + \frac{1 - c_{H_2O}^{melt}}{W}} \tag{10}$$

$$X_{Cl}^{melt} = \frac{\frac{c_{Cl}^{melt}}{35.45}}{\frac{c_{H_2O}^{melt}}{18} + \frac{1 - c_{H_2O}^{melt}}{W}} \tag{11}$$

$$X_{OH}^{melt} = \frac{\frac{1}{2} - \sqrt{\frac{1}{4} - \left\{ \left(\frac{K_2 - 4}{K_2} \right) \left([H_2O_t] - [H_2O_t]^2 \right) \right\}}}{\left(\frac{K_2 - 4}{2K_2} \right)} \tag{12}$$

where W is the molar fraction of dry silicate melts, which was taken here as 33 g/mol [75]; $[H_2O_t]$ is the molar fraction of total water in melts; K_2 is the equilibrium constant for the conversion from H_2O_m to OH:

$$[H_2O_t] = \frac{\frac{c_{H_2O}^{melt}}{18}}{\frac{c_{H_2O}^{melt}}{18} + \frac{1 - c_{H_2O}^{melt}}{W}} \tag{13}$$

$$\ln K_2 = a + b/T \tag{14}$$

According to Liu et al. [88], for dacite, a equals 1.49 and b equals −2634. Concentrations of H_2O in the Eastern Manus Basin dacite obtained by this calculation program ranged from 1.4−2.1 wt. % (based on F concentrations in melts, error ±30−40%) and 1.2−1.5 wt.% (based on Cl concentrations in melts, error ±30−40%), respectively (Table S4 in Supplementary Materials). Sun et al. [15] have measured H_2O contents of volcanic glasses in the East Manus Basin, and the results ranged from 1.11−1.67 wt.%, which is generally consistent with the range we estimated, indicating that the results obtained here should be reliable.

Previous studies have revealed that the H_2O contents of MORB magmas are generally in the range of 100−450 ppm [89], while the primary melts of basalts in subduction zones generally contain higher H_2O contents, indicating the influence of water as a fluid phase transferred from the subduction slab to the mantle wedge [90]. The H_2O contents of dacitic magmas in the Eastern Manus Basin are obviously higher than those of MORB magmas, again indicating the addition of slab fluids. To be noted, in previous linear regression calculations of K_d, the low F contents in experimental melts caused a larger analysis error than that of Cl [75]. Thus, the H_2O contents calculated from Cl concentrations may be closer to the true value, which indicates that the magma chamber of dacitic melts in the Eastern Manus Basin is located about 2 km below the seafloor at a pressure of 0.05 GPa−0.1 GPa [15].

5.4. Sources of F- and Cl-Rich Components

The F and Cl concentrations in melts of the Eastern Manus Basin are significantly higher than those in the primitive mantle (25 ppm F and 17 ppm Cl, [91]). Moreover, since halogens are incompatible with mantle minerals, their concentrations will be further lost in the mantle during its partial melt [92]. Subduction is an important mechanism for the circulation of halogens from surface reservoirs to the mantle [9,92,93]. According to the above discussion, magmas in the Eastern Manus Basin are influenced by the incorporation of subduction components during subduction. Moreover, numerous previous studies have also proven that magmas in this region are obviously influenced by subduction components (mainly by fluids released from the subducting altered oceanic crust, as well as a little sediment melt, e.g. [16,18,39,85] and references therein), yet no significant magmatic mixing and crustal contamination occurred during magma evolution. Hence, the high F and Cl concentrations of our samples may be related to the pollution in the mantle source region.

Seawater has a high Cl, but low F, abundance, which is 19,500 ppm and 1.3 ppm respectively [93]. Therefore, the direct addition of seawater is too low to explain the enrichment of F in samples. In contrast, sediments near subduction zones may be a reasonable source of F; for example, up to 1300 ppm in pelagic clay [94]. The repeated leaching of overlying sediments and assimilation of oceanic crust near the subduction zones by seawater may cause an increase of F contents in the oceanic crust before its subduction [92]. Since F is an incompatible element with fluids [79], the enrichment of F caused by the above process may not be enough. According to Pagé et al. [92], apatite is the mineral phase that most preferentially accommodates F in subduction zones. F may be redistributed to phengite and lawsonite as apatite eventually decomposes at $\sim$200 km, then lawsonite and phengite decomposition releases F to the deeper upper mantle at $\sim$280–300 km in cold subduction zones [35,92], which could be another reasonable source of F-rich components. Additionally, the F-rich seafloor sediments could also contribute F efficiently to the mantle source region as melts during subduction.

Studies of back-arc magmatism have indicated that slab fluids contributing to magmatism in back-arc basins are Cl-rich, as shown by off-axis seamount lavas of the Lau Basin and submarine basaltic glasses of the Mariana back-arc trough, both of which exhibit Cl-rich characteristics due to subduction [6,10,95]. Although the contribution of slab fluids can be identified with Cl/K_2O and Cl/TiO_2 to Ba/Nb ratios, the mobility of K differs from that of Cl, and the value of Cl/TiO_2 changes during magma evolution, since Cl is more incompatible than Ti [10]. So neither Cl/K_2O nor Cl/TiO_2 is appropriate for quantifying the contribution of subduction-released fluids to Cl. Sun et al. [15] indicated that the plots of Cl/K_2O and Cl/TiO_2 versus Ba/Nb can be refined with the plots of Cl/Nb versus Ba/Nb and U/Nb. Cl, Ba, U, and Nb are incompatible elements, where Ba and U are fluid-mobile LILE, and the behavior of Cl resembles them, while Nb is one of the most inactive elements during subduction [15,48,96,97]. The ratios of Cl, Ba, U to Nb can maximally offset the partial melting of the mantle, as well as the crystallization and differentiation of magmas, thereby providing a useful indication of the mobility of these elements during subduction. Here, we plot the Cl/Nb of our samples and the collected volcanic rocks of other subduction systems in the western Pacific against Ba/Nb and U/Nb, respectively (Figure 11; Cl in our samples was calculated from apatite components, and the rest of data are shown in Table S4).

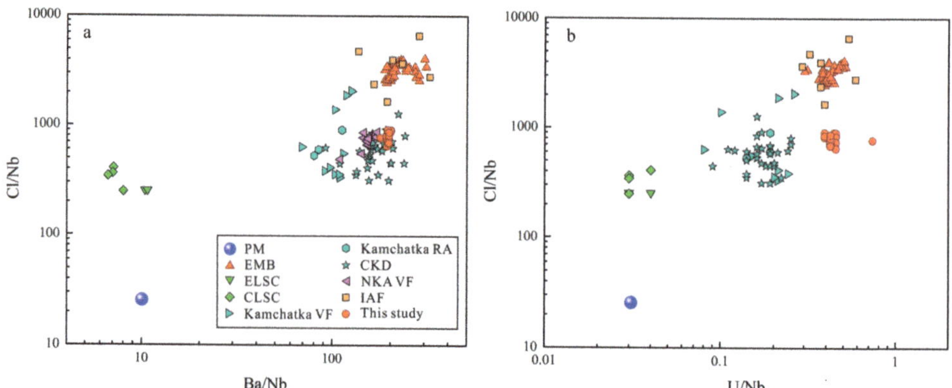

Figure 11. Plots of (**a**) Cl/Nb versus Ba/Nb and (**b**) Cl/Nb versus U/Nb in subduction-relate volcanic rocks (data from: PM, Primitive Mantle [91]; EMB, Eastern Manus Basin [15]; ELSC, Easte Lau Spreading Center [10,98]; CLSC, Central Lau Spreading Center [10,98]; Kamchatka VF, volca front of Kamchatka [99]; Kamchatka RA, rear-arc of Kamchatka [99]; CKD, The Central Kamch Depression [99]; NKA VF, volcanic front of Northern Kurile Arc [99]; IAF, Izu arc front [93]).

Figure 11 indicates that, compared to the primitive mantle, the mantle source of study exhibited intense enrichment of Cl, Ba, and U, implying that it has been influen by Cl-, Ba-, and U-rich components. Such components are most probably the subducti released fluids [10]. Moreover, the elemental ratios of samples in this study and volca rocks from other subduction-related systems did not exhibit a single linear trend with t primitive mantle. At the same Cl/Nb values, volcanic rocks from the Kamchatka, Izu ai and Lau basin exhibit lower Ba/Nb and U/Nb values relative to those of this study, which appears to indicate the influence of other Cl-rich and Ba-, U-poor components.

Kent et al. [10] indicated that shallow assimilation of Cl-rich seawater derived compo- nents during magma transport, storage, or eruption in submarine environments can also result in increased Cl concentrations in magmas, and, compared to subduction-released fluids, such materials are characterized as Ba- and U-poor. Sun et al. [15] also found that the samples were commonly influenced by Cl-rich seawater derived components through the analysis of matrix glasses in several back-arc basins and active spreading ridges, including the Eastern Manus Basin. Compared to the matrix glasses of the Eastern Manus Basin analyzed by Sun et al. [15], samples in this study had similar Ba/Nb and U/Nb values, but lower Cl/Nb values, indicating that samples in this study were relatively poorly influenced by Cl-rich seawater derived components. Kent et al. [10] analyzed matrix glasses of the Eastern Lau Spreading Center and Central Lau Spreading Center, finding that their Cl/Nb ratios are significantly higher than that of the primitive mantle, while the Ba/Nb values are low to the MORB, which indicated the minimal subduction components input, hence they considered that the high Cl/Nb ratios in magmas are the result of assimilation b Cl-rich seawater-derived components. As shown in Table S4, the Cl/Nb values of ou dacite samples ranged from 649 to 909, while the contribution of the primitive mantle lay around 26 only [91]. Taking the average of the Cl/Nb values in magmas from the Eastern Lau Spreading Center and Central Lau Spreading Center (about 400 [15]) as the influence from Cl-rich seawater derived components, then approximately 14−21% of Cl in the Eastern Manus Basin magma source region is contributed by subduction-released fluids, and mainly by fluids released from subducting altered oceanic crust. However, as discussed above, the mantle source region of this study has been less influenced by Cl-rich seawater derived components, so the contribution of subduction-released fluids to Cl may be higher.

6. Conclusions

In this study, dacite from the Eastern Manus Basin was selected as the object for study, and apatites in dacite were analyzed by EMPA. The results were as follows:

(1) Apatites in dacite samples crystallized under a volatile-undersaturated condition with temperatures ranging from 935 to 952 °C. Compared with apatites in dacite of other subduction systems in the western Pacific, they exhibited rich in Cl but poor in F and SO_3.

(2) The minimum S contents of the host melts have a narrow range of 2−65 ppm and 8−11 ppm. Additionally, the host magma was under weak reduction conditions, and the relatively low oxygen fugacity could be caused by reduction brought from crystallization and separation of Fe^{3+}-rich magnetite, which also caused the low SO_3 contents in apatites.

(3) The F and Cl contents of melts in this study were within the range of subduction-related dacite, which is 185−448 ppm and 1059−1588 ppm, respectively, and such relatively high Cl concentrations of melts may have an impact on the compositions and exsolution of magmatic fluids, as well as on magmatic rheology. Further, the H_2O contents of melts were deduced to be in the range of 1.4−2.1% and 1.2−1.5% based on the F and Cl contents of melts, respectively, which indicates the location of the magma chamber at a depth of ~2 km. The high Cl/F ratio and H_2O contents of samples indicated the addition of slab-derived fluids in the parent magmas.

(4) The high F and Cl values exhibited in melts may come from both pollution in the magma source region, where the F may be influenced by F-rich seafloor sediments, as well as the release of F from lawsonite and phengite decomposition; high Cl may come from slab fluids and shallow assimilation of Cl-rich seawater derived components, where the Cl contributed by slab fluids to the mantle source region may account for 14−21% of the total Cl contents in melts, and possibly even higher.

Supplementary Materials: The following supporting information can be downloaded at: https://www.mdpi.com/article/10.3390/jmse10050698/s1, Tables S1–S4 and Text S1. Table S1: Abundances of major elements (wt.%) in apatites from the dacite; Table S2: Estimations of S concentrations and magmatic redox states from compositions in apatite and whole-rock; Table S3: Calculations of F, Cl and H_2O contents in melts from apatite compositions; Table S4: Cl and representative trace element concentrations of volcanic rocks from this study and other subduction related systems (ppm); Text S1 include the formulas used for calculations and the references of Tables S1–S4.

Author Contributions: Methodology, X.D. and Z.C.; software, X.D.; formal analysis, X.D.; validation, X.D.; investigation, X.D.; resources, Z.Z.; data curation, X.D.; writing—original draft preparation, X.D.; writing—review and editing, X.D., Z.C. and Z.Z.; visualization, X.D.; supervision, Z.Z.; project administration, Z.Z.; funding acquisition, Z.Z. All authors have read and agreed to the published version of the manuscript.

Funding: This work was supported by the National Natural Science Foundation of China (Grant No. 91958213), Strategic Priority Research Program (B) of the Chinese Academy of Sciences (Grant No. XDB42020402), National Basic Research Program of China (Grant No. 2013CB429700), and Special Fund for the Taishan Scholar Program of Shandong Province (Grant No. ts201511061).

Informed Consent Statement: Not applicable.

Data Availability Statement: All the data are given in the Supporting Information.

Acknowledgments: We are grateful for the valuable comments and suggestions from the anonymous reviewers and editors.

Conflicts of Interest: The authors declare no conflict of interest.

References

1. Stock, M.J.; Humphreys, M.C.S.; Smith, V.C.; Isaia, R.; Brooker, R.A.; Pyle, D.M. Tracking volatile behaviour in sub-volcanic plumbing systems using apatite and glass: Insights into pre-eruptive processes at Campi Flegrei, Italy. *J. Petrol.* **2018**, *59*, 2463–2491. [CrossRef]
2. Howarth, G.H.; Pernet-Fisher, J.F.; Bodnar, R.J.; Taylor, L.A. Evidence for the exsolution of cl-rich fluids in martian magmas: Apatite petrogenesis in the enriched lherzolitic shergottite northwest Africa 7755. *Geochim. Cosmochim. Acta* **2015**, *166*, 234–248. [CrossRef]
3. Annen, C.; Blundy, J.D.; Sparks, R.S.J. The genesis of intermediate and silicic magmas in deep crustal hot zones. *J. Petrol.* **2006**, *47*, 505–539. [CrossRef]
4. Wallace, P.J. Volatiles in subduction zone magmas: Concentrations and fluxes based on melt inclusion and volcanic gas data. *J. Volcanol. Geotherm. Res.* **2005**, *140*, 217–240. [CrossRef]
5. Anderson, A. Parental basalts in subduction zones—Implications for continental evolution. *J. Geophys. Res.* **1982**, *87*, 7047–7060. [CrossRef]
6. Stolper, E.; Newman, S. The role of water in the petrogenesis of mariana trough magmas. *Earth Planet. Sci. Lett.* **1994**, *121*, 293–325. [CrossRef]
7. Grove, T.L.; Parman, S.W.; Bowring, S.A.; Price, R.C.; Baker, M.B. The role of an H_2O-rich fluid component in the generation of primitive basaltic andesites and andesites from the Mt. Shasta region, N California. *Contrib. Mineral. Petrol.* **2002**, *142*, 375–396. [CrossRef]
8. Manning, C.E. The chemistry of subduction-zone fluids. *Earth Planet. Sci. Lett.* **2004**, *223*, 1–16. [CrossRef]
9. Barnes, J.D.; Manning, C.E.; Scambelluri, M.; Selverstone, J. The behavior of halogens during subduction-zone processes. In *The Role of Halogens in Terrestrial and Extraterrestrial Geochemical Processes*; Springer: Berlin/Heidelberg, Germany, 2018. [CrossRef]
10. Kent, A.J.R.; Peate, D.W.; Newman, S.; Stolper, E.M.; Pearce, J.A. Chlorine in submarine glasses from the Lau Basin: Seawater contamination and constraints on the composition of slab-derived fluids. *Earth Planet. Sci. Lett.* **2002**, *202*, 361–377. [CrossRef]
11. Kamenetsky, V.S.; Binns, R.A.; Gemmell, J.B.; Crawford, A.J.; Mernagh, T.P.; Maas, R.; Steele, D. Parental basaltic melts and fluids in eastern Manus backarc Basin: Implications for hydrothermal mineralisation. *Earth Planet. Sci. Lett.* **2001**, *184*, 685–702. [CrossRef]
12. Miller, D.J.; Vanko, D.A.; Paulick, H. Data report: Petrology and geochemistry of fresh, recent dacite lavas at Pual Ridge, Papua New Guinea, from an active, felsic-hosted seafloor hydrothermal system. *Proc. Ocean Drill. Program Sci. Results* **2007**, *193*, 1–31. [CrossRef]
13. Yang, K.; Scott, S.D. Magmatic degassing of volatiles and ore metals into a hydrothermal system on the modern sea floor of the eastern Manus Back-Arc Basin, western Pacific. *Econ. Geol.* **2002**, *97*, 1079–1100. [CrossRef]
14. Beier, C.; Bach, W.; Turner, S.; Niedermeier, D.; Woodhead, J.; Erzinger, J.; Krumm, S. Origin of silicic magmas at spreading centres—An example from the south east Rift, Manus Basin. *J. Petrol.* **2015**, *56*, 255–272. [CrossRef]
15. Sun, W.D.; Binns, R.A.; Fan, A.C.; Kamenetsky, V.S.; Wysoczanski, R.; Wei, G.J.; Hu, Y.H.; Arculus, R.J. Chlorine in submarine volcanic glasses from the eastern Manus Basin. *Geochim. Cosmochim. Acta* **2007**, *71*, 1542–1552. [CrossRef]
16. Martinez, F.; Taylor, B. Controls on back-arc crustal accretion: Insights from the Lau, Manus and Mariana Basins. *Geol. Soc. Lond. Spec. Publ.* **2003**, *219*, 19–54. [CrossRef]
17. Siegburg, M.; Klügel, A.; Rocholl, A.; Bach, W. Magma plumbing and hybrid magma formation at an active back-arc basin volcano: North Su, Eastern Manus Basin. *J. Volcanol. Geotherm. Res.* **2018**, *362*, 1–16. [CrossRef]
18. Yao, M.A.; Zeng, Z.; Chen, S.; Yin, X.; Wang, X. Origin of the volcanic rocks erupted in the eastern Manus Basin: Basaltic andesite-andesite-dacite associations. *J. Ocean Univ. China* **2017**, *16*, 389–402. [CrossRef]
19. Zhao, X.; Tian, L.; Sun, J.; Huang, P.; Yan, L.I.; Gao, Y. Petrogenesis of volcanic rocks from eastern Manus Basin: Indications in mineralogy and geochemistry. *J. Oceanol. Limnol.* **2021**, *39*, 21. [CrossRef]
20. Seewald, J.S.; Reeves, E.P.; Bach, W.; Saccocia, P.J.; Craddock, P.R.; Iii, W.; Sylva, S.P.; Pichler, T.; Rosner, M.; Walsh, E. Submarine venting of magmatic volatiles in the eastern Manus Basin, Papua New Guinea. *Geochim. Cosmochim. Acta* **2015**, *163*, 178–199. [CrossRef]
21. Martinez, F.; Taylor, B. Backarc Spreading, Rifting, and Microplate Rotation, between Transform Faults in the Manus Basin. *Mar. Geophys. Res.* **1996**, *18*, 203–224. [CrossRef]
22. Piccoli, P.M.; Candela, P.A. Apatite in igneous systems. In *Phosphates: Geochemical, Geobiological, and Materials Importance*; Kohn, M.J., Rakovan, J., Hughes, J.M., Eds.; Mineralogical Society of America: Washington, DC, USA, 2002; Volume 48, pp. 255–292, ISBN 978-0-939950-60-7.
23. Moore, P.B. *Apatite. Its Crystal Chemistry, Mineralogy, Utilization and Geologic and Biologic Occurrences*; McConnell, D., Ed.; Applied Mineralogy; Springer: New York, NY, USA, 1973; Volume 5, p. 111.
24. Wyllie, P.J.; Cox, K.G.; Biggar, G.M. The habit of apatite in synthetic systems and igneous rocks. *J. Petrol.* **1962**, *3*, 238–243. [CrossRef]
25. Webster, J.D.; Piccoli, P.M. Magmatic apatite: A powerful, yet deceptive, mineral. *Elements* **2015**, *11*, 177–182. [CrossRef]
26. Elliott, J.C.; Wilson, R.; Dowker, S.E.P. Apatite structures. *Adv. X-ray Anal.* **2002**, *45*, 172–181.
27. Zhang, Y.; Zeng, Z.; Yin, X.; Qi, H.; Chen, S.; Wang, X.; Shu, Y.; Chen, Z. Petrology and Mineralogy of Pumice from the Iheya North Knoll, Okinawa Trough: Implications for the Differentiation of Crystal-Poor and Volatile-Rich Melts in the Magma Chamber. *Geol. J.* **2018**, *53*, 2732–2745. [CrossRef]
28. Parat, F.; Holtz, F.; Kluegel, A. S-rich apatite-hosted glass inclusions in Xenoliths from La Palma: Constraints on the volatile partitioning in evolved alkaline magmas. *Contrib. Mineral. Petrol.* **2011**, *162*, 463–478. [CrossRef]

29. Peng, G.Y.; Luhr, J.F.; McGee, J.J. Factors controlling sulfur concentrations in volcanic apatite. *Am. Miner.* **1997**, *82*, 1210–1224. [CrossRef]
30. Kusebauch, C.; John, T.; Whitehouse, M.J.; Engvik, A.K. Apatite as probe for the halogen composition of metamorphic fluids (Bamble Sector, SE Norway). *Contrib. Mineral. Petrol.* **2015**, *170*, 34. [CrossRef]
31. Webster, J.D.; Goldoff, B.A.; Flesch, R.N.; Nadeau, P.A.; Silbert, Z.W. Hydroxyl, Cl, and F partitioning between high-silica rhyolitic melts-apatite-fluid(s) at 50–200 MPa and 700–1000 degrees C. *Am. Miner.* **2017**, *102*, 61–74. [CrossRef]
32. Boudreau, A.E.; McCallum, I.S. Investigations of the stillwater complex: Part V. apatites as indicators of evolving fluid composition. *Contrib. Mineral. Petrol.* **1989**, *102*, 138–153. [CrossRef]
33. Kusebauch, C.; John, T.; Whitehouse, M.J.; Klemme, S.; Putnis, A. Distribution of halogens between fluid and apatite during fluid-mediated replacement processes. *Geochim. Cosmochim. Acta* **2015**, *170*, 225–246. [CrossRef]
34. Zafar, T.; Leng, C.-B.; Zhang, X.-C.; Rehman, H.U. Geochemical attributes of magmatic apatite in the kukaazi granite from western Kunlun Orogenic Belt, NW China: Implications for granite petrogenesis and Pb-Zn (-Cu-W) mineralization. *J. Geochem. Explor.* **2019**, *204*, 256–269. [CrossRef]
35. Zafar, T.; Rehman, H.U.; Mahar, M.A.; Alam, M.; Oyebamiji, A.; Rehman, S.U.; Leng, C.-B. A critical review on petrogenetic, metallogenic and geodynamic implications of granitic rocks exposed in north and east China: New insights from apatite geochemistry. *J. Geodyn.* **2020**, *136*, 101723. [CrossRef]
36. Lee, S.-M.; Ruellan, E. Tectonic and magmatic evolution of the Bismarck Sea, Papua New Guinea: Review and new synthesis. *Geophys. Monogr.* **2006**, *166*, 263–286. [CrossRef]
37. Fourre, E.; Jean-Baptiste, P.; Charlou, J.L.; Donval, J.P.; Ishibashi, J.I. Helium isotopic composition of hydrothermal fluids from the Manus Back-Arc Basin, Papua New Guinea. *Geochem. J.* **2006**, *40*, 245–252. [CrossRef]
38. Tregoning, P. Plate kinematics in the western Pacific derived from geodetic observations. *J. Geophys. Res. Solid Earth* **2002**, *107*, ECV-7. [CrossRef]
39. Park, S.-H.; Lee, S.-M.; Kamenov, G.D.; Kwon, S.-T.; Lee, K.-Y. Tracing the origin of subduction components beneath the south east rift in the Manus Basin, Papua New Guinea. *Chem. Geol.* **2010**, *269*, 339–349. [CrossRef]
40. Bach, W.; Roberts, S.; Vanko, D.A.; Binns, R.A.; Yeats, C.J.; Craddock, P.R.; Humphris, S.E. Controls of fluid chemistry and complexation on rare-earth element contents of anhydrite from the pacmanus subseafloor hydrothermal system, Manus Basin, Papua New Guinea. *Miner. Depos.* **2003**, *38*, 916–935. [CrossRef]
41. Yang, K.H.; Scott, S.D. Vigorous exsolution of volatiles in the magma chamber beneath a hydrothermal system on the modern sea floor of the eastern Manus Back-Arc Basin, western Pacific: Evidence from melt inclusions. *Econ. Geol.* **2005**, *100*, 1085–1096. [CrossRef]
42. Chen, Z.; Zeng, Z.; Wang, X.; Yin, X.; Chen, S.; Guo, K.; Lai, Z.; Zhang, Y.; Ma, Y.; Qi, H. U-Th/He dating and chemical compositions of apatite in the dacite from the southwestern Okinawa trough: Implications for petrogenesis. *J. Asian Earth Sci.* **2018**, *161*, 1–13. [CrossRef]
43. Ketcham, R.A. Technical note: Calculation of stoichiometry from EMP data for apatite and other phases with mixing on monovalent anion sites. *Am. Miner.* **2015**, *100*, 1620–1623. [CrossRef]
44. O'Reilly, S.Y.; Griffin, W.L. Apatite in the mantle: Implications for metasomatic processes and high heat production in phanerozoic mantle. *Lithos* **2000**, *53*, 217–232. [CrossRef]
45. Imai, A. Variation of Cl and SO_3 contents of microphenocrystic apatite in intermediate to silicic igneous rocks of cenozoic Japanese Island Arcs: Implications for porphyry Cu metallogenesis in the western Pacific Island Arcs. *Resour. Geol.* **2004**, *54*, 357–372. [CrossRef]
46. Zheng, Y.; Zhang, C.; Jia, F.; Liu, H.; Yan, Q. Apatite and zircon geochemistry in Yao'an alkali-rich porphyry gold deposit, southwest China: Implications for petrogenesis and mineralization. *Minerals* **2021**, *11*, 1293. [CrossRef]
47. Harrison, T.M.; Watson, E.B. The behavior of apatite during crustal anatexis: Equilibrium and kinetic considerations. *Geochim. Cosmochim. Acta* **1984**, *48*, 1467–1477. [CrossRef]
48. Straub, S.M.; Layne, G.D. Decoupling of fluids and fluid-mobile elements during shallow subduction. Evidence from halogen-rich andesite melt inclusions from the Izu arc volcanic front. *Geochem. Geophys. Geosyst.* **2003**, *4*, 9003. [CrossRef]
49. Parat, F.; Holtz, F.; Streck, M.J. Sulfur-bearing magmatic accessory minerals. In *Sulfur in Magmas and Melts: Its Importance for Natural and Technical Processes*; Behrens, H., Webster, J.D., Eds.; Mineralogical Society of America & Geochemical Society: Chantilly, France, 2011; Volume 73, pp. 285–314, ISBN 978-0-939950-87-4.
50. Zhu, J.-J.; Richards, J.P.; Rees, C.; Creaser, R.; DuFrane, S.A.; Locock, A.; Petrus, J.A.; Lang, J. Elevated magmatic sulfur and chlorine contents in ore-forming magmas at the red chris porphyry Cu-Au deposit, northern British Columbia, Canada. *Econ. Geol.* **2018**, *113*, 1047–1075. [CrossRef]
51. Johnson, E.R.; Wallace, P.J.; Cashman, K.V.; Delgado Granados, H. Degassing of volatiles (H_2O, CO_2, S, Cl) during ascent, crystallization, and eruption at mafic monogenetic volcanoes in central Mexico. *J. Volcanol. Geotherm. Res.* **2010**, *197*, 225–238. [CrossRef]
52. Fleet, M.E. Xanes spectroscopy of sulfur in earth materials. *Can. Mineral.* **2005**, *43*, 1811–1838. [CrossRef]
53. Chelle-Michou, C.; Chiaradia, M. Amphibole and apatite insights into the evolution and mass balance of Cl and S in magmas associated with porphyry copper deposits. *Contrib. Mineral. Petrol.* **2017**, *172*, 105. [CrossRef]

54. Richards, J.P.; Lopez, G.P.; Zhu, J.-J.; Creaser, R.A.; Locock, A.J.; Mumin, A.H. Contrasting tectonic settings and sulfur contents of magmas associated with cretaceous porphyry Cu +/− Mo +/− Au and intrusion-related iron oxide Cu-Au deposits in northern chile. *Econ. Geol.* **2017**, *112*, 295–318. [CrossRef]

55. Pan, Y.M.; Fleet, M.E. Compositions of the apatite-group minerals: Substitution mechanisms and controlling factors. In *Phosphates: Geochemical, Geobiological, and Materials Importance*; Kohn, M.J., Rakovan, J., Hughes, J.M., Eds.; Mineralogical Society of America & Geochemical Society: Chantilly, France, 2002; Volume 48, pp. 13–49. ISBN 978-0-939950-60-7.

56. Miles, A.J.; Graham, C.M.; Hawkesworth, C.J.; Gillespie, M.R.; Hinton, R.W.; Bromiley, G.D. Apatite: A new redox proxy for silicic magmas? *Geochim. Cosmochim. Acta* **2014**, *132*, 101–119. [CrossRef]

57. Myers, J.; Eugster, H. The system Fe-Si-O—Oxygen buffer calibrations to 1500k. *Contrib. Mineral. Petrol.* **1983**, *82*, 75–90. [CrossRef]

58. Kress, V.; Carmichael, I. The compressibility of silicate liquids containing Fe_2O_3 and the effect of composition, temperature, oxygen fugacity and pressure on their redox states. *Contrib. Mineral. Petrol.* **1991**, *108*, 82–92. [CrossRef]

59. Parkinson, I.J.; Arculus, R.J. The redox state of subduction zones: Insights from arc-peridotites. *Chem. Geol.* **1999**, *160*, 409–423. [CrossRef]

60. Wood, B.; Bryndzia, L.; Johnson, K. Mantle oxidation-state and its relationship to tectonic environment and fluid speciation. *Science* **1990**, *248*, 337–345. [CrossRef] [PubMed]

61. Arculus, R. Aspects of magma genesis in arcs. *Lithos* **1994**, *33*, 189–208. [CrossRef]

62. Kleinsasser, J.M.; Simon, A.C.; Konecke, B.A.; Kleinsasser, M.J.; Beckmann, P.; Holtz, F. Sulfide and sulfate saturation of dacitic melts as a function of oxygen fugacity. *Geochim. Cosmochim. Acta* **2022**, *326*, 1–16. [CrossRef]

63. Kelley, K.A.; Cottrell, E. Water and the oxidation state of subduction zone magmas. *Science* **2009**, *325*, 605–607. [CrossRef]

64. Meng, X.; Kleinsasser, J.M.; Richards, J.P.; Tapster, S.R.; Jugo, P.J.; Simon, A.C.; Kontak, D.J.; Robb, L.; Bybee, G.M.; Marsh, J.H.; et al. Oxidized sulfur-rich arc magmas formed porphyry Cu deposits by 1.88 Ga. *Nat. Commun.* **2021**, *12*, 2189. [CrossRef]

65. Richards, J.P. The oxidation state, and sulfur and Cu contents of arc magmas: Implications for metallogeny. *Lithos* **2015**, *233*, 27–45. [CrossRef]

66. Crabtree, S.M.; Lange, R.A. An evaluation of the effect of degassing on the oxidation state of hydrous andesite and dacite magmas: A comparison of pre- and post-eruptive Fe^{2+} concentrations. *Contrib. Mineral. Petrol.* **2012**, *163*, 209–224. [CrossRef]

67. Waters, L.E.; Lange, R.A. No effect of H_2O degassing on the oxidation state of magmatic liquids. *Earth Planet. Sci. Lett.* **2016**, *447*, 48–59. [CrossRef]

68. de Hoog, J.C.M.; Hattori, K.H.; Hoblitt, R.P. Oxidized sulfur-rich mafic magma at Mount Pinatubo, Philippines. *Contrib. Mineral. Petrol.* **2004**, *146*, 750–761. [CrossRef]

69. Grocke, S.B.; Cottrell, E.; de Silva, S.; Kelley, K.A. The role of crustal and eruptive processes versus source variations in controlling the oxidation state of iron in central andean magmas. *Earth Planet. Sci. Lett.* **2016**, *440*, 92–104. [CrossRef]

70. Jenner, F.E.; Hauri, E.H.; Bullock, E.S.; Koenig, S.; Arculus, R.J.; Mavrogenes, J.A.; Mikkelson, N.; Goddard, C. The competing effects of sulfide saturation versus degassing on the behavior of the chalcophile elements during the differentiation of hydrous melts. *Geochem. Geophys. Geosyst.* **2015**, *16*, 1490–1507. [CrossRef]

71. Jenner, F.E.; O'Neill, H.S.C.; Arculus, R.J.; Mavrogenes, J.A. The magnetite crisis in the evolution of arc-related magmas and the initial concentration of Au, Ag and Cu. *J. Petrol.* **2010**, *51*, 2445–2464. [CrossRef]

72. Webster, J.D.; Tappen, C.M.; Mandeville, C.W. Partitioning behavior of chlorine and fluorine in the system apatite-melt-fluid. II: Felsic silicate systems at 200 MPa. *Geochim. Cosmochim. Acta* **2009**, *73*, 559–581. [CrossRef]

73. Brenan, J. Partitioning of fluorine and chlorine between apatite and aqueous fluids at high-pressure and temperature—Implications for the F and Cl content of high P-T fluids. *Earth Planet. Sci. Lett.* **1993**, *117*, 251–263. [CrossRef]

74. Tacker, R.; Stormer, J. A Thermodynamic model for apatite solid-solutions, applicable to high-temperature geologic problems. *Am. Miner.* **1989**, *74*, 877–888.

75. Li, W.; Costa, F. A thermodynamic model for F-Cl-OH partitioning between silicate melts and apatite including non-ideal mixing with application to constraining melt volatile budgets. *Geochim. Cosmochim. Acta* **2020**, *269*, 203–222. [CrossRef]

76. Boyce, J.W.; Tomlinson, S.M.; McCubbin, F.M.; Greenwood, J.P.; Treiman, A.H. The lunar apatite paradox. *Science* **2014**, *344*, 400–402. [CrossRef] [PubMed]

77. Li, H.; Hermann, J. Apatite as an indicator of fluid salinity: An experimental study of chlorine and fluorine partitioning in subducted sediments. *Geochim. Cosmochim. Acta* **2015**, *166*, 267–297. [CrossRef]

78. Li, H.; Hermann, J. Chlorine and fluorine partitioning between apatite and sediment melt at 2.5 GPa, 800 degrees C: A new experimentally derived thermodynamic model. *Am. Miner.* **2017**, *102*, 580–594. [CrossRef]

79. Aiuppa, A.; Baker, D.R.; Webster, J.D. Halogens in volcanic systems. *Chem. Geol.* **2009**, *263*, 1–18. [CrossRef]

80. Wallace, P.J.; Plank, T.; Edmonds, M.; Hauri, E.H. Volatiles in magmas. In *The Encyclopedia of Volcanoes*; Academic Press: Cambridge, MA, USA, 2015; pp. 163–183. [CrossRef]

81. Zimova, M.; Webb, S.L. The combined effects of chlorine and fluorine on the viscosity of aluminosilicate melts. *Geochim. Cosmochim. Acta* **2007**, *71*, 1553–1562. [CrossRef]

82. Zimova, M.; Webb, S. The effect of chlorine on the viscosity of Na_2O-Fe_2O_3-Al_2O_3-SiO_2 melts. *Am. Miner.* **2006**, *91*, 344–352. [CrossRef]

83. Webster, J.D. The exsolution of magmatic hydrosaline chloride liquids. *Chem. Geol.* **2004**, *210*, 33–48. [CrossRef]

84. Signorelli, S.; Carroll, M.R. Experimental constraints on the origin of chlorine emissions at the soufriere hills volcano, montserrat. *Bull. Volcanol.* **2001**, *62*, 431–440. [CrossRef]

85. Kamenov, G.D.; Perfit, M.R.; Mueller, P.A.; Jonasson, I.R. Controls on magmatism in an island arc environment: Study of lavas and sub-arc xenoliths from the Tabar-Lihir-Tanga-Feni island chain, Papua New Guinea. *Contrib. Mineral. Petrol.* **2008**, *155*, 635–656. [CrossRef]

86. Boyce, J.W.; Hervig, R.L. Apatite as a monitor of late-stage magmatic processes at volcan irazA(o), Costa Rica. *Contrib. Mineral. Petrol.* **2009**, *157*, 135–145. [CrossRef]

87. Zhang, Y.X. H_2O in rhyolitic glasses and melts: Measurement, speciation, solubility, and diffusion. *Rev. Geophys.* **1999**, *37*, 493–516. [CrossRef]

88. Liu, Y.; Behrens, H.; Zhang, Y.X. The speciation of dissolved H_2O in dacitic melt. *Am. Miner.* **2004**, *89*, 277–284. [CrossRef]

89. Danyushevsky, L.V.; Eggins, S.M.; Falloon, T.J.; Christie, D.M. H_2O abundance in depleted to moderately enriched mid-ocean ridge magmas; part I: Incompatible behaviour, implications for mantle storage, and origin of regional variations. *J. Petrol.* **2000**, *41*, 1329–1364. [CrossRef]

90. Sobolev, A.V.; Chaussidon, M. H_2O concentrations in primary melts from supra-subduction zones and mid-ocean ridges: Implications for H_2O storage and recycling in the mantle. *Earth Planet. Sci. Lett.* **1996**, *137*, 45–55. [CrossRef]

91. Mcdonough, W.; Sun, S. The composition of the earth. *Chem. Geol.* **1995**, *120*, 223–253. [CrossRef]

92. Page, L.; Hattori, K.; de Hoog, J.C.M.; Okay, A.I. Halogen (F, Cl, Br, I) behaviour in subducting slabs: A study of lawsonite blueschists in western Turkey. *Earth Planet. Sci. Lett.* **2016**, *442*, 133–142. [CrossRef]

93. Straub, S.M.; Layne, G.D. The systematics of chlorine, fluorine, and water in izu arc front volcanic rocks: Implications for volatile recycling in subduction zones. *Geochim. Cosmochim. Acta* **2003**, *67*, 4179–4203. [CrossRef]

94. Li, Y.H. A brief discussion on the mean oceanic residence time of elements. *Geochim. Cosmochim. Acta* **1982**, *46*, 2671–2675. [CrossRef]

95. Kamenetsky, V.S.; Crawford, A.J.; Eggins, S.; Muhe, R. Phenocryst and melt inclusion chemistry of near-axis seamounts, valu fa ridge, Lau Basin: Insight into mantle wedge melting and the addition of subduction components. *Earth Planet. Sci. Lett.* **1997**, *151*, 205–223. [CrossRef]

96. Mc, A.; Jag, B. Geochemical and geodynamical constraints on subduction zone magmatism. *Earth Planet. Sci. Lett.* **1991**, *102*, 358–374. [CrossRef]

97. Pearce, J.A.; Peate, D.W. Tectonic implications of the composition of volcanic arc magmas. *Ann. Rev. Earth Planet. Sci.* **1995**, *23*, 251–285. [CrossRef]

98. Sun, W.D.; Bennett, V.C.; Eggins, S.M.; Arculus, R.J.; Perfit, M.R. Rhenium systematics in submarine MORB and back-arc basin glasses: Laser ablation ICP-MS results. *Chem. Geol.* **2003**, *196*, 259–281. [CrossRef]

99. Portnyagin, M.; Hoernle, K.; Plechov, P.; Mironov, N.; Khubunaya, S. Constraints on mantle melting and composition and nature of slab components in volcanic arcs from volatiles (H_2O, S, Cl, F) and trace elements in melt inclusions from the kamchatka arc. *Earth Planet. Sci. Lett.* **2007**, *255*, 53–69. [CrossRef]

Journal of
*Marine Science
and Engineering*

Article

Lithium, Oxygen and Magnesium Isotope Systematics of Volcanic Rocks in the Okinawa Trough: Implications for Plate Subduction Studies

Zhigang Zeng [1,2,3,*], Xiaohui Li [1], Yuxiang Zhang [1,4] and Haiyan Qi [1,4]

1   Seafloor Hydrothermal Activity Laboratory, CAS Key Laboratory of Marine Geology and Environment, Institute of Oceanology, Chinese Academy of Sciences, Qingdao 266071, China; xiaohuili0526@foxmail.com (X.L.); yxzhang@qdio.ac.cn (Y.Z.); qihaiyan@qdio.ac.cn (H.Q.)
2   Laboratory for Marine Mineral Resources, Qingdao National Laboratory for Marine Science and Technology, Qingdao 266071, China
3   College of Marine Science, University of Chinese Academy of Sciences, Beijing 100049, China
4   Center for Ocean Mega-Science, Chinese Academy of Sciences, Qingdao 266071, China
*   Correspondence: zgzeng@qdio.ac.cn

Abstract: Determining the influence of subduction input on back-arc basin magmatism is important for understanding material transfer and circulation in subduction zones. Although the mantle source of Okinawa Trough (OT) magmas is widely accepted to be modified by subducted components, the role of slab-derived fluids is poorly defined. Here, major element, trace element, and Li, O and Mg isotopic compositions of volcanic lavas from the middle OT (MOT) and southern OT (SOT) were analyzed. Compared with the MOT volcanic lavas, the T9-1 basaltic andesite from the SOT exhibited positive Pb anomalies, significantly lower Nd/Pb and Ce/Pb ratios, and higher Ba/La ratios, indicating that subducted sedimentary components affected SOT magma compositions. The δ^7Li, δ^{18}O, and δ^{26}Mg values of the SOT basaltic andesite ($-5.05‰$ to $4.98‰$, $4.83‰$ to $5.80‰$ and $-0.16‰$ to $-0.09‰$, respectively) differed from those of MOT volcanic lavas. Hence, the effect of the Philippine Sea Plate subduction component, (low δ^7Li and δ^{18}O and high δ^{26}Mg) on magmas in the SOT was clearer than that in the MOT. This contrast likely appears because the amounts of fluids and/or melts derived from altered oceanic crust (AOC, lower δ^{18}O) and/or subducted sediment (lower δ^7Li, higher δ^{18}O and δ^{26}Mg) injected into magmas in the SOT are larger than those in the MOT and because the injection ratio between subducted AOC and sediment is always >1 in the OT. The distance between the subducting slab and overlying magma may play a significant role in controlling the differences in subduction components injected into magmas between the MOT and SOT.

Keywords: Li-O-Mg isotopes; magma; plate subduction; Okinawa Trough; western Pacific

Citation: Zeng, Z.; Li, X.; Zhang, Y.; Qi, H. Lithium, Oxygen and Magnesium Isotope Systematics of Volcanic Rocks in the Okinawa Trough: Implications for Plate Subduction Studies. *J. Mar. Sci. Eng.* **2022**, *10*, 40. https://doi.org/10.3390/jmse10010040

Academic Editor: Anabela Oliveira

Received: 14 November 2021
Accepted: 14 December 2021
Published: 31 December 2021

Publisher's Note: MDPI stays neutral with regard to jurisdictional claims in published maps and institutional affiliations.

1. Introduction

The Okinawa Trough is a young back-arc basin in the western Pacific, and its magmas have been affected by the subducted Philippine Sea plate (PSP) [1]. The chemical and stable isotope (Li, O, and Mg) compositions of back-arc volcanic lavas are conventionally used to study the contributions of subducted slabs to magmas [1–11]. However, magmas produced by plate subduction are relatively enriched in large ion lithophile elements and light rare earth elements, but are depleted in high field strength elements; these magmas usually have low Ce/Pb ratios since Pb is preferentially extracted from subducted oceanic crust and/or sediments during plate subduction and dehydration processes [8–10,12]. As incompatible elements share similar partition coefficients in magmas, their ratios are rarely modified by the partial melting or crystallization processes. Thus, the incompatible element ratios can be used to trace the compositions of mantle sources. For example, Ba/La ratio is a good indicator of subduction components [13]. Ce/Pb ratios of sediments exhibit a distinctive range, below the values in mantle rocks, and are not significantly affected by fractional crystallization and partial melting processes [14]; thus, these ratios can be used to estimate the contributions of plate subduction components. Th concentrations of marine sediments are higher than the mantle values, and Th is fluid immobile relative to large ion lithophile elements. A high Th/Rb ratio in magma indicates that a sediment melt was added to the magma source [15]. In addition, high Th/La, Th/Nb, and Th/Nd ratios of the volcanic rocks indicate the contribution of sediment melts to the mantle wedge [16–19]. Ba/La, Cs/La, B/Nb, Pb/Ce, B/La, and Li/Y ratios can also serve as proxies for fluids dehydrated from subducted sediments [20].

Moreover, lithium (Li) is a fluid mobile element, and considerable Li isotope fractionation occurs during oceanic crust alteration and plate subduction [21–24]. Subducted sediments and altered oceanic crust (AOC) have high Li contents (up to 80 ppm) and a broad range of δ^7Li values ($-12‰$ to $+21‰$) [4,25–28]. During the metamorphism and dehydration of subducted slabs, fluids carrying heavy Li isotopes are released from the AOC and metasomatize the overlying mantle, resulting in a high-δ^7Li mantle wedge and a low-δ^7Li residual plate [29–31]. However, Guo, et al. [1] studied Li isotope data from middle Okinawa Trough (MOT) volcanic lavas and found that the MOT magmas were affected by subduction components (with AOC:sediment = 96:4). The δ^7Li values of volcanic lavas from the Izu Arc in Japan vary across the arc, decreasing with increasing depth to the subducting plate, indicating that the amount of slab fluids decreases with increasing depth [32–34]. Furthermore, Benton, et al. [35] analyzed the Li isotope compositions of volcanic lavas in the Mariana Arc front seamounts and found that they have heterogeneous δ^7Li signatures, which reflect a complex history of exchange between the slab fluids and the forearc mantle. These results indicate that there may be uncertainties in using Li isotope data for back-arc volcanic lavas to uncover magma sources and plate subduction and implies that other isotopic tools should be used to clarify magmatic processes in western Pacific plate subduction zones.

However, the oxygen (O) isotopic compositions of the MOT volcanic lavas reveal that the magma source is mantle peridotites modified by subducted slab components [36]. The O isotopic compositions of volcanic glasses obtained from the Manus Basin and the Mariana Trough indicate that the Mariana Trough magma was affected by subduction-modified mantle [37] and that the Manus Basin magma had a $\delta^{18}O$-depleted mantle reservoir, wherein $\delta^{18}O$ was increased by recent subduction and a sediment component with low $\delta^{18}O$ and high $^3He/^4He$ values was derived from the Manus Basin plume [37]. Moreover, the O isotope compositions of volcanic glasses and phenocrysts from the Lau Basin suggest that a subduction cycle involving oxygen-rich mantle materials altered the original $\delta^{18}O$ of the magma [38]. All these results indicate that the O isotope data for volcanic lavas may not completely reflect plate subduction processes.

Magnesium (Mg) isotopes are important geochemical tracers of both magmatic sources containing recycled crustal materials [39–41] and sources of fluids dehydrated from hydrous minerals in subducting plate [5,40,42–44]. However, the Mg isotope compositions of white schist in the western Alps reveal that the dehydration of serpentinites formed in the mantle wedge during plate subduction and exhumation can release Mg-rich fluids to the subduction channel [5]. Furthermore, a study of the Mg isotope compositions of prograde metamorphic rocks from eastern China revealed that Mg isotope fractionation was limited during continental subduction [41]. All these results also indicate that the Mg isotope data for volcanic lavas may not completely determine magma sources and plate subduction processes.

Therefore, we sought to combine the Li isotope composition of volcanic lavas with their O and Mg isotope compositions to understand the magmatic processes involved in forming volcanic lavas and the influence of plate subduction on magmas in the Okinawa Trough (OT). However, previous studies have shown that subducted sediments significantly influence the mantle source of magmas beneath the OT [2,10,19,45,46]. Although slab-derived fluids are also considered to contribute to the OT magmas [47–50], the evidence for this contribution is insufficient. Furthermore, Li and O isotope studies have been conducted in the OT, but the data remain limited to only three Li isotope analyses in the MOT. O isotope studies have primarily focused on crustal contamination as a source of magma formation [36,51–53], and the study of the Mg isotopic compositions of volcanic lavas has not yet been applied to the OT. Here, we investigate the major and trace elements as well as the Sr, Nd, Li, O, and Mg isotopic compositions of volcanic lavas from the MOT and SOT, with the goal of establishing how OT volcanic lavas form and what they represent for understanding the effects of subduction components on back-arc magmas.

2. Geologic Setting

The OT is a young back-arc basin developed in response to the subduction of the PSP (Figure 1) [11,54–56] and provides a window into understanding the influences of plate subduction on the back-arc basin evolution, mantle melting, crust-mantle interactions, and seafloor hydrothermal activity [1,11,46,57,58]. According to the prevailing tectonics, the OT can be divided into three parts: northern OT (NOT), MOT, and SOT, bounded by the Tokara Fault and the Kerama Fault, respectively [1,11,50]. The crustal thicknesses of the NOT, MOT, and SOT segments are 15 to 23 km, 12 to 18 km, and 10 to 16 km, respectively [54,59–64]. The depths to the Mohorovičić discontinuity (Moho) are 27–30 km and 16–22 km in the NOT and SOT, respectively [59,65,66]. The subduction direction of PSP is nearly perpendicular to the axis of the MOT, with a slab depth of 150–200 km, and becomes progressively more oblique to the south, with a slab depth of ~150 km [11,47,50,67,68]. The subduction rates of PSP increase from 4.9 to 7.3 cm/a from north to south [11,56]. In addition, the convergence rate gradually increases from ~6–7 cm/a in central Ryukyu to ~7–13 cm/a in southern Ryukyu [11,56,69], and the back-arc extension rate increases from 2 cm/a in the NOT to 5 cm/a in the SOT [65,69,70].

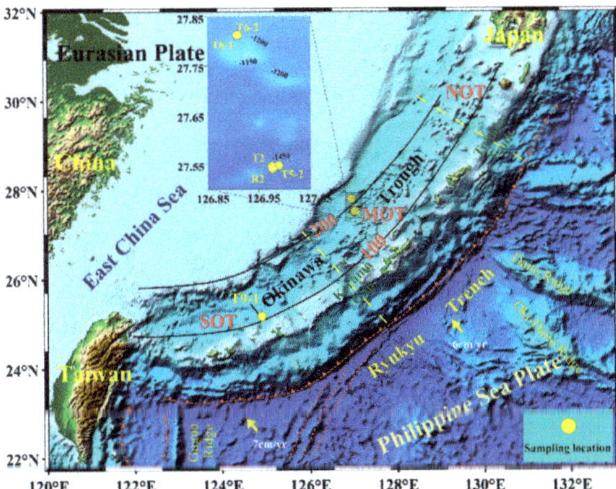

Figure 1. Regional geologic map of the Okinawa Trough (OT) showing the sampling locations. MOT, middle Okinawa Trough; SOT, southern Okinawa Trough; and NOT, northern Okinawa Trough. Black solid lines represent the depth contours of the subducting plate (Wadati-Benioff zone) [11,67,71]. Yellow dotted lines mark the Tokara Fault and the Kerama Fault.

The OT has undergone substantial magmatic activity that has generated abundant volcanic lavas. The NOT is dominated by rhyolites and dacites [6,53]. The magma in the NOT could be generated by the differentiation of basaltic melt contaminated by an enriched crustal component [53]. The MOT volcanic lavas exhibit bimodal compositional distribution, forming a basaltic-rhyolitic dominant suite with scarce andesitic lavas [6,72,73]. The Sr-Nd-Pb isotope compositions suggest that the magma source of the MOT volcanic lavas resembles the depleted mantle (DM) but also shows signatures of the enriched mantle II (EMII), indicating that the mantle source for the MOT volcanic lavas is a hybrid of the DM and EMII endmembers [74]. Geochemical modelling shows that the magma source for the MOT volcanic lavas could be generated by hybridizing approximately 0.8–2.0% subducted sediments with 98.0–99.2% mantle rocks [74]. The SOT is dominated by rhyolites, basalts and basaltic andesites [50,72,75,76]. In the SOT, the petrogenesis of rhyolites is due to the ascent of basaltic magma from a deep magma reservoir to a shallow depth where fractional crystallization and crustal assimilation occurred [75]. The decrease in the DUPAL-like anomaly from the rhyolites sampled in the western part of the MOT to the volcanic lavas in the MOT axial zone can be explained by the injection of asthenospheric mantle into the pre-existing mantle with DUPAL-like signature during back-arc extension. The Tl-Pb-Sr-Nd isotopic compositions of volcanic lavas from the NOT to the SOT can be accounted for by the sediment inputs of <1%, 0.1–1% and 0.3–2%, respectively, by weight to the depleted mantle source [46]. The magma sources of the SOT basalts are mainly affected by slab fluids and bulk sediments [47]. The magma sources of the MOT basalts are affected by fluids derived from both sediment and AOC [47]. Based on the above findings, we aim to use the geochemical and Li-O-Mg isotope compositions of OT volcanic lavas to understand the origins of magmas and the effects of plate subduction on magmas in this study.

3. Sampling and Methods

3.1. Sample Collection

The volcanic lavas were collected from the MOT and SOT using a TV grab sampler in 2014 and 2016 during HOBAB (Hello Back Arc Basin) 2 and 4 cruises, respectively. The R2, T5-2, and T2 samples were obtained from the Iheya Ridge (MOT), samples T6-1 and T6-2 were obtained from the western slope of the Iheya Ridge in the MOT, and sample T9-1 was obtained from the Yaeyama Graben (SOT) (Table S1 and Figure 1).

T9-1 basaltic andesite from the SOT is a black vesicular lava sample with a compact, massive structure. It contains few phenocrysts, which predominantly consist of clinopyroxene (~5%), orthopyroxene (~5%), plagioclase (~10%), a small amount of olivine (~1%), and accessory minerals (magnetite and ilmenite). The groundmass of the T9-1 basaltic andesite is mostly composed of plagioclase and clinopyroxene microlites (Figure 2a). The phenocrysts in the T9-1 basaltic andesite are euhedral to subhedral and range from ~0.1 to ~1.0 mm in size.

R2 basalt from the MOT is a black lava with a moderately porphyritic massive structure. It contains olivine (~5%), plagioclase (~5%), and clinopyroxene (~1%) phenocrysts. The phenocrysts in the R2 basalt are euhedral to subhedral, with sizes ranging from ~0.05 to 6 mm. The groundmass of the R2 basalt is dominated by clinopyroxene and plagioclase microlites (Figure 2b). Both the T5-2 trachyandesite and T2 andesite from the MOT are black lavas with compact, massive structures. Thin sections reveal that the T5-2 trachyandesite and T2 andesite are porphyritic and that ~15% of the phenocrysts are plagioclase with minor clinopyroxene and orthopyroxene. The phenocrysts in the T5-2 trachyandesite and T2 andesite samples are euhedral to subhedral, with sizes ranging from ~40 to ~300 μm (Figure 2c,d). The T6-1 pumice from the MOT (Figure 2e) is a light, white, vesicular lava. The phenocrysts (<5%) in the T6-1 pumice are primarily clinopyroxene, orthopyroxene, and plagioclase and range from ~0.2 to 0.5 mm in size. The groundmass in the T6-1 pumice is mainly composed of plagioclase microlites (Figure 2e). The T6-2 pumice from the MOT is a gray-black vesicular lava. The phenocrysts (<5%) in the T6-2 pumice mainly consist of clinopyroxene, plagioclase, and orthopyroxene, with sizes ranging from ~0.2 to 0.4 mm in size. The groundmass in the T6-2 pumice is also mainly composed of plagioclase microlites (Figure 2f).

Details about the sample processing, electron microprobe, major element and trace element, and Sr and Nd isotope analytic methods are given in the Supplementary Texts, and the sampling locations and sample analytic results can be found in Tables S1–S8.

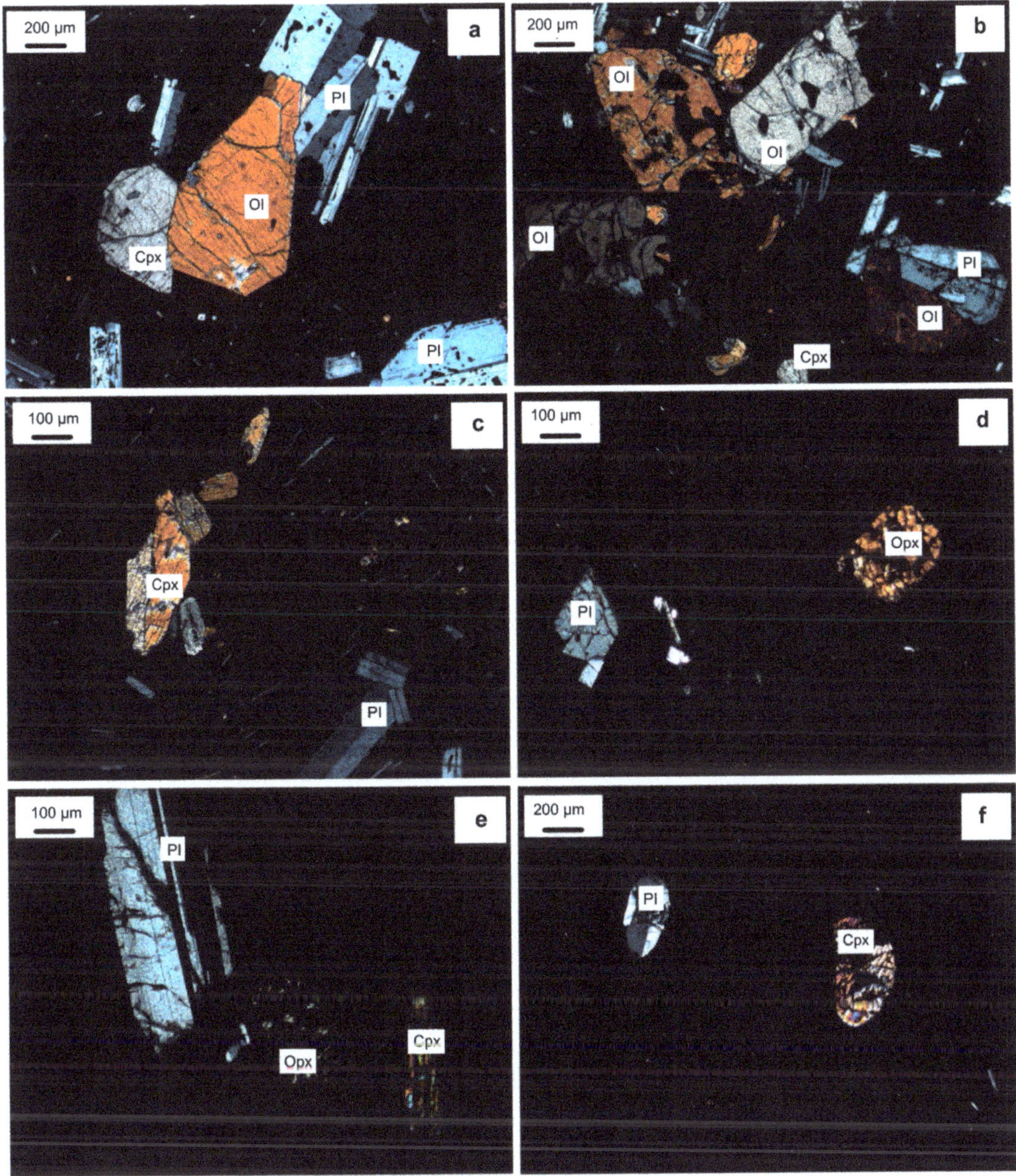

Figure 2. Representative photomicrographs of the MOT and SOT volcanic lava samples. (**a**) T9-1 is from the SOT and (**b**) R2, (**c**) T5-2, (**d**) T2, (**e**) T6-1, and (**f**) T6-2 are from the MOT. Abbreviations: olivine (Ol); magnetite (Mt); clinopyroxene (Cpx); orthopyroxene (Opx); and plagioclase (Pl).

3.2. Li, O, and Mg Isotope Analyses

The whole-rock powders and minerals were fully digested. All reagents (HF, HNO_3, and HCl) were doubly distilled, and Milli-Q® water (18.2 MΩ·cm) (Merck Millipore Inc., Billerica, MA, USA) was used. For the isotopic analyses, Li was separated using organic solvent-free two-step liquid chromatography in a clean laboratory at the University of

Science and Technology of China (USTC). The procedure used was described by [77]. All separations were monitored using inductively coupled plasma mass spectrometry (ICP-MS) analysis to guarantee both a high Li yield (>99.8% recovery) and a low Na/Li ratio (<0.5). The final Li concentrations of the solutions used for the multi-collector ICP-MS (MC-ICP-MS) analyses were targeted in the range of 50–100 ppb to ensure the precision and accuracy of the results. The total procedural blanks for column chromatography alone and for sample digestion and column chromatography combined were <~0.03 ng of Li. Compared with the ~200–5000 ng of Li used for our analyses, the blank correction was negligible at the uncertainty levels achieved ($\leq$0.2‰, see below).

The lithium isotope compositions were analyzed at the USTC using a Neptune Plus MC-ICP-MS system (Thermo Fisher Scientific Inc., Waltham, MA, USA) operating in wet plasma mode using an X skimmer cone and jet sample cones (Table S9). The samples were introduced through a low-flow PFA nebulizer (~50 µL/min) coupled with a quartz spray chamber. The ^{7}Li and ^{6}Li isotopes were measured simultaneously by two opposing Faraday cups. Each sample analysis was bracketed before and after by analyses of the reference material L-SVEC. For a solution containing 100 ppb of Li and an uptake rate of 50 µL/min, the typical ^{7}Li intensity was ~8 V. The in-run precision of the ^{7}Li/^{6}Li measurements was $\leq$0.2‰ for one block of 60 ratios. The external precision was based on the long-term analysis of an in-house standard (Li-QCUSTC = +8.8‰ $\pm$ 0.2‰ (2 standard deviations (SD), n = 161)). For international rock standards, repeat analyses at the USTC yielded values of +4.4‰ $\pm$ 0.3‰ (2SD, n = 8) for BHVO-2, −0.8‰ $\pm$ 0.3‰ (2SD, n = 29) for GSP-2, and +5.9‰ $\pm$ 0.5‰ (2SD, n = 9) for AGV-1, which were within the uncertainty limits of previously published results [77,78]. The results are reported in the delta notation [δ^7Li = ((^{7}Li/^{6}Li)$_{sample}$/(^{7}Li/^{6}Li)$_{standard}$ −1) × 1000] relative to the L-SVEC Li isotope standard [79].

The whole-rock and mineral oxygen isotope compositions were measured using laser fluorination with a 25 W MIR-10 carbon dioxide (CO_2) laser at the CAS Key Laboratory of Crust-Mantle Materials and Environments at the USTC, Hefei. The oxygen isotope analyses and the data acquisition followed the methods described by [80,81]. O_2 was extracted from the samples through reaction with bromine pentafluoride (BrF_5) in nickel (Ni) bombs, and it was then converted to CO_2 through reaction with a hot carbon rod. The δ^{18}O of CO_2 was measured using a Delta+ mass spectrometer. Reference minerals GBW04409 quartz (δ^{18}O = 11.11‰ $\pm$ 0.1‰) [81] and in-house standard 04BXL07 garnet (δ^{18}O$_{V\text{-}SMOW}$ = 3.70‰ $\pm$ 0.1‰) [82] were analyzed during each run. We analyzed the O isotope compositions of the whole rocks and the minerals from each sample twice, except for R2-2-Cpx and T2-Opx (Table S9). The data are reported using the usual δ^{18}O notation relative to Vienna standard mean ocean water (V-SMOW). On a given day, the reproducibility of each standard was better than $\pm$0.2‰ (2σ) for δ^{18}O.

The Mg isotopes were measured using a Thermo Scientific Neptune Plus MC-ICP-MS system following the methods of [83] at the CAS Laboratory of Crust-Mantle Materials and Environments at the USTC, Hefei. The whole-rock powders and minerals were fully digested to obtain ~20 µg of Mg for chemical purification. A mixture of concentrated HF-HNO$_3$ was used for digestion. Mg was purified in Savillex columns loaded with 2 mL of Bio-Rad AG®50W-X12 resin (Bio-Rad Laboratories Inc., Hercules, CA, USA). For the Mg isotope analyses, sample-standard bracketing was used, wherein DSM-3 was the bracketing standard. The Mg isotope compositions are reported using the standard δ notation relative to DSM-3. The uncertainties in the δ^{25}Mg and δ^{26}Mg values of the standards and the samples are reported as 2SD based on repeated measurements (Table S9). The data quality was carefully controlled through the repeated analysis of multiple Mg isotope standards (Table S9). The long-term external precision was better than $\pm$0.05‰. Furthermore, each sample was processed three times under the same conditions. During the analyses, the δ^{26}Mg values obtained for BCR-2 (−0.215‰ $\pm$ 0.017‰; n = 3) and BHVO-2 (−0.205‰ $\pm$ 0.020‰; n = 3) were all identical within the established uncertainty limits (−0.162‰ $\pm$ 0.014‰ for BCR-2 and −0.216‰ $\pm$ 0.035‰ for BHVO-2) [83].

4. Results

4.1. Clinopyroxene and Orthopyroxene

The clinopyroxene contents in R2 basalt, the orthopyroxene contents in T5-2 trachyandesite, T2 andesite, T6-1 and T6-2 rhyolites from the MOT, and the orthopyroxene content in T9-1 basaltic andesite from the SOT, were analyzed using an electron microprobe. The data are presented in Tables S2–S6. The clinopyroxene and orthopyroxene were unzoned, and their compositions were homogeneous (Figure 2 and Tables S2–S6). There was no notable difference between the Mg#s of clinopyroxene and orthopyroxene [80–84 in R2 ($n = 11$), 67–76 in T9-1 ($n = 18$), 59–72 in T2 ($n = 14$), 34–48 in T6-1 ($n = 29$), and 47–52 in T6-2 ($n = 6$)] (Tables S2–S6).

However, assuming that the mineral phases represent equilibrium compositions, the OT clinopyroxene and orthopyroxene composition data (Tables S2–S6) allowed the use of the clinopyroxene and orthopyroxene-liquid geothermobarometer of [84,85] to calculate the crystallization temperatures and pressures of the magma along its ascent path. The standard errors of the temperature and pressure calibrations were 39 °C and 2.1 kbar, respectively [85]. According to the clinopyroxene-liquid geothermobarometer (Table S4) [84], the crystallization temperature, pressure, and magma depth for formation of the clinopyroxene phenocrysts in the R2 basalt from the MOT were estimated to be in the ranges of 1112–1137 °C, 0.14–0.42 GPa, and ~8.0–16.9 km, respectively. From the orthopyroxene-liquid geothermobarometers (Table S3) [85], the crystallization temperature, pressure, and magma depth for formation of the orthopyroxene phenocrysts in the T2 andesite from the MOT were calculated to be in the ranges of 1021–1095 °C, 0.03–0.37 GPa, and 4.3–15.3 km, respectively (Table S3). The crystallization temperature, pressure, and magma depth for formation of the orthopyroxene phenocrysts in the T6-1 and T6-2 rhyolites from the MOT were estimated to be 821–884 °C and 844–858 °C, 0.11–0.59 GPa and 0.01–0.11 GPa, and 7.1–22.2 km and 3.7–7.1 km, respectively (Tables S5 and S6). The crystallization temperature, pressure, and magma source depth for forming the orthopyroxene phenocrysts in the T9-1 basaltic andesite from the SOT were estimated to be 1095–1138 °C, 0.16–0.40 GPa, and 8.4–16.1 km, respectively (Table S2).

4.2. Major Element and Trace Element Compositions of the Volcanic Lavas

The major element concentrations of the volcanic lavas obtained from the MOT and SOT are presented in Table S7. No increase in $^{87}Sr/^{86}Sr$ ratios was observed with increasing LOI (Figure S1), indicating that the samples were free of seawater alteration. On the total alkalis-silica (TAS) diagram, the volcanic lavas plot in the R2 basalt, T9-1 basaltic andesite, T5-2 trachyandesite, T2 andesite, T6-1 and T6-2 rhyolite fields, indicate they can be classified as subalkaline ($SiO_2 = 51.62 - 73.85$ wt.%; $Na_2O + K_2O = 2.63 - 8.16$ wt.%) (Figure 3) [86–88]. On the K_2O vs. SiO_2 diagram, the T9-1 basaltic andesite from the SOT and the R2 basalt from the MOT plot in the low-K arc tholeiitic field, the T5-2 trachyandesite and T2 andesite from the MOT plot in the medium-K calc-alkaline field, and the T6-1 and T6-2 rhyolites from the MOT, are in the high-K calc-alkaline field (Figure 3). On the Harker diagrams (Figure S2), the major element oxides (except for Na_2O, TiO_2, and P_2O_5) of the MOT volcanic lavas exhibit first-order trends, indicating that they evolved through the fractional crystallization of magma [89]. In addition, the T9-1 basaltic andesite from the SOT had lower Na_2O (2.20 wt.%) and higher Al_2O_3 (18.55 wt.%) contents than the MOT volcanic lavas ($Na_2O = 2.77$–5.43 wt.%, $Al_2O_3 = 12.23$–17.08 wt.%) (Figure S2).

The trace element concentrations of the OT lava samples are presented in Table S8. The primitive mantle-normalized spider diagrams for the MOT and SOT volcanic lavas reveal obvious enrichment in large ion lithophile elements relative to the high field strength elements and rare earth elements (Figure 4). All the samples had negative niobium (Nb), tantalum (Ta), and titanium (Ti) anomalies and distinctly positive Pb anomalies (Figure 4). Moreover, the MOT rhyolites exhibited significant Sr, P, and Ti depletions (Figure 4e,f), suggesting mineral fractionation. The R2 basalt from the MOT and the T9-1 basaltic andesite from the SOT had slight Sr enrichments (Figure 4a). The chondrite-normalized REE diagrams of the

MOT and SOT volcanic lavas reveal light REE (LREE) enrichment relative to heavy REEs (HREEs) (Figure 4), and the MOT volcanic lava samples (($La/Yb)_N$ = 2.23–3.68) were more fractionated than the T9-1 basaltic andesite sample (($La/Yb)_N$ = 1.66) from the SOT. The MOT rhyolites exhibited significant negative europium (Eu) anomalies (Eu/Eu^*_{T6-1} = 0.55, and Eu/Eu^*_{T6-2} = 0.36, where Eu/Eu^* = $2Eu_N/(Sm_N + Gd_N)$), the MOT trachyandesite and andesite exhibited small negative Eu anomalies (Eu/Eu^*_{T5-2} = 0.85, and Eu/Eu^*_{T2} = 0.86), and the MOT basalt and the SOT basaltic andesite exhibited negligible Eu anomalies (Eu/Eu^*_{R2} = 0.99, and Eu/Eu^*_{T9-1} = 1.05). However, fractionation between the LREEs and HREEs was higher in the MOT volcanic lavas (($La/Yb)_N$ = 2.23−3.68) than in the SOT lava (($La/Yb)_N$ = 1.66).

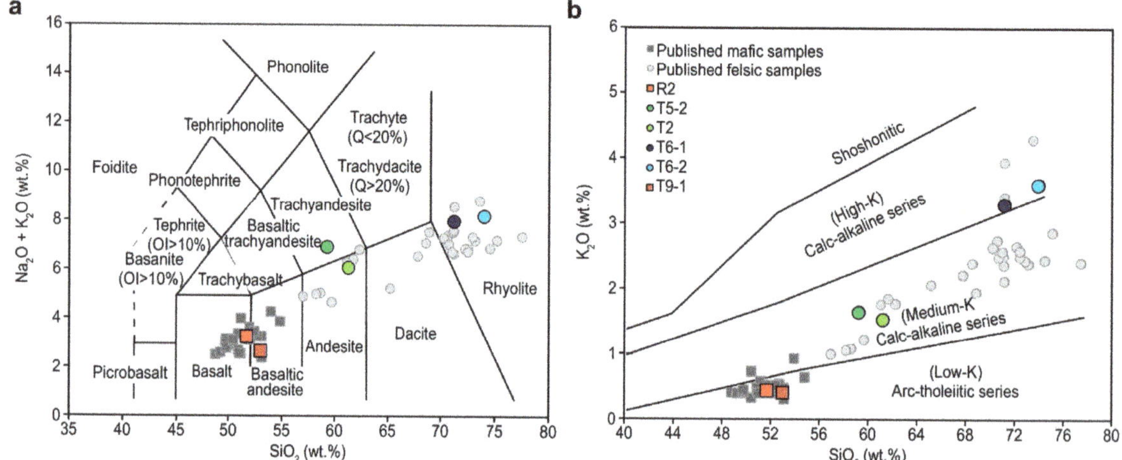

Figure 3. Classification diagrams for the MOT and SOT volcanic lavas. (**a**) Plots of $Na_2O + K_2O$ vs. SiO_2. The base diagram is from [87], and the boundary between the alkaline and subalkaline rocks is from [86]. (**b**) K_2O (wt.%) vs. SiO_2 (wt.%). Boundaries are from [88]. Data for published mafic samples are from [1,36,50,74]. Data for published felsic samples from [1,73,74]. Data for samples R2, T5-2, T6-1, T2, T6-2, and T9-1 are from this study.

4.3. Li–O–Mg Isotope Compositions of the Volcanic Lavas and Minerals

The Li isotopic compositions of the OT volcanic lavas and minerals are reported in Table S9 and are plotted in Figure 5a,b. The δ^7Li values of all samples vary from −5.05‰ to +5.61‰ (Figure 5a). The highly fractionated rhyolites (T6-1 and T6-2) have higher δ^7Li values than the mafic and intermediate lavas (Figure S3b), probably as a result of late-stage magma evolution; thus, these data are not included in the following discussion. The T5-2 trachyandesite from the MOT had lower δ^7Li values (−0.83‰ to +2.90‰) than the other MOT samples (Figure 5a). The majority of the δ^7Li values in the MOT volcanic lavas fell within the range of mid-ocean ridge basalts (MORBs) (+1.50‰ to +6.85‰) (Figure 5a,b) [22,33,91–93], indicating that these samples originated from similar magmatic sources. The glass in T9-1 basaltic andesite from the SOT had the highest δ^7Li value (+4.98‰) (Table S9), while the δ^7Li values of clinopyroxene (+0.49‰), orthopyroxene (+1.04‰), and plagioclase (−5.05‰) in the T9-1 basaltic andesite sample from the SOT were all lower than those of MORBs (Table S9 and Figure 5a). The δ^7Li values of the MOT and SOT plagioclase and clinopyroxene phenocrysts were lower than those of olivine, orthopyroxene, and glass (Table S9 and Figure 5a). Furthermore, most of the δ^7Li values obtained in this study were lower than those obtained for the MOT volcanic lavas by [1] (δ^7Li = +2.6‰ to +6.9‰, *n* = 3). The δ^7Li values of the OT basaltic lavas (δ^7Li_{R2-1} = +3.33‰, and δ^7Li_{9-1} = +3.45‰) were lower than those reported for the Lau Basin basalts (δ^7Li = +4.32‰ to +4.82‰) [4].

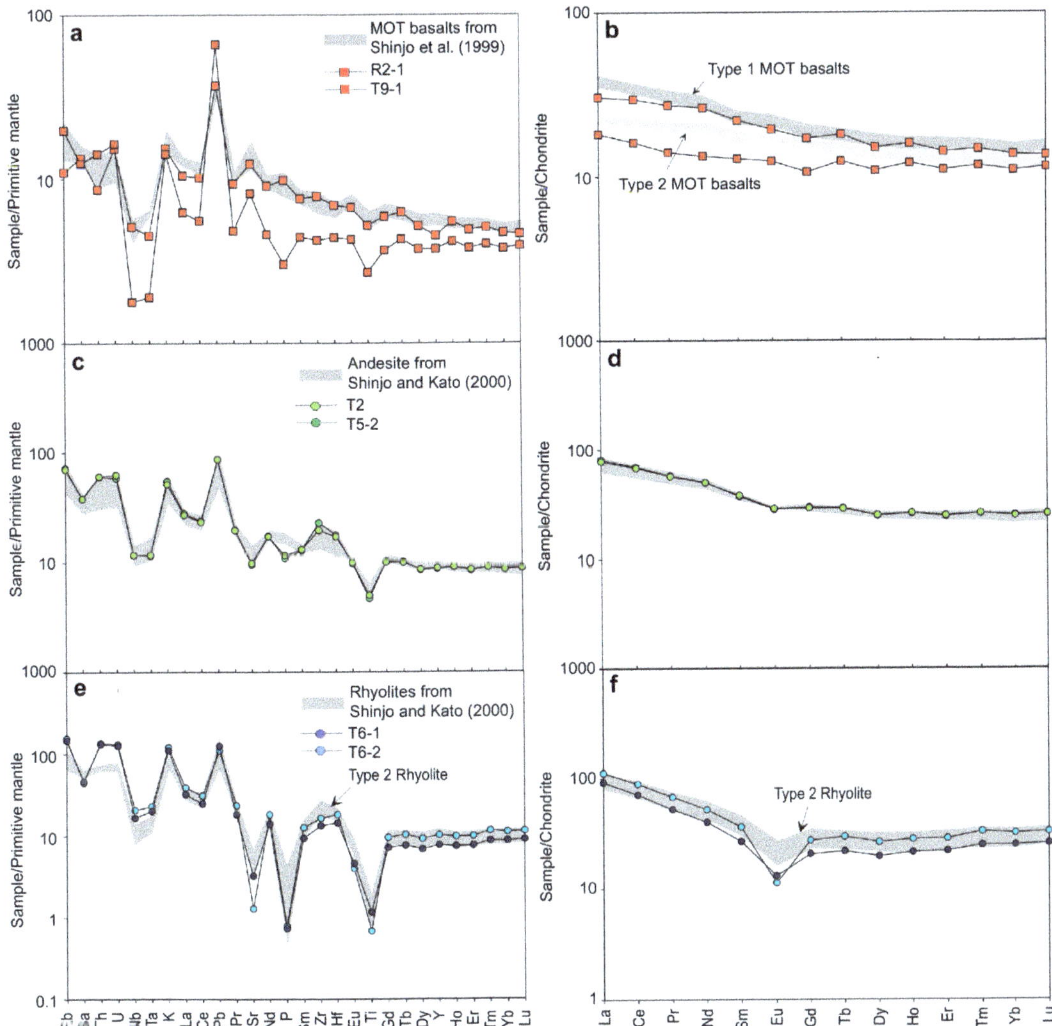

Figure 4. (**a,c,e**) Trace element patterns normalized to primitive mantle and (**b,d,f**) corresponding rare earth element (REE) patterns normalized to chondritic contents for the MOT and SOT volcanic lavas. Primitive mantle and chondrite contents are from [90]. (**a,b**) show the trace element distributions of the MOT basalt and the SOT basaltic andesite, respectively. (**c,d**) show the trace element distributions of the MOT andesite and trachyandesite, respectively. (**e,f**) show the trace element distributions of the MOT rhyolites. Data are from [50,73], and this study.

In addition, the Li isotope fractionations ($\Delta^7\mathrm{Li}_{x-y} - \delta^7\mathrm{Li}_x - \delta^7\mathrm{Li}_y$) between coexisting mineral phases (x-y) were also calculated. The $\Delta^7\mathrm{Li}_{cpx\text{-}ol}$ (Li isotope fractionation between coexisting clinopyroxene and olivine) and $\Delta^7\mathrm{Li}_{pl\text{-}ol}$ (Li isotope fractionation between coexisting plagioclase and olivine) values of the R2 basalt from the MOT were −3.76 and −3.77, respectively. The $\Delta^7\mathrm{Li}_{cpx\text{-}ol}$, $\Delta^7\mathrm{Li}_{opx\text{-}ol}$, and $\Delta^7\mathrm{Li}_{pl\text{-}ol}$ values of the T5-2 trachyandesite from the MOT were −3.24, −0.07, and −0.55, respectively. The $\Delta^7\mathrm{Li}_{cpx\text{-}ol}$, $\Delta^7\mathrm{Li}_{opx\text{-}ol}$, and $\Delta^7\mathrm{Li}_{pl\text{-}ol}$ values of the T9-1 basaltic andesite from the SOT were −2.35, −1.80, and −7.88, respectively. The $\Delta^7\mathrm{Li}$ values of all samples were negative. This cannot be explained by closed system equilibrium between the different minerals and olivine.

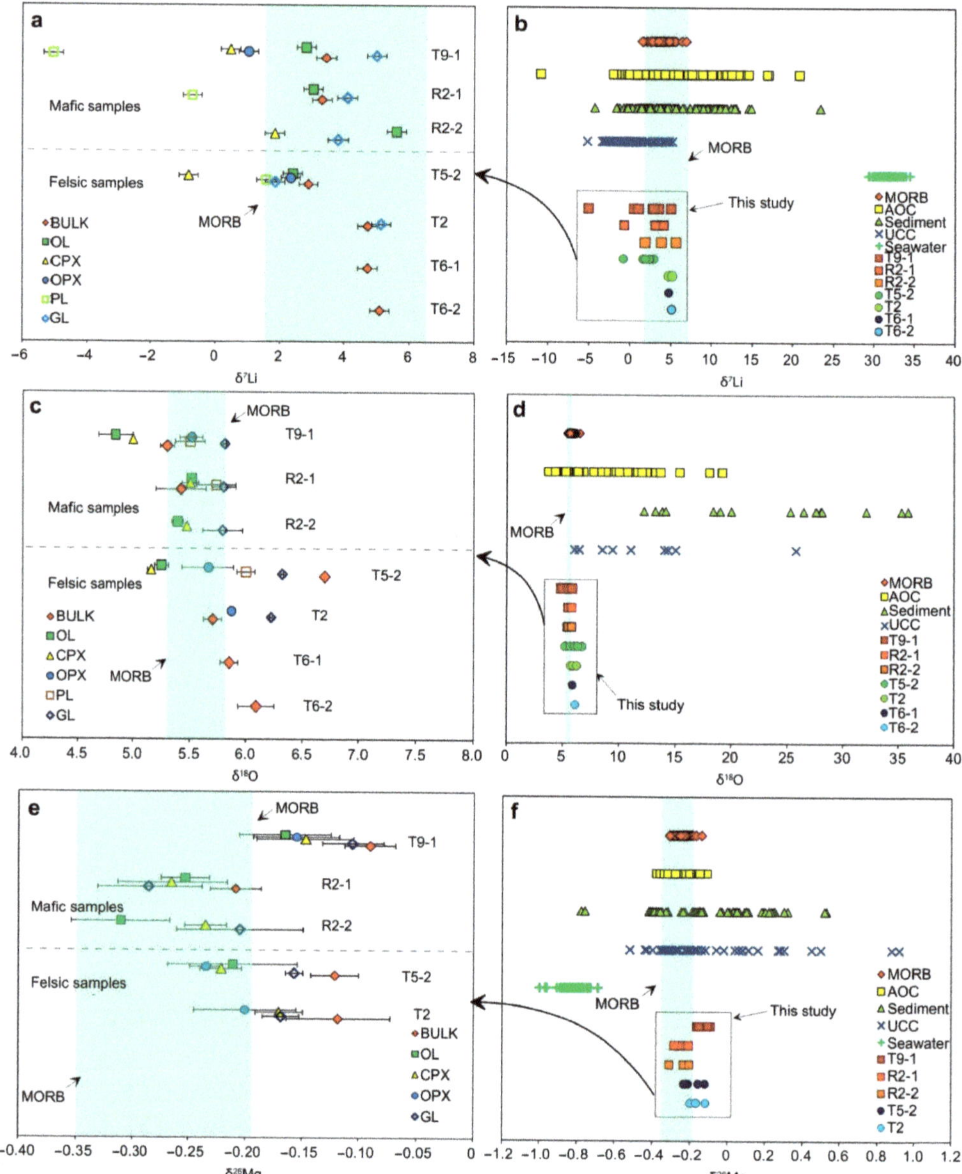

Figure 5. The δ^7Li distributions of (**a**) the different whole-rock and mineral separates from the Okinawa Trough (OT) and (**b**) mid-ocean ridge basalts (MORBs), AOC, sediment, upper continental crust, and seawater. The δ^{18}O distributions of (**c**) the different whole-rock and mineral separates from the OT and (**d**) MORBs, AOC, sediment, and upper continental crust. The δ^{26}Mg distributions of (**e**) the different whole-rock and mineral separates from the OT and (**f**) MORBs, AOC, sediment, upper continental crust, and seawater. The MORB δ^7Li range is from [22,33,91–93]. The AOC δ^7Li data are from [4,22,25,94]. The sediment δ^7Li data are from [3,4,25,27,95–97]. Upper continental crust (UCC) δ^7Li data are from [98]. The MORB δ^{18}O range is from [99–101]. AOC δ^{18}O data are from [102,103]. Sediment δ^{18}O data are from [104]. UCC δ^{18}O data are from [105]. The MORB δ^{26}Mg range is from [106,107]. AOC δ^{26}Mg data are from [39]. Sediment δ^{26}Mg data are from [108]. UCC δ^{26}Mg data are from [109].

The oxygen isotopic compositions of the OT volcanic lavas and minerals varied from +4.83‰ to +6.69‰ (Table S9 and Figure 5c). The $\delta^{18}O$ values showed no significant increase with increasing SiO_2 concentrations for whole-rock samples (Figure S3a). The $\delta^{18}O$ values (+5.38‰ to +5.79‰) of R2-1 and R2-2 basalts from the MOT fell within the MORB range (Figure 5c,d) (+5.30‰ to +5.80‰) [110,111]. The $\delta^{18}O$ values of the T5-2 trachyandesite exhibited a broader range (+5.15‰ to +6.69‰) than those of the other MOT samples (Figure 5c). The majority of the $\delta^{18}O$ values (+4.83‰ to +5.80‰) of the T9-1 basaltic andesite and its mineral samples from the SOT were lower than those of the MOT volcanic lava samples (+5.15‰ to +6.69‰) (Table S9 and Figure 5c). Moreover, the $\delta^{18}O$ values of all the OT volcanic lava and mineral samples analyzed in this study overlapped with the $\delta^{18}O$ range previously reported for volcanic lavas from the Manus Basin (+4.96‰ to +6.68‰) [37,100]. The whole-rock $\delta^{18}O$ values of T5-2 (+6.69‰) and T6-2 (+6.08‰) fell within the range of the MOT volcanic lavas (+6.0‰ to +7.6‰) analyzed by [36]. The $\delta^{18}O$ values of the T9-1 (+5.29‰), R2-1 (+5.41‰), R2-2 (+5.38‰), T2 (+5.69‰), and T6-1 (+5.84‰) whole-rock samples (Table S9) were lower than those of the volcanic lavas (+6.6‰ to +8.8‰) analyzed by [36,53]. The majority of the $\delta^{18}O$ values of the T9-1 basaltic andesite from the SOT and the R2-1 and R2-2 basalts from the MOT were lower than those measured for the volcanic lavas of the North Fiji Basin ($\delta^{18}O$ = +5.78‰ to +6.06‰) [38] and the Mariana back-arc basin ($\delta^{18}O$ = +5.8‰ to +6.0‰) [7].

In addition, the $\delta^{18}O$ values of plagioclase and glass were higher than those of olivine, clinopyroxene, and orthopyroxene in the MOT and SOT volcanic lavas (Table S9), and the $\delta^{18}O$ values of olivine, clinopyroxene, and orthopyroxene exhibited smaller variations than those of volcanic glass (Figure 5c). The $\Delta^{18}O_{cpx-ol}$ and $\Delta^{18}O_{pl-ol}$ values calculated for the R2 basalt from the MOT were 0.08 and 0.22, respectively. The $\Delta^{18}O_{cpx-ol}$, $\Delta^{18}O_{opx-ol}$, and $\Delta^{18}O_{pl-ol}$ values calculated for the T5-2 trachyandesite from the MOT were −0.09, 0.41, and 0.75, respectively. The $\Delta^{18}O_{cpx-ol}$, $\Delta^{18}O_{opx-ol}$, and $\Delta^{18}O_{pl-ol}$ values calculated for the T9-1 basaltic andesite from the SOT were 0.16, 0.67, and 0.66, respectively. These data imply that the crystallization of the magma and minerals preferentially incorporated isotopically heavy O into the orthopyroxene and plagioclase phenocrysts, resulting in the relative enrichment of lighter O isotope in the olivine phenocrysts.

The Mg isotope compositions of the OT volcanic lavas and their minerals varied from −0.31‰ to −0.09‰ (Table S9 and Figure 5e). No correlations existed between Mg isotopes and SiO_2 concentrations for whole-rock samples (Figure S3c). The $\delta^{26}Mg$ values of the R2 basalt from the MOT ranged from −0.31 to −0.20‰, and all the values fell within the MORB range (−0.26‰ ± 0.07‰) [107] (Figure 5e,f and Table S9). The whole-rock $\delta^{26}Mg$ values (−0.25‰ to −0.11‰) of the MOT andesite (T2) and trachyandesite (T5-2) were slightly higher than those of MORBs. All the $\delta^{26}Mg$ values (−0.16‰ to −0.09‰) of the T9-1 basaltic andesite from the SOT were higher than those of MORBs (Figure 5e,f). The $\delta^{26}Mg$ values of the clinopyroxene, orthopyroxene, and glass in the T9-1 basaltic andesite from the SOT were higher than the $\delta^{26}Mg$ values of those from the MOT volcanic lavas (Table S9 and Figure 5c). Furthermore, the $\Delta^{26}Mg_{cpx-ol}$ value calculated for the R2 basalt from the MOT was −0.01. The $\Delta^{26}Mg_{cpx-ol}$ and $\Delta^{26}Mg_{opx-ol}$ values calculated for the T5-2 trachyandesite from the MOT were −0.01 and −0.02, respectively. The $\Delta^{26}Mg_{cpx-ol}$ and $\Delta^{26}Mg_{opx-ol}$ values calculated for the T9-1 basaltic andesite from the SOT were 0.02 and 0.01, respectively.

However, only the δ^7Li (+3.33‰ to +4.09‰), $\delta^{18}O$ (+5.41‰ to +5.78‰), and $\delta^{26}Mg$ (−0.28‰ to −0.20‰) values of the glass or whole rock from the mafic R2 basalt might be considered indicative of magma source compositions.

5. Discussion

5.1. Genesis and Evolution of Magmas

The crystallization temperatures and pressure ranges (Tables S2–S6) of the corresponding ascending magmas were estimated from the pyroxene compositions of the volcanic lavas (Section 4.1) [84,85,112,113]. The crystallization temperature, pressure, and magma depth for forming the orthopyroxene phenocrysts in T6-2 rhyolites from the MOT were lower and shallower than those for forming the clinopyroxene phenocrysts in the R2 basalt, the orthopyroxene phenocrysts in the T2 andesite from the MOT, and the orthopyroxene phenocrysts in the T9-1 basaltic andesite from the SOT (Tables S2–S6, Figure 6). The average depths of ascending magmas for forming the clinopyroxene phenocrysts in the R2 basalt, orthopyroxene phenocrysts in the T2 andesite, orthopyroxene phenocrysts in T6-1 rhyolites from the MOT, and orthopyroxene phenocrysts in the T9-1 basaltic andesite from the SOT were estimated to be 11.3 km (n = 11), 10.9 km (n = 14), 12.7 km (n = 28), and 12.6 km (n = 13), respectively. These results suggest that the ascending magmas for forming the R2 basalt, T2 andesite, and T6-1 rhyolites in the MOT and the T9-1 basaltic andesite in the SOT originated from near the crust-mantle transition zone (13–14 km) [62].

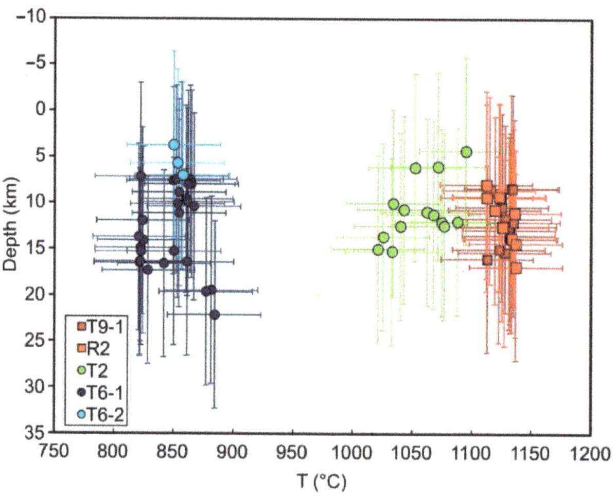

Figure 6. Crystallization temperatures and magma source depths for the MOT and SOT volcanic lavas.

The LILE, Pb, and Sr enrichments and the Nb, Ta, and Ti depletions in the MOT and SOT lavas (Figure 4) produced trace element distribution patterns similar to those of the continental crust [114]. If considerable crustal contamination or magma mixing occurred, the $^{87}Sr/^{86}Sr$ and $^{143}Nd/^{144}Nd$ of the volcanic lavas would then increase and decrease, respectively, with increasing SiO_2 [12,89]. However, except for the T6-1 and T6-2 rhyolites, the $^{87}Sr/^{86}Sr$ and $^{143}Nd/^{144}Nd$ of the MOT and SOT volcanic lavas remained nearly constant with increasing SiO_2 content (Figure 7), which is unlikely to suggest mixing or assimilation between variably evolved OT magmas, indicating that crustal contamination had a minimal effect on the evolution of MOT and SOT magmas. However, the $^{87}Sr/^{86}Sr$ ratios of MOT volcanic lavas could be strongly affected by the involvement of AOC or sediments, resulting in elevated $^{87}Sr/^{86}Sr$ ratios in the R2 and T2 volcanic lavas. Furthermore, as observed in the Harker diagrams (Figure S2), the significant negative correlations among SiO_2 and Al_2O_3, CaO, FeO_t, and MgO highlight the fractional crystallization of olivine, clinopyroxene, and plagioclase [115]. P_2O_5 and TiO_2 initially increased and then decreased with increasing SiO_2 content, suggesting that apatite fractionally crystallized after the SiO_2 content reached >60%. Magnetite fractionally crystallized when the SiO_2 content was >55% (Figure S2) [115,116].

The sample locations, trace element patterns, and Sr-Nd isotopic compositions of the T5-2 trachyandesite (27°32.86′ N, 126°59.36′ E, 1283 m) and T2 andesite in the MOT (27°32.76′ N, 126°58.52′ E, 1240 m; $^{87}Sr/^{86}Sr$ = 0.704522, $^{143}Nd/^{144}Nd$ = 0.512865) [74] were similar to those previously reported for MOT andesite (287-2: 27°29.50′ N, 126°50.00′ E, 1380 m; $^{87}Sr/^{86}Sr$ = 0.704252, $^{143}Nd/^{144}Nd$ = 0.512806) [73] (Figure 4c), indicating that samples T5-2, T2, and 287-2 may be derived from a similar magma source [117]. Furthermore,

Li, et al. [74] and Shinjo and Kato [73] suggested that the 287-2 and T2 samples were formed by fractional crystallization of basaltic magmas, implying that the T5-2 sample was also produced in this way. Similarly, the sampling location (Table S1), trace element distribution pattern, and Sr-Nd isotope compositions of the R2 basalt from the MOT (27°32.47′ N, 126°58.62′ E, 1309.7 m; $^{87}Sr/^{86}Sr$ = 0.704188, $^{143}Nd/^{144}Nd$ = 0.512763) [74] were similar to those of Type 1 basalt (A6: 27°31.33′ N, 126°56.60′ E, 967 m; $^{87}Sr/^{86}Sr$ = 0.704044, $^{143}Nd/^{144}Nd$ = 0.512827) (Figure 4b) previously reported by [50], indicating that the R2 sample and Type 1 sample A6 from [50] were both produced by the crystallization of a similar magma source and did not suffer notable crustal contamination or undergo any assimilation and fractional crystallization (AFC) processes [50].

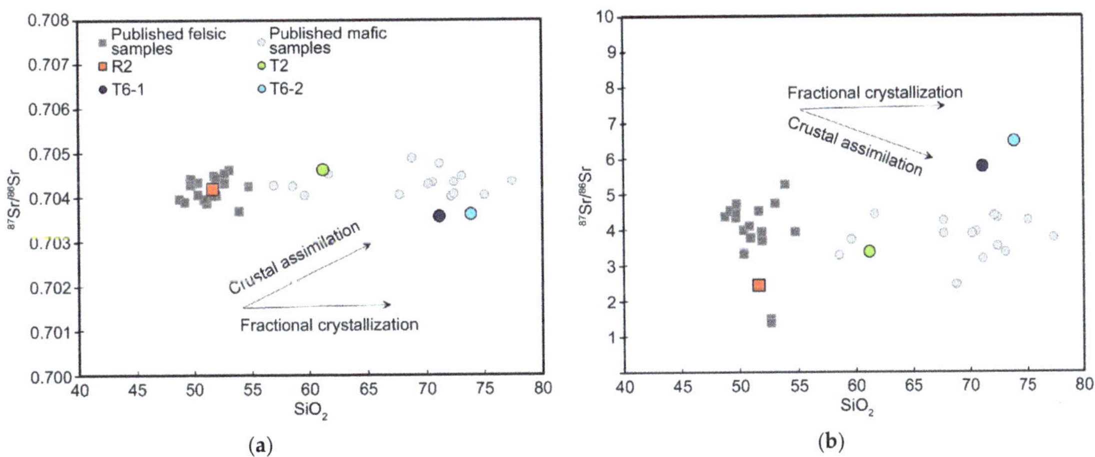

Figure 7. (a) SiO_2 vs. $^{87}Sr/^{86}Sr$ and (b) SiO_2 vs. $^{143}Nd/^{144}Nd$ plots showing that crustal assimilation slightly influenced the magmatic evolution of the MOT and SOT volcanic lavas. Data for published mafic samples are from [1,36,50,74]. Data for published felsic samples are from [1,73,74]. Data for samples T6-1, T2, and T6-2 are from this study.

However, the T9-1 basaltic andesite from the SOT had lower trace element and REE concentrations than the R2 basalt from the MOT (Table S8, Figure 4a,b), implying that the T9-1 basaltic andesite and the R2 basalt originated from different magmatic sources and evolved through different processes [50]. Furthermore, the trace element distribution pattern of the T9-1 basaltic andesite from the SOT differed from those of other SOT samples previously reported by [50], suggesting that the source of the SOT magma was heterogeneous and that fractional crystallization of a primitive magma was insufficient to form the different compositions of the SOT magmas [50].

In addition, the T6-1 and T6-2 samples were obtained from the western slope of the trough (Figure 1), and they exhibited the lowest and highest $^{87}Sr/^{86}Sr$ and $^{143}Nd/^{144}Nd$, respectively, among all the OT samples (Figure 7). These pumices all had lower crystallization temperatures and magma depths than the T2 samples obtained from the Iheya Ridge (Figure 6). Zhang, et al. [118] analyzed the geochemistry of the T6-1 and T6-2 pumices and concluded that they did not originate from MOT basaltic rocks through either partial melting or fractional crystallization processes; rather, they were generated from a potassium-rich magma source in the OT. The trace element and REE patterns of the T6-1 and T6-2 rhyolites from the MOT resemble those of Type 2 rhyolites previously reported by [73] (Table S8 and Figure 4e), which can be explained by AFC processes in the MOT basaltic magma [73]. Therefore, the T6-1 and T6-2 rhyolites from the MOT originated from different magma sources and underwent magmatic evolution histories different from those of other MOT samples.

5.2. Influence of Subduction Components Inferred from Trace Elements

The MOT basalts and SOT basaltic andesite were notably enriched in LILEs and Pb and were depleted in HFSEs (Figure 4 and Table S8), indicating that the magma from which these lavas formed had been affected by subduction component input [9]. On the Ba/La vs. Th/Rb and Ce/Pb vs. Ba/La diagrams (Figure 8), all the MOT and SOT whole-rock samples plot are within the subducted sediment zone, indicating that all these volcanic lavas have been affected by subducted sediments. The Ba/La vs. Th/Yb and Ba/Th vs. Th/Nb diagrams (Figure 8) also show that the MOT and SOT trace element ratios trend toward sediment assimilation. All these diagrams suggest that the OT volcanic rocks have been affected by subducted sediment. However, whether slab-derived fluids influenced the mantle beneath the back-arc is still uncertain.

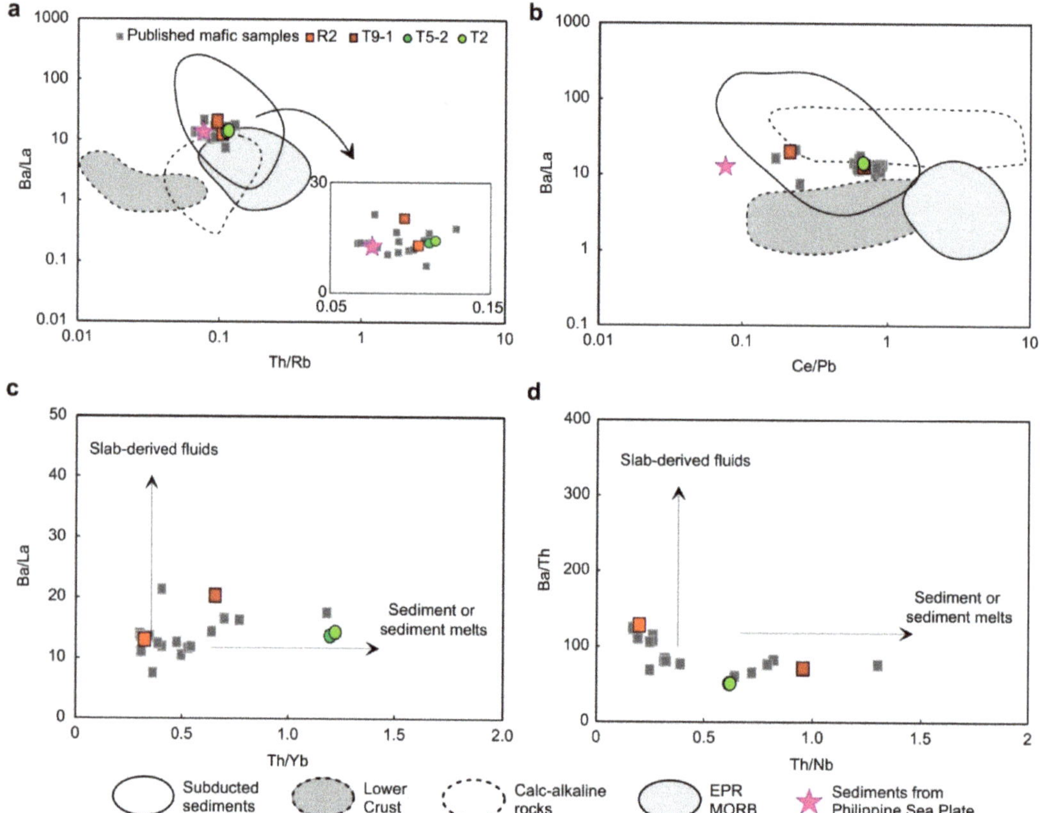

Figure 8. Trace element ratio plots: (**a**) Th/Rb vs. Ba/La, (**b**) Ce/Pb vs. Ba/La, (**c**) Th/Yb vs. Ba/La, and (**d**) Th/Nb vs. Ba/Th for the MOT and SOT volcanic lavas. Compositions of the EPR MORB data are from PetDB database (www.earthchem.org/petdb accessed on 10 December 2021). Subducted sediment compositions are from [10]. Canary ocean island basalt (OIB) contents are from [119]. Calc-alkaline basalt contents are from [120,121], and the values for the lower crust in the North China Craton are from [122]. Data for published mafic samples are from [1,36,50,74]. Data for published felsic samples from [1,73,74]. Data for samples R2, T5-2, T6-1, T2, and T6-2 in the MOT and T9-1 in the SOT are from this study.

The Pb content of crustal sediments (19.9 ppm) [10] is two orders of magnitude greater than that of the mantle (0.6 ppm) [12,90], and the addition of dehydration fluids from the subducting plate can increase the contents of fluid-active elements [108]. However, the ratios of fluid-inactive elements to active elements, such as the Nd/Pb ratio (1.33) of the T9-1 basaltic andesite from the SOT, were notably lower than those of MOT samples (basalt = 4.65, andesite = 3.80, trachyandesite = 3.79, and rhyolites = 2.15–3.16), and the Nd/Pb ratios of all MOT and SOT samples were lower than that of the primitive mantle (~7.3) [90], suggesting that a Pb-rich component (i.e., subducted sediment) was present in the OT magma source [12].

The Ba/La vs. Th/Yb and Ba/Th vs. Th/Nb diagrams (Figure 8) show that the MOT and SOT samples trend toward sediment assimilation, which suggests that the OT volcanic lavas were affected mainly by subducted sediment-derived melts and that subduction fluids had less influence on the magma. Furthermore, the Ba/La ratio of the SOT basaltic andesite (20.32) was considerably higher than those of MOT basalt (13.05), andesite (13.79), trachyandesite (14.35), and rhyolites (11.76–14.81). The Ce/Pb ratio of the T9-1 basaltic andesite (2.11) from the SOT was significantly lower than those of MOT samples (4.99–7.05). Compared to the Ba/La ratios of MOT volcanic rocks (11.76–14.81), the Ba/La ratios (20.32) of T9-1 basaltic andesite from the SOT were closer to the Ba/La ratios of sediment in the Philippines Sea (25.11) [10] (Figure 8), all of which indicate that the SOT magma from which the T9-1 basaltic andesite formed was affected by subducted sediments [8,12–14,74,115].

5.3. Subduction Input of Low-δ^7Li Components to the OT Magmas

The Li isotope characteristics of back-arc volcanic lavas are known to be potentially influenced by subducted slab-derived components [1,123,124], which include AOC-derived hydrous fluids, subducted sediments, and oceanic crust- and/or sediment-derived silicate melts [125,126]. Although the low δ^7Li values are not consistent with subduction fluids or the asthenosphere (i.e., MORBs), they may be related to the melting of a dehydrated slab [4,96,124,127]. Previous research has demonstrated that low δ^7Li values can be interpreted as dehydration signatures resulting from the hydrothermal alteration of oceanic crust (−10.90‰ to +20.80‰) [26,94]. During high-temperature plate subduction processes, low-δ^7Li components are released and interact with the upper mantle, thereby resulting in low δ^7Li values in the range of −6‰ to +10‰ [35]. Other studies have shown that alteration of the upper oceanic crust at high-temperature can also result in low δ^7Li values (e.g., −1.7‰) [26,128], and forearc serpentinites exhibit δ^7Li values of less than −6‰ [35]. Additionally, eclogites, which represent dehydrated oceanic crust, have low δ^7Li values (−11‰ to +0.3‰), and the direct melting of such siliceous eclogites could result in low-δ^7Li melts [28]. Some sediments, especially young oceanic sediments that have not been altered by seawater, also exhibit low δ^7Li values (−4.31‰ to +23.33‰) [27]. Thus, fluids with relatively low δ^7Li values may also be released from subducted sediments [3].

The δ^7Li values of glass and whole-rock samples from the T5-2 trachyandesite, T2 andesite, R2 basalt in the MOT and the T9-1 basaltic andesite in the SOT were significantly lower than those of whole-rock data for T6-1 and T6-2 rhyolites in the MOT (Table S9). This contrast implies that the magmas from which the T5-2 trachyandesite, T2 andesite, and R2 basalt in the MOT and the T9-1 basaltic andesite in the SOT formed could have been affected by a low-δ^7Li component that may be released from oceanic sediments or subducted AOC.

Furthermore, previous research has shown that Li isotopes are not fractionated in closed systems at temperatures greater than 350 °C [33,129,130]. Thus, Li isotopes are not expected to have fractionated during OT magma evolution [33,131,132]. However, the δ^7Li values of clinopyroxene in the T5-2 trachyandesite (−0.83‰) from the MOT and T9-1 basaltic andesite (0.49‰) from the SOT and of plagioclase in the R2-1 basalt (−0.71‰) from the MOT and T9-1 basaltic andesite (−5.05‰) from the SOT were lower than those of olivine, orthopyroxene, glass (Table S9), and MORBs (+1.50‰ to +6.85‰) [22,33,91–93] (Figure 5a,b). This difference suggests that the Li isotope compositions of plagioclase and pyroxene may have been modified by a low-δ^7Li fluid or melt during magma evolution and that the ^{6}Li content was enriched in clinopyroxene and plagioclase during mineral crystallization and mantle metasomatism. Overall, the negative Li isotope fractionation

($\Delta^7 Li_{x\text{-}ol} < 0$) between silicate minerals and olivine during mineral crystallization may have been due to the low-$\delta^7 Li$ components that are released from oceanic sediments or AOC.

5.4. Subduction of Low-and High-$\delta^{18}O$ Components

For volcanic rocks, magma differentiation from mafic to felsic compositions at high temperatures does not result in significant oxygen isotope fractionation (usually less than 0.3–0.4‰) (e.g., from 5.8‰ for basalts to 6.1‰ for rhyolites at ~90% differentiation) [133,134]. The $\delta^{18}O$ values of glass (6.31‰) and plagioclase (5.99‰) in the T5-2 trachyandesite, glass (6.21‰) and orthopyroxene (5.86‰) in the T2 andesite, the whole-rock samples of T5-2 (6.69‰) trachyandesite, T6-1 (5.84‰), and T6-2 (6.08‰) rhyolites from the MOT (Table S9) were all higher than those of MORBs (+5.30 to +5.80‰) [110,111] (Figure 5c); this feature also indicates that these volcanic lava samples were produced by high-$\delta^{18}O$ magmas [135,136]. The high $\delta^{18}O$ values of the evolved magmas may be related to crustal contamination; however, no correlations were observed between the Sr-Nd isotopes and SiO_2 concentrations in OT lavas (Figure 7), suggesting that little crustal contamination occurred [11,36,50,73,74]. Alternatively, the high-$\delta^{18}O$ signatures could be due to the subduction of $\delta^{18}O$-rich components [110,137]. Marine carbonates ($\delta^{18}O$ = 25–32‰), siliceous oozes ($\delta^{18}O$ = 35–42‰), and pelagic clays ($\delta^{18}O$ = 15–25‰), which are major components of sediments subducting beneath the volcanic arc, all have higher $\delta^{18}O$ values than MORBs [2]. Therefore, the addition of subducted sediments to the magma could explain the higher $\delta^{18}O$ values of MOT volcanic rock samples.

However, O isotopes can be used to constrain the contribution of slab melts to island arc magmas, thereby indicating the origin of a melt [2]. The $\delta^{18}O$ values of the R2-1 (5.41–5.79‰) and R2-2 (5.38–5.78‰) basalts from the MOT fell within the MORB range (5.3–5.8‰), implying that the volcanic lava samples in the MOT originated from mantle-derived magma (Figure 5c) [111,138]. The whole-rock and mineral $\delta^{18}O$ values (5.15–6.31‰) of the T5-2 trachyandesite varied significantly (Figure 5c) and may have been affected by different magmatic components [2] during magma ascent. Moreover, the $\delta^{18}O$ values of whole-rock samples (5.29‰), olivine (4.83‰) and clinopyroxene (4.99‰) from the T9-1 basaltic andesite in the SOT were lower than those (5.30–5.80‰) [110,111] of MORBs (Figure 5c). This result indicates the presence of a low-$\delta^{18}O$ component during magma evolution [37,139,140]. Furthermore, AOC had a large range of $\delta^{18}O$ values (+2‰ to +14‰) [141], and the low $\delta^{18}O$ values of ocean island basalts (OIBs) and MORBs can be explained by the addition of subducted oceanic crust to the magma source region [142–145]. Therefore, the low $\delta^{18}O$ values of the T9-1 basaltic andesite from the SOT are attributed to siliceous melts or fluids derived from subducted AOC [143,145]. Furthermore, the $\delta^{18}O$ value (5.50‰) of orthopyroxene in the T9-1 basaltic andesite from the SOT was lower than that (5.86‰) of orthopyroxene in the T2 andesite from the MOT, which was consistently related to the varying crystallization temperatures (Tables S2 and S3), suggesting that the high-temperature (1095–1138 °C) orthopyroxenes were characterized by ^{16}O enrichment in the SOT, whereas the low-temperature (1020–1095 °C) orthopyroxenes were characterized by ^{16}O depletion in the MOT.

As with the Li isotopes, the $\delta^{18}O$ values of plagioclase and glass were higher than those of MOT and SOT olivine, clinopyroxene, and orthopyroxene (Table S9 and Figure 5c), and all the samples exhibited positive O isotopic fractionations between other minerals and olivine ($\Delta^{18}O_{x\text{-}ol} > 0$). This feature suggests that compared to olivine, plagioclase and pyroxene preferentially incorporated ^{18}O rather than ^{16}O during silicate mineral crystallization, which indicates that the O isotope compositions of plagioclase and pyroxene may have been modified by a high-$\delta^{18}O$ fluid or melt [137].

5.5. Contribution of High-$\delta^{26}Mg$ Slab Fluids

The $\delta^{26}Mg$ values of MOT volcanic samples (−0.21‰ to −0.28‰) fell within the range of MORBs (−0.26‰ ± 0.07‰) [107] (Figure 5e). The $\delta^{26}Mg$ values of the T2 andesite (−0.20 to −0.12‰) were similar to those of the T5-2 trachyandesite (−0.23‰ to −0.12‰) (Figure 5e), both of which were slightly higher than those of MORBs (Figure 5e). The $\delta^{26}Mg$

values ($-0.16‰$ to $-0.09‰$) of the T9-1 basaltic andesite in the SOT, which were slightly higher than those in MOT volcanic lavas (Table S9), were also higher than those of MORBs (Figure 5e), all of which indicates that these lavas may have originated from high-δ^{26}Mg magmatic components [5,39,146].

Previous studies have shown that higher Mg isotopes in subduction zones may originate from seawater alteration, chemical weathering, dissolution of continental crust, partial melting of oceanic crust, and/or sediment and slab dehydration [107,108,147–151]. Dissolution of continental crust can be ruled out due to the absence of correlations between the Sr-Nd isotopic compositions and the indices of magma differentiation (e.g., SiO_2; Figure 7). Furthermore, volcanic lavas can have isotopically heavier δ^{26}Mg values than MORBs if they are generated by the melting of subducted oceanic crust containing a garnet residual phase because garnet has much lower δ^{26}Mg values [41,42,108,152]. However, because the chemical compositions of our samples were quite different from those of slab products (e.g., adakites) (Figure 4), the possibility of slab melting can also be excluded. Therefore, either subducted sediments or slab dehydration may explain the high δ^{26}Mg values of volcanic lava samples analyzed in this study.

On average, forearc sediments have heavy Mg isotope compositions (-0.10 ± 0.61) that could provide a source for volcanic lavas with heavy Mg isotope compositions [108]. Moreover, the δ^{26}Mg values of the Avacha volcanic lavas from the Kamchatka Peninsula range from $-0.25‰$ to $-0.06‰$ (average $= -0.18‰ \pm 0.10‰$ (2SD)) [153], which has been interpreted as the result of the lower mantle being affected by subduction fluids released by the Pacific slab [148]. Similarly, the δ^{26}Mg values of the Avacha volcanic lavas are consistent with those of whole-rock samples from MOT andesite ($-0.17‰$ to $-0.12‰$), trachyandesite ($-0.16‰$ to $-0.12‰$), and basalt ($-0.28‰$ to $-0.20‰$) and SOT basaltic andesite ($-0.11‰$ to $-0.09‰$), implying that fluids released from subducted sediment or oceanic crust contributed to the OT magma source.

5.6. Mixing of Subduction Components

The above discussion suggests that the OT magma source was influenced by both subducted sediments and AOC [1,47,50]. A residual slab endmember, represented by eclogite, and an AOC endmember interacting in different proportions with a MORB-producing magma could produce lavas with different δ^7Li, δ^{18}O, and δ^{26}Mg values [7,20,33]. According to the δ^{18}O vs. δ^7Li and δ^{26}Mg vs. δ^7Li diagrams (Figure 9), OT volcanic lavas with different δ^7Li, δ^{18}O, and δ^{26}Mg values can be produced by interactions between MORB and different proportions of subducted sediment (i.e., low δ^7Li and high δ^{18}O and δ^{26}Mg) and AOC (i.e., high δ^7Li and low δ^{18}O), wherein the AOC, oceanic sediments, and mantle wedge (i.e., MORB) are the endmembers contributing to the MOT and SOT magmas. The whole-rock and glass data from the R2 basalt in the MOT and the T9-1 basaltic andesite occur in the SOT plot between subducted sediments and AOC, indicating that the contributions from AOC and sediments were 80% to 92% and 20% to 8%, respectively; therefore, the mixing of different subduction components can produce Li, O, and Mg isotope characteristics in the R2 basalt from the MOT and the T9-1 basaltic andesite from the SOT (Figure 9 and Table S9).

5.7. Implications of Differences in Plate Subduction

As discussed above, the SOT magma from which the T9-1 basaltic andesite formed was influenced by more slab-derived fluids, although the supporting data are limited. However, the thermal structure of a subduction zone is the key to determining the depth at which the subducted slab dehydrates [154–157], and this thermal structure is mainly determined by the depth of the subducted slab, age of the subducting plate, convergence rate, subduction zone geometry (especially the subduction angle), subduction zone shear heating rate, and nature of the mantle wedge [11,65,123,155,157]. Furthermore, dehydration usually decreases with increasing subduction depth [47,123,158], and most subducted slabs dehydrate considerably at a depth of ~80 km, leaving <1% of the water to be carried deeper by the subducting slab [154]. The subducting plates lose notable volumes of water at relatively shallow depth (modeled here at a depth of 80 km), and the further dehydration of most slabs is only minor (e.g., Kamchatka and Calabria) (Figure S4).

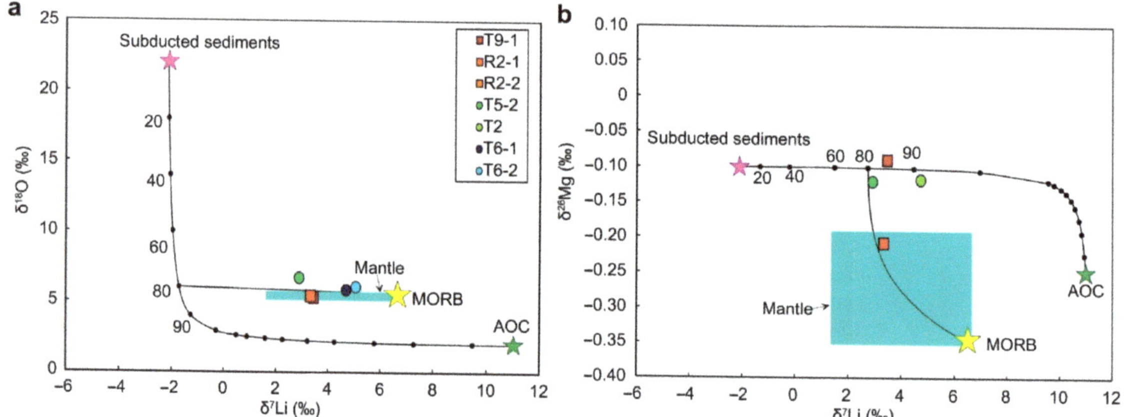

Figure 9. Plots of (**a**) $\delta^{18}O$ vs. δ^7Li and (**b**) $\delta^{26}Mg$ vs. δ^7Li for the MOT and SOT samples. Isotopic compositions ($\delta^{18}O = 2\text{‰}$, $\delta^7Li = 11\text{‰}$, and $\delta^{26}Mg = -0.25\text{‰}$) of AOC are from [33,39,137], respectively. Isotopic compositions ($\delta^{18}O = 22\text{‰}$, $\delta^7Li = -2.1\text{‰}$, and $\delta^{26}Mg = -0.1\text{‰}$) of subducted sediments are from [7,33,108], respectively. Isotope compositions ($\delta^{18}O = 5.5\text{‰}$, $\delta^7Li = 6.5\text{‰}$, and $\delta^{26}Mg = -0.35\text{‰}$) of mid-ocean ridge basalts (MORBs) are from [22,107,110], respectively.

The OT is influenced by the subduction of the PSP, and the difference between the subduction rates of the SOT and the MOT is relatively small [11,50,56,69]. Therefore, the difference between the subduction rates likely had only a limited impact on plate dehydration [47,50,157]. Furthermore, the subduction direction of the PSP is nearly perpendicular to the axis of the MOT and becomes oblique to the SOT [11,50,65]. Although the slab subduction angle beneath the SOT is higher compared to that beneath the MOT, this small difference does not significantly influence slab material transport [11,47,50]. Moreover, the difference between the crustal thicknesses of the MOT and the SOT (i.e., ~16 and ~14 km, respectively) is negligible when compared to the subduction depths of the PSP in the MOT and SOT (~200 and ~150 km, respectively), suggesting that the slight difference between the crustal thicknesses of the MOT and the SOT is not enough to cause different degrees of slab dehydration. However, this difference may reflect the degree of crustal contamination when the magma ascends from a deep magma chamber.

The Ryukyu subduction zone has a fairly rapid subduction rate (~82 mm/a) [159], which is in line with that of cold subduction zone structures [47], suggesting that large degrees of dehydration occurred when the slab reached depths of ~80 to 100 km, at which point sediment fluid/melt entered the mantle wedge [157]. Moreover, the ascending magma from which the T9-1 basaltic andesite in the SOT formed corresponds to a Wadati–Benioff zone depth of ~150 km, and the MOT has Wadati–Benioff zone depths of ~150 to 200 km (Figure 1). However, compared with the PSP subducting slab (depths: ~150 to 200 km) in the MOT, the PSP subducting slab in the SOT exhibits a significantly shallower subduction depth (~150 km), which is consistent with the variation in the subduction angle, i.e., nearly perpendicular beneath the MOT and becoming progressively more oblique beneath the SOT [11,50,65]. This variation indicates that the subducting slab (depth of ~150 km) beneath the SOT was closer to the ascending magma (avg. depth = 12.6 km) that formed the T9-1 basaltic andesite in the SOT than the subducting slab (depth of ~200 km) beneath the MOT. The deep ascending magmas (depth: 8.4–16.1 km and ~8.0–16.9 km) that formed the clinopyroxene and orthopyroxene phenocrysts with high crystallization temperatures (1095–1138 °C and 1111–1137 °C, respectively) in the mafic volcanic lavas

(SOT basaltic andesite T9-1 and MOT basalt R2) were closer to the subducted slab than the shallow ascending magma (depth: 3.7–7.1 km) that formed the orthopyroxene phenocrysts with low crystallization temperatures (844–858 °C) in the felsic volcanic lavas (MOT rhyolites T6-2). These data imply that the deeper the ascending magma is, the closer it is to the subducted slab, and the stronger the influence of the subducted components on the ascending magma (Figure 10).

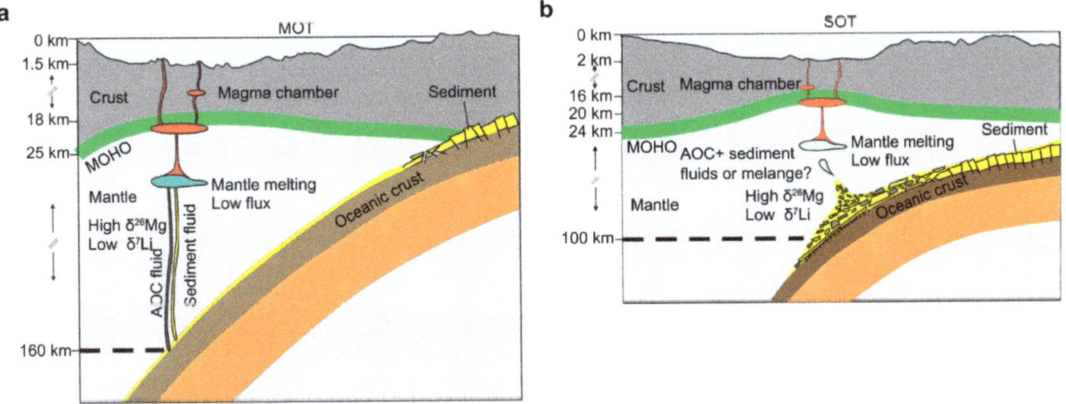

Figure 10. Schematic diagram showing the different contributions of subduction components to the (**a**) MOT and (**b**) SOT.

6. Conclusions

(1) The T9-1 basaltic andesite in the SOT had lower trace element contents than the MOT samples and exhibited stronger Nb, Ta, and Ti depletions and a positive Pb anomaly. The Nd/Pb and Ce/Pb ratios of the SOT basaltic andesite were significantly lower than those of the MOT volcanic lavas, while the Ba/La ratio of the SOT basaltic andesite was significantly higher, indicating that the magma from which the T9-1 basaltic andesite from the SOT formed was influenced by subduction components and experienced an injection of sediment components.

(2) The δ^7Li values of the whole-rock samples from the T6-1 and T6-2 rhyolites in the MOT were higher than those of whole-rock and glass separates from the T5-2 trachyandesite, T2 andesite, and R2 basalt in the MOT and the T9-1 basaltic andesite in the SOT. These results imply that the magmas that formed the R2 basalt in the MOT and the T9-1 basaltic andesite in the SOT could have been affected by a low-δ^7Li component that may have been released from oceanic sediments or subducted AOC. The δ^7Li values of plagioclase and clinopyroxene phenocrysts were lower than those of glass and olivine in the MOT and SOT volcanic lavas, suggesting that ^{6}Li was preferentially removed from the magma and incorporated into plagioclase and clinopyroxene, resulting in relative enrichment in heavy Li isotopes in olivine and glass.

(3) The SOT and MOT magmas were influenced by low-δ^{18}O AOC fluids or melts and high-δ^{18}O sediment components, respectively. The δ^{18}O values of plagioclase and glass were higher than those of olivine, clinopyroxene, and orthopyroxene phenocrysts in the MOT and SOT volcanic lavas, indicating that ^{16}O was preferentially removed from the magma and incorporated into olivine, clinopyroxene, and orthopyroxene, resulting in the relative enrichment in heavy ^{18}O isotopes in plagioclase phenocrysts and glass.

(4) The δ^{26}Mg values of clinopyroxene, orthopyroxene, and glass were higher in the T9-1 basaltic andesite from the SOT than those in the MOT volcanic lavas, and the δ^{18}O values of the T9-1 basaltic andesite from the SOT were lower than the δ^{18}O of whole-rock volcanic samples from the MOT. These differences imply that the influence of subduction components with high δ^{26}Mg and low δ^{18}O values was stronger in the SOT than in the

MOT, which may have occurred because the amount of subducted sediment and AOC dehydration fluids injected into the magma in the SOT was larger than that in the MOT.

(5) The $\delta^{18}O$ and $\delta^{26}Mg$ values of pyroxenes in the SOT were lower and higher than those in the MOT, respectively, which is consistent with the variations from the SOT to the MOT in the crystallization temperatures, pressures, and depths of the ascending magmas from which volcanic lavas formed. These results suggest that high-temperature pyroxenes originating from a deep magma (avg. depth = 12.6 km, n = 13), which was located near the crust-mantle transition zone (13–14 km), are characterized by ^{16}O and ^{26}Mg enrichments in the SOT. Low-temperature pyroxenes originating from a shallow magma (avg. depth = 10.9 km, n = 14) are characterized by ^{16}O and ^{26}Mg depletions in the MOT. However, the distance between the subducting slab and the overlying magma may have played a significant part in controlling the differences between the MOT and SOT in the amounts of subduction components injected into the magma. The deeper the magma is, the closer it is to the subducted slab, and the stronger the effects of the subducted inputs on the magma.

Supplementary Materials: The following are available online at https://www.mdpi.com/article/10.3390/jmse10010040/s1, Figures S1–S4, Tables S1–S9, and Text S1. These materials include (1) additional analytical methods, (2) major elements, trace elements, and isotopes data obtained in this study, and (3) additional plots for major elements and isotopes.

Author Contributions: Conceptualization, Z.Z.; methodology, X.L.; investigation, Z.Z., X.L., Y.Z. and H.Q.; resources, Z.Z.; data curation, H.Q.; writing—original draft preparation, Z.Z. and X.L.; writing—review and editing, Y.Z.; supervision, Z.Z.; project administration, Z.Z.; funding acquisition, Z.Z. All authors have read and agreed to the published version of the manuscript.

Funding: This work was supported by the NSFC Major Research Plan on West-Pacific Earth System Multispheric Interactions (project number: 91958213), the National Program on Global Change and Air-Sea Interaction (Grant No. GASI-GEOGE-02), the Taishan Scholar Foundation of Shandong Province (Grant No. ts201511061), and the National Basic Research Program of China (Grant No. 2013CB429700).

Institutional Review Board Statement: Not applicable.

Data Availability Statement: All the data that support the findings in this study are given in the Supporting Information or by contacting the corresponding author.

Acknowledgments: We would like to thank the crews of the R/V KEXUE during the HOBAB 2 and 4 cruises for their help with the sample collection. We thank Professor Yilin Xiao and Professor Fang Huang for performing the lithium, oxygen, and magnesium isotope analyses. We are most grateful for the detailed and constructive comments and suggestions provided by Professor David Selby of the Department of Earth Sciences, University of Durham, which greatly improved an earlier version of the manuscript.

Conflicts of Interest: The authors declare no conflict of interest.

References

1. Guo, K.; Zhai, S.; Yu, Z.; Wang, S.; Zhang, X.; Wang, X. Geochemical and Sr-Nd-Pb-Li isotopic characteristics of volcanic rocks from the Okinawa Trough: Implications for the influence of subduction components and the contamination of crustal materials. *J. Mar. Syst.* **2018**, *180*, 140–151. [CrossRef]
2. Bindeman, I.N.; Eiler, J.M.; Yogodzinski, G.M.; Tatsumi, Y.; Stern, C.R.; Grove, T.L.; Portnyagin, M.; Hoernle, K.; Danyushevsky, L.V. Oxygen isotope evidence for slab melting in modern and ancient subduction zones. *Earth Planet. Sci. Lett.* **2005**, *235*, 480–496. [CrossRef]
3. Chan, L.H.; Kastner, M. Lithium isotopic compositions of pore fluids and sediments in the Costa Rica subduction zone: Implications for fluid processes and sediment contribution to the arc volcanoes. *Earth Planet. Sci. Lett.* **2000**, *183*, 275–290. [CrossRef]
4. Chan, L.H.; Leeman, W.P.; You, C.F. Lithium isotopic composition of Central American volcanic arc lavas: Implications for modification of subarc mantle by slab-derived fluids: Correction. *Chem. Geol.* **2002**, *182*, 293–300. [CrossRef]
5. Chen, Y.X.; Schertl, H.P.; Zheng, Y.F.; Huang, F.; Zhou, K.; Gong, Y.Z. Mg–O isotopes trace the origin of Mg-rich fluids in the deeply subducted continental crust of Western Alps. *Earth Planet. Sci. Lett.* **2016**, *456*, 157–167. [CrossRef]

6.   Huang, P.; Li, A.; Jiang, H. Geochemical features and their geological implications of volcanic rocks from the northern and middle Okinawa Trough. *Acta Petrol. Sin.* **2006**, *22*, 1703–1712.

7.   Ito, E.; Stern, R.J. Oxygen- and strontium-isotopic investigations of subduction zone volcanism: The case of the Volcano Arc and the Marianas Island Arc. *Earth Planet. Sci. Lett.* **1986**, *76*, 312–320. [CrossRef]

8.   Niu, Y.; Wilson, M.; Humphreys, E.R.; O'Hara, M.J. A trace element perspective on the source of ocean island basalts (OIB) and fate of subducted ocean crust (SOC) and mantle lithosphere (SML). *Episodes* **2012**, *35*, 310–327. [CrossRef]

9.   Pearce, J.A.; Peate, D.W. Tectonic implications of the composition of volcanic ARC magmas. *Annu. Rev. Earth Planet. Sci.* **1995**, *23*, 251–285. [CrossRef]

10.   Plank, T.; Langmuir, C.H. The chemical composition of subducting sediment and its consequences for the crust and mantle. *Chem. Geol.* **1998**, *145*, 325–394. [CrossRef]

11.   Shinjo, R. Geochemistry of high Mg andesites and the tectonic evolution of the Okinawa Trough–Ryukyu arc system. *Chem. Geol.* **1999**, *157*, 69–88. [CrossRef]

12.   Yang, Y.Z.; Wang, Y.; Ye, R.S.; Li, S.Q.; He, J.F.; Siebel, W.; Chen, F. Petrology and geochemistry of Early Cretaceous A-type granitoids and late Mesozoic mafic dikes and their relationship to adakitic intrusions in the lower Yangtze River belt, Southeast China. *Int. Geol. Rev.* **2017**, *59*, 62–79. [CrossRef]

13.   Taylor, B.; Martinez, F. Back-arc basin basalt systematics. *Earth Planet. Sci. Lett.* **2003**, *210*, 481–497. [CrossRef]

14.   White, W.M.; Duncan, R.A. Geochemistry and geochronology of the Society Islands: New evidence for deep mantle recycling. In *Earth Processes: Reading the Isotopic Code*; Basu, A., Hart, S., Eds.; American Geophysical Union: Washington, DC, USA, 1996; pp. 183–206.

15.   Duan, X.; Sun, H.; Yang, W.; Su, B.; Xiao, Y.; Hou, Z.; Shi, H. Melt–peridotite interaction in the shallow lithospheric mantle of the North China Craton. Evidence from melt inclusions in the quartz-bearing orthopyroxene-rich websterite from Hannuoba. *Int. Geol. Rev.* **2014**, *56*, 448–472. [CrossRef]

16.   Class, C.; Miller, D.M.; Goldstein, S.L.; Langmuir, C.H. Distinguishing melt and fluid subduction components in Umnak Volcanics, Aleutian Arc. *Geochem. Geophys Geosyst* **2000**, *1*, 1004. [CrossRef]

17.   Elliott, T.; Plank, T.; Zindler, A.; White, W.; Bourdon, B. Element transport from slab to volcanic front at the Mariana arc. *J. Geophys. Res. Solid Earth* **1997**, *102*, 14991–15019. [CrossRef]

18.   George, R.; Turner, S.; Hawkesworth, C.; Morris, J.; Nye, C.; Ryan, J.; Zheng, S.H. Melting processes and fluid and sediment transport rates along the Alaska-Aleutian arc from an integrated U-Th-Ra-Be isotope study. *J. Geophys. Res. Solid Earth* **2003**, *108*, 2252. [CrossRef]

19.   Plank, T. Constraints from thorium/lanthanum on sediment recycling at subduction zones and the evolution of the continents. *J. Pet.* **2005**, *46*, 921–944. [CrossRef]

20.   Singer, B.S.; Jicha, B.R.; Leeman, W.P.; Rogers, N.W.; Thirlwall, M.F.; Ryan, J.; Nicolaysen, K.E. Along-strike trace element and isotopic variation in Aleutian Island arc basalt: Subduction melts sediments and dehydrates serpentine. *J. Geophys. Res.* **2007**, *112*, B06206. [CrossRef]

21.   Brenan, J.M.; Ryerson, F.J.; Shaw, H.F. The role of aqueous fluids in the slab-to-mantle transfer of boron, beryllium, and lithium during subduction: Experiments and models. *Geochim. Cosmochim. Acta* **1998**, *62*, 3337–3347. [CrossRef]

22.   Chan, L.; Edmond, J.; Thompson, G.; Gillis, K. Lithium isotopic composition of submarine basalts: Implications for the lithium cycle in the oceans. *Earth Planet. Sci. Lett.* **1992**, *108*, 151–160. [CrossRef]

23.   Teng, F.Z.; Rudnick, R.L.; McDonough, W.F.; Gao, S.; Tomascak, P.B.; Liu, Y. Lithium isotopic composition and concentration of the deep continental crust. *Chem. Geol.* **2008**, *255*, 47–59. [CrossRef]

24.   Wunder, B.; Meixner, A.; Romer, R.L.; Heinrich, W. Temperature-dependent isotopic fractionation of lithium between clinopyroxene and high-pressure hydrous fluids. *Contrib Miner. Pet.* **2006**, *151*, 112–120. [CrossRef]

25.   Bouman, C.; Elliott, T.; Vroon, P.Z. Lithium inputs to subduction zones. *Chem. Geol.* **2004**, *212*, 59–79. [CrossRef]

26.   Chan, L.H.; Alt, J.C.; Teagle, D.A.H. Lithium and lithium isotope profiles through the upper oceanic crust: A study of seawater–basalt exchange at ODP Sites 504B and 896A. *Earth Planet. Sci. Lett.* **2002**, *201*, 187–201. [CrossRef]

27.   Chan, L.H.; Leeman, W.P.; Plank, T. Lithium isotopic composition of marine sediments. *Geochem. Geophys Geosyst* **2006**, *7*, Q06005. [CrossRef]

28.   Zack, T.; Tomascak, P.B.; Rudnick, R.L.; Dalpé, C.; McDonough, W.F. Extremely light Li in orogenic eclogites: The role of isotope fractionation during dehydration in subducted oceanic crust. *Earth Planet. Sci. Lett.* **2003**, *208*, 279–290. [CrossRef]

29.   Marschall, H.R.; von Strandmann, P.A.E.P.; Seitz, H.M.; Elliott, T.; Niu, Y. The lithium isotopic composition of orogenic eclogites and deep subducted slabs. *Earth Planet. Sci. Lett.* **2007**, *262*, 563–580. [CrossRef]

30.   Xiao, Y.; Hoefs, J.; Hou, Z.; Simon, K.; Zhang, Z. Fluid/rock interaction and mass transfer in continental subduction zones: Constraints from trace elements and isotopes (Li, B, O, Sr, Nd, Pb) in UHP rocks from the Chinese Continental Scientific Drilling Program, Sulu, East China. *Contrib. Miner. Pet.* **2011**, *162*, 797–819. [CrossRef]

31.   Xiao, Y.; Sun, H.; Gu, H.; Huang, J.; Li, W.; Liu, L. Fluid/melt in continental deep subduction zones: Compositions and related geochemical fractionations. *Sci. China Earth Sci.* **2015**, *58*, 1457–1476. [CrossRef]

32.   Ishikawa, T.; Nakamura, E. Origin of the slab component in arc lavas from across-arc variation of B and Pb isotopes. *Nature* **1994**, *370*, 205–208. [CrossRef]

33. Moriguti, T.; Nakamura, E. Across-arc variation of Li isotopes in lavas and implications for crust/mantle recycling at subduction zones. *Earth Planet. Sci. Lett.* **1998**, *163*, 167–174. [CrossRef]

34. Nakamura, E.; Campbell, I.H.; Sun, S.S. The influence of subduction processes on the geochemistry of Japanese alkaline basalts. *Nature* **1985**, *316*, 55–58. [CrossRef]

35. Benton, L.D.; Ryan, J.G.; Savov, I.P. Lithium abundance and isotope systematics of forearc serpentinites, Conical Seamount, Mariana forearc: Insights into the mechanics of slab-mantle exchange during subduction. *Geochem. Geophys. Geosyst.* **2004**, *5*, Q08J12. [CrossRef]

36. Honma, H.; Kusakabe, M.; Kagami, H.; Iizumi, S.; Sakai, H.; Kodama, Y.; Kimura, M. Major and trace element chemistry and D/H, $^{18}O/^{16}O$, $^{87}Sr/^{86}Sr$ and $^{143}Nd/^{144}Nd$ ratios of rocks from the spreading center of the Okinawa Trough, a marginal back-arc basin. *Geochem. J.* **1991**, *25*, 121–136. [CrossRef]

37. Macpherson, C.G.; Hilton, D.R.; Mattey, D.P.; Sinton, J.M. Evidence for an ^{18}O-depleted mantle plume from contrasting $^{18}O/^{16}O$ ratios of back-arc lavas from the Manus Basin and Mariana Trough. *Earth Planet. Sci. Lett.* **2000**, *176*, 171–183. [CrossRef]

38. Macpherson, C.G.; Mattey, D.P. Oxygen isotope variations in Lau Basin lavas. *Chem. Geol.* **1998**, *144*, 177–194. [CrossRef]

39. Huang, J.; Ke, S.; Gao, Y.; Xiao, Y.; Li, S. Magnesium isotopic compositions of altered oceanic basalts and gabbros from IODP site 1256 at the East Pacific Rise. *Lithos* **2015**, *231*, 53–61. [CrossRef]

40. Li, S.G.; Yang, W.; Ke, S.; Meng, X.; Tian, H.; Xu, L.; He, Y.; Huang, J.; Wang, X.C.; Xia, Q.; et al. Deep carbon cycles constrained by a large-scale mantle Mg isotope anomaly in eastern China. *Natl. Sci. Rev.* **2017**, *4*, 111–120. [CrossRef]

41. Wang, S.J.; Teng, F.Z.; Li, S.G.; Hong, J.A. Magnesium isotopic systematics of mafic rocks during continental subduction. *Geochim Cosmochim Acta* **2014**, *143*, 34–48. [CrossRef]

42. Wang, S.J.; Teng, F.Z.; Williams, H.M.; Li, S.G. Magnesium isotopic variations in cratonic eclogites: Origins and implications. *Earth Planet. Sci. Lett.* **2012**, *359-360*, 219–226. [CrossRef]

43. Xiao, Y.; Teng, F.Z.; Zhang, H.F.; Yang, W. Large magnesium isotope fractionation in peridotite xenoliths from eastern North China craton: Product of melt–rock interaction. *Geochim Cosmochim Acta* **2013**, *115*, 241–261. [CrossRef]

44. Yang, W.; Teng, F.Z.; Zhang, H.F.; Li, S.G. Magnesium isotopic systematics of continental basalts from the North China craton: Implications for tracing subducted carbonate in the mantle. *Chem. Geol.* **2012**, *328*, 185–194. [CrossRef]

45. Shimoda, G.; Tatsumi, Y.; Nohda, S.; Ishizaka, K.; Jahn, B.M. Setouchi high-Mg andesites revisited: Geochemical evidence for melting of subducting sediments. *Earth Planet. Sci. Lett.* **1998**, *160*, 479–492. [CrossRef]

46. Shu, Y.; Nielsen, S.G.; Zeng, Z.; Shinjo, R.; Blusztajn, J.; Wang, X.; Chen, S. Tracing subducted sediment inputs to the Ryukyu arc-Okinawa Trough system: Evidence from thallium isotopes. *Geochim Cosmochim Acta* **2017**, *217*, 462–491. [CrossRef]

47. Guo, K.; Zeng, Z.G.; Chen, S.; Zhang, Y.X.; Qi, H.Y.; Ma, Y. The influence of a subduction component on magmatism in the Okinawa Trough: Evidence from thorium and related trace element ratios. *J. Asian Earth Sci.* **2017**, *145*, 205–216. [CrossRef]

48. Li, Z.G.; Chu, F.Y.; Dong, Y.H.; Liu, J.Q.; Chen, L. Geochemical constraints on the contribution of Louisville seamount materials to magmagenesis in the Lau back-arc basin, SW Pacific. *Int. Geol. Rev.* **2015**, *57*, 978–997. [CrossRef]

49. Pi, J.L.; You, C.F.; Wang, K.L. The influence of Ryukyu subduction on magma genesis in the Northern Taiwan Volcanic Zone and Middle Okinawa Trough—Evidence from boron isotopes. *Lithos* **2016**, *260*, 242–252. [CrossRef]

50. Shinjo, R.; Chung, S.L.; Kato, Y.; Kimura, M. Geochemical and Sr-Nd isotopic characteristics of volcanic rocks from the Okinawa Trough and Ryukyu Arc: Implications for the evolution of a young, intracontinental back arc basin. *J. Geophys Res. Solid Earth* **1999**, *104*, 10591–10608. [CrossRef]

51. Chang-Hwa, C.; Typhoon, L.; Yuch-Ning, S.; Cheng-Hong, C.; Wen-Yu, H. Magmatism at the onset of back-arc basin spreading in the Okinawa Trough. *J. Volcanol. Geotherm. Res.* **1995**, *69*, 313–322. [CrossRef]

52. Zengqian, H.; Zaw, K.; Yanhe, L.; Qiling, Z.; Zhigang, Z.; Urabe, T. Contribution of magmatic fluid to the active hydrothermal system in the JADE Field, Okinawa trough: Evidence from fluid inclusions, oxygen and helium isotopes. *Int. Geol. Rev.* **2010**, *47*, 420–437. [CrossRef]

53. Zeng, Z.; Yu, S.; Wang, X.; Fu, Y.; Yin, X.; Zhang, G.; Wang, X.; Chen, S. Geochemical and isotopic characteristics of volcanic rocks from the northern East China Sea shelf margin and the Okinawa Trough. *Acta Oceanol. Sin.* **2010**, *29*, 48–61. [CrossRef]

54. Lee, C.S.; Shor, G.G.; Bibee, L.D.; Lu, R.S.; Hilde, T.W.C. Okinawa Trough: Origin of a back-arc basin. *Mar. Geol.* **1980**, *35*, 219–241. [CrossRef]

55. Seno, T.; Maruyama, S. Paleogeographic reconstruction and origin of the Philippine Sea. *Tectonophysics* **1984**, *102*, 53–84. [CrossRef]

56. Seno, T.; Stein, S.; Gripp, A.E. A model for the motion of the Philippine Sea Plate consistent with NUVEL-1 and geological data. *J. Geophys. Res. Solid Earth* **1993**, *98*, 17941–17948. [CrossRef]

57. Hoang, N.; Uto, K. Upper mantle isotopic components beneath the Ryukyu arc system: Evidence for 'back-arc' entrapment of Pacific MORB mantle. *Earth Planet. Sci. Lett.* **2006**, *249*, 229–240. [CrossRef]

58. Ishizuka, H.; Kawanobe, Y.; Sakai, H. Petrology and geochemistry of volcanic rocks dredged from the Okinawa Trough, an active back-arc basin. *Geochem. J.* **1990**, *24*, 75–92. [CrossRef]

59. Han, B.; Zhang, X.H.; Pei, J.X.; Zhang, W.G. Characteristics of crust-mantle in the East China Sea and adjacent regions. *Prog. Geophys.* **2007**, *22*, 376–382.

60. Iwasaki, T.; Hirata, N.; Kanazawa, T.; Melles, J.; Suyehiro, K.; Urabe, T.; Möller, L.; Makris, J.; Shimamura, H. Crustal and upper mantle structure in the Ryukyu Island Arc deduced from deep seismic sounding. *Geophys. J. Int.* **1990**, *102*, 631–651. [CrossRef]

61. Klingelhoefer, F.; Lee, C.S.; Lin, J.Y.; Sibuet, J.C. Structure of the southernmost Okinawa Trough from reflection and wide-angle seismic data. *Tectonophys* **2009**, *466*, 281–288. [CrossRef]
62. Liu, B.; Li, S.Z.; Suo, Y.H.; Li, G.X.; Dai, L.M.; Somerville, I.D.; Guo, L.L.; Zhao, S.J.; Yu, S. The geological nature and geodynamics of the Okinawa Trough, Western Pacific. *Geol. J.* **2016**, *51*, 416–428. [CrossRef]
63. Nakahigashi, K.; Shinohara, M.; Suzuki, S.; Hino, R.; Shiobara, H.; Takenaka, H.; Nishino, M.; Sato, T.; Yoneshima, S.; Kanazawa, T. Seismic structure of the crust and uppermost mantle in the incipient stage of back arc rifting—Northernmost Okinawa Trough. *Geophys Res. Lett.* **2004**, *31*, L02614. [CrossRef]
64. Sibuet, J.C.; Hsu, S.K.; Shyu, C.T.; Liu, C.S. Structural and kinematic evolutions of the Okinawa Trough backarc basin. In *Backarc Basins: Tectonics and Magmatism*; Taylor, B., Ed.; Springer: Boston, MA, USA, 1995; pp. 343–379.
65. Arai, R.; Kodaira, S.; Yuka, K.; Takahashi, T.; Miura, S.; Kaneda, Y. Crustal structure of the southern Okinawa Trough: Symmetrical rifting, submarine volcano, and potential mantle accretion in the continental back-arc basin. *J. Geophys Res. Solid Earth* **2017**, *122*, 622–641. [CrossRef]
66. Furukawa, M.; Tokuyama, H.; Abe, S.; Nishizawa, A.; Kinoshita, H. Report on DELP 1988 cruises in the Okinawa Trough: Part 2. seismic reflection studies in the southwestern part of the Okinawa Trough. *Bull. Earthq Res. Inst. Univ. Tokyo* **1991**, *66*, 17–36.
67. Letouzey, J.; Kimura, M. Okinawa Trough genesis: Structure and evolution of a backarc basin developed in a continent. *Mar. Pet. Geol.* **1985**, *2*, 111–130. [CrossRef]
68. Sibuet, J.C.; Deffontaines, B.; Hsu, S.K.; Thareau, N.; Le Formal, J.P.; Liu, C.S. Okinawa trough backarc basin: Early tectonic and magmatic evolution. *J. Geophys Res. Solid Earth* **1998**, *103*, 30245–30267. [CrossRef]
69. Argus, D.F.; Gordon, R.G.; DeMets, C. Geologically current motion of 56 plates relative to the no-net-rotation reference frame. *Geochem. Geophys Geosyst* **2011**, *12*, Q11001. [CrossRef]
70. Kubo, A.; Fukuyama, E. Stress field along the Ryukyu Arc and the Okinawa Trough inferred from moment tensors of shallow earthquakes. *Earth Planet. Sci. Lett.* **2003**, *210*, 305–316. [CrossRef]
71. Pezzopane, S.K.; Wesnousky, S.G. Large earthquakes and crustal deformation near Taiwan. *J. Geophys Res.* **1989**, *94*, 7250–7264. [CrossRef]
72. Li, W.; Yang, Z.; Wang, Y.; Zhang, B. The petrochemical features of the volcanic rocks in the Okinawa Trough and their geological significance. *Acta Petrol. Sin.* **1997**, *13*, 538–550.
73. Shinjo, R.; Kato, Y. Geochemical constraints on the origin of bimodal magmatism at the Okinawa Trough, an incipient back-arc basin. *Lithos* **2000**, *54*, 117–137. [CrossRef]
74. Li, X.; Zeng, Z.; Chen, S.; Ma, Y.; Yang, H.; Zhang, Y.; Chen, Z. Geochemical and Sr-Nd-Pb isotopic compositions of volcanic rocks from the Iheya Ridge, the middle Okinawa Trough: Implications for petrogenesis and a mantle source. *Acta Oceanol Sin.* **2018**, *37*, 73–88. [CrossRef]
75. Chen, Z.; Zeng, Z.; Wang, X.; Zhang, Y.; Yin, X.; Chen, S.; Ma, Y.; Li, X.; Qi, H. Mineral chemistry indicates the petrogenesis of rhyolite from the southwestern Okinawa Trough. *J. Ocean. Univ. China* **2017**, *16*, 1097–1108. [CrossRef]
76. Ishikawa, M.; Sato, H.; Furukawa, M.; Kimura, M.; Kato, Y.; Tsugaru, R.; Shimamura, K. Report on DELP 1988 cruises in the Okinawa Trough: Part 6. petrology of volcanic rocks. *Bull. Earthq. Res. Inst. Univ. Tokyo* **1991**, *66*, 151–177.
77. Gao, Y.; Casey, J.F. Lithium isotope composition of ultramafic geological reference materials JP-1 and DTS-2. *Geostand Geoanal Res.* **2012**, *36*, 75–81. [CrossRef]
78. Rudnick, R.L.; Tomascak, P.B.; Njo, H.B.; Gardner, L.R. Extreme lithium isotopic fractionation during continental weathering revealed in saprolites from South Carolina. *Chem. Geol.* **2004**, *212*, 45–57. [CrossRef]
79. Flesch, G.D.; Anderson, A.R.; Svec, H.J. A secondary isotopic standard for $^6Li/^7Li$ determinations. *Int. J. Mass Spectrom. Ion. Phys.* **1973**, *12*, 265–272. [CrossRef]
80. Zheng, Y.F.; Wang, Z.R.; Li, S.G.; Zhao, Z.F. Oxygen isotope equilibrium between eclogite minerals and its constraints on mineral Sm-Nd chronometer. *Geochim. Cosmochim. Acta* **2002**, *66*, 625–634. [CrossRef]
81. Zheng, Y.F.; Fu, B.; Li, Y.; Xiao, Y.; Li, S. Oxygen and hydrogen isotope geochemistry of ultrahigh-pressure eclogites from the Dabie Mountains and the Sulu terrane. *Earth Planet. Sci. Lett.* **1998**, *155*, 113–129. [CrossRef]
82. Gong, B.; Zheng, Y.F.; Chen, R.X. TC/EA-MS online determination of hydrogen isotope composition and water concentration in eclogitic garnet. *Phys. Chem. Min.* **2007**, *34*, 687–698. [CrossRef]
83. An, Y.; Wu, F.; Xiang, Y.; Nan, X.; Yu, X.; Yang, J.; Yu, H.; Xie, L.; Huang, F. High-precision Mg isotope analyses of low-Mg rocks by MC-ICP-MS. *Chem. Geol.* **2014**, *390*, 9–21. [CrossRef]
84. Putirka, K.; Condit, C.D. Cross section of a magma conduit system at the margin of the Colorado Plateau. *Geology* **2003**, *31*, 701–704. [CrossRef]
85. Putirka, K.D. Thermometers and barometers for volcanic systems. *Rev. Mineral. Geochem.* **2008**, *69*, 61–120. [CrossRef]
86. Irvine, T.N.; Baragar, W.R.A. A guide to the chemical classification of the common volcanic rocks. *Can. J. Earth Sci.* **1971**, *8*, 523–548. [CrossRef]
87. Le Maitre, R.W. *A classification of Igneous Rocks and Glossary of Terms: Recommendations of the International Union of Geological Sci.ences Subcommission on the Systematics of Igneous Rocks*; Blackwell: Oxford, UK, 1989.
88. Roberts, M.P.; Clemens, J.D. Origin of high-potassium, talc-alkaline, I-type granitoids. *Geology* **1993**, *21*, 825–828. [CrossRef]
89. Hu, Y.; Niu, Y.; Li, J.; Ye, L.; Kong, J.; Chen, S.; Zhang, Y.; Zhang, G. Petrogenesis and tectonic significance of the late Triassic mafic dikes and felsic volcanic rocks in the East Kunlun Orogenic Belt, Northern Tibet Plateau. *Lithos* **2016**, *245*, 205–222. [CrossRef]

90. Sun, S.S.; McDonough, W.F. Chemical and isotopic systematics of oceanic basalts: Implications for mantle composition and processes. *Geol. Soc. Lond. Spec. Publ.* **1989**, *42*, 313–345. [CrossRef]
91. Elliott, T.; Thomas, A.; Jeffcoate, A.; Niu, Y. Lithium isotope evidence for subduction-enriched mantle in the source of mid-ocean-ridge basalts. *Nature* **2006**, *443*, 565–568. [CrossRef]
92. Nishio, Y.; Nakai, S.i.; Ishii, T.; Sano, Y. Isotope systematics of Li, Sr, Nd, and volatiles in Indian Ocean MORBs of the Rodrigues Triple Junction: Constraints on the origin of the DUPAL anomaly. *Geochim. Cosmochim. Acta* **2007**, *71*, 745–759. [CrossRef]
93. Tomascak, P.B.; Langmuir, C.H.; le Roux, P.J.; Shirey, S.B. Lithium isotopes in global mid-ocean ridge basalts. *Geochim. Cosmochim. Acta* **2008**, *72*, 1626–1637. [CrossRef]
94. Brant, C.; Coogan, L.A.; Gillis, K.M.; Seyfried, W.E.; Pester, N.J.; Spence, J. Lithium and Li-isotopes in young altered upper oceanic crust from the East Pacific Rise. *Geochim. Cosmochim. Acta* **2012**, *96*, 272–293. [CrossRef]
95. James, R.H.; Rudnicki, M.D.; Palmer, M.R. The alkali element and boron geochemistry of the Escanaba Trough sediment-hosted hydrothermal system. *Earth Planet. Sci. Lett.* **1999**, *171*, 157–169. [CrossRef]
96. Leeman, W.P.; Tonarini, S.; Chan, L.H.; Borg, L.E. Boron and lithium isotopic variations in a hot subduction zone—the southern Washington Cascades. *Chem. Geol.* **2004**, *212*, 101–124. [CrossRef]
97. Zhang, L.; Chan, L.H.; Gieskes, J.M. Lithium isotope geochemistry of pore waters from ocean drilling program Sites 918 and 919, Irminger Basin. *Geochim. Cosmochim. Acta* **1998**, *62*, 2437–2450. [CrossRef]
98. Teng, F.Z.; McDonough, W.F.; Rudnick, R.L.; Dalpé, C.; Tomascak, P.B.; Chappell, B.W.; Gao, S. Lithium isotopic composition and concentration of the upper continental crust. *Geochim. Cosmochim. Acta* **2004**, *68*, 4167–4178. [CrossRef]
99. Cooper, K.M.; Eiler, J.M.; Sims, K.W.W.; Langmuir, C.H. Distribution of recycled crust within the upper mantle: Insights from the oxygen isotope composition of MORB from the Australian-Antarctic Discordance. *Geochem. Geophys. Geosyst.* **2009**, *10*, Q12004. [CrossRef]
100. Eiler, J.M.; Schiano, P.; Kitchen, N.; Stolper, E.M. Oxygen-isotope evidence for recycled crust in the sources of mid-ocean-ridge basalts. *Nature* **2000**, *403*, 530–534. [CrossRef]
101. Ito, E.; White, W.M.; Göpel, C. The O, Sr, Nd and Pb isotope geochemistry of MORB. *Chem. Geol.* **1987**, *62*, 157–176. [CrossRef]
102. Gregory, R.T.; Taylor, H.P. An oxygen isotope profile in a section of Cretaceous oceanic crust, Samail Ophiolite, Oman: Evidence for $\delta 18O$ buffering of the oceans by deep (>5 km) seawater-hydrothermal circulation at mid-ocean ridges. *J. Geophys Res. Solid Earth* **1981**, *86*, 2737–2755. [CrossRef]
103. Staudigel, H.; Davies, G.R.; Hart, S.R.; Marchant, K.M.; Smith, B.M. Large scale isotopic Sr, Nd and O isotopic anatomy of altered oceanic crust: DSDP/ODP sites417/418. *Earth Planet. Sci. Lett.* **1995**, *130*, 169–185. [CrossRef]
104. Alt, J.; Shanksiii, W. Stable isotope compositions of serpentinite seamounts in the Mariana forearc: Serpentinization processes, fluid sources and sulfur metasomatism. *Earth Planet. Sci. Lett.* **2006**, *242*, 272–285. [CrossRef]
105. Simon, L.; Lécuyer, C. Continental recycling: The oxygen isotope point of view. *Geochem. Geophys Geosyst.* **2005**, *6*, Q08004. [CrossRef]
106. Bourdon, B.; Tipper, E.T.; Fitoussi, C.; Stracke, A. Chondritic Mg isotope composition of the Earth. *Geochim. Cosmochim. Acta* **2010**, *74*, 5069–5083. [CrossRef]
107. Teng, F.Z.; Li, W.Y.; Ke, S.; Marty, B.; Dauphas, N.; Huang, S.; Wu, F.Y.; Pourmand, A. Magnesium isotopic composition of the Earth and chondrites. *Geochim. Cosmochim. Acta* **2010**, *74*, 4150–4166. [CrossRef]
108. Teng, F.Z.; Hu, Y.; Chauvel, C. Magnesium isotope geochemistry in arc volcanism. *Proc. Natl. Acad. Sci. USA* **2016**, *113*, 7082–7087. [CrossRef]
109. Li, W.Y.; Teng, F.Z.; Ke, S.; Rudnick, R.L.; Gao, S.; Wu, F.Y.; Chappell, B.W. Heterogeneous magnesium isotopic composition of the upper continental crust. *Geochim. Cosmochim. Acta* **2010**, *74*, 6867–6884. [CrossRef]
110. Eiler, J.M. Oxygen isotope variations of basaltic lavas and upper mantle rocks. *Rev. Miner. Geochem.* **2001**, *43*, 319–364. [CrossRef]
111. Mattey, D.; Lowry, D.; Macpherson, C. Oxygen isotope composition of mantle peridotite. *Earth Planet. Sci. Lett.* **1994**, *128*, 231–241. [CrossRef]
112. Beattie, P. Olivine-melt and orthopyroxene-melt equilibria. *Contrib. Miner. Pet.* **1993**, *115*, 103–111. [CrossRef]
113. Davis, B.T.C.; Boyd, F.R. The join Mg2Si2O6-CaMgSi2O6at 30 kilobars pressure and its application to pyroxenes from kimberlites. *J. Geophys Res.* **1966**, *71*, 3567–3576. [CrossRef]
114. Rudnick, R.L.; Gao, S. Composition of the continental crust. In *Treatise on Geochemistry*; Holland, H.D., Turekian, K.K., Eds.; Elsevier: Amsterdam, The Netherlands, 2003; Volume 3, pp. 1–64.
115. Park, S.H.; Lee, S.M.; Kamenov, G.D.; Kwon, S.T.; Lee, K.Y. Tracing the origin of subduction components beneath the South East rift in the Manus Basin, Papua New Guinea. *Chem. Geol.* **2010**, *269*, 339–349. [CrossRef]
116. Hong, L.B.; Zhang, Y.H.; Qian, S.P.; Liu, J.Q.; Ren, Z.Y.; Xu, Y.G. Constraints from melt inclusions and their host olivines on the petrogenesis of Oligocene-Early Miocene Xindian basalts, Chifeng area, North China Craton. *Contrib. Miner. Pet.* **2013**, *165*, 305–326. [CrossRef]
117. Sun, H.; Xiao, Y.; Gao, Y.; Lai, J.; Hou, Z.; Wang, Y. Fluid and melt inclusions in the Mesozoic Fangcheng basalt from North China Craton: Implications for magma evolution and fluid/melt-peridotite reaction. *Contrib. Miner. Pet.* **2013**, *165*, 885–901. [CrossRef]
118. Zhang, Y.; Zeng, Z.; Li, X.; Yin, X.; Wang, X.; Chen, S.; Li, S. High-potassium volcanic rocks from the Okinawa Trough: Implications for a cryptic potassium-rich and DUPAL-like source. *Geol. J.* **2018**, *53*, 1755–1766. [CrossRef]

119. Prægel, N.O.; Holm, P.M. Lithospheric contributions to high-MgO basanites from the Cumbre Vieja Volcano, La Palma, Canary Islands and evidence for temporal variation in plume influence. *J. Volcanol. Geotherm. Res.* **2006**, *149*, 213–239. [CrossRef]

120. Zhang, H.F.; Sun, M. Geochemistry of Mesozoic basalts and mafic dikes, southeastern North China Craton, and tectonic implications. *Int. Geol. Rev.* **2002**, *44*, 370–382. [CrossRef]

121. Zhang, J.J.; Zheng, Y.F.; Zhao, Z.F. Geochemical evidence for interaction between oceanic crust and lithospheric mantle in the origin of Cenozoic continental basalts in east-central China. *Lithos* **2009**, *110*, 305–326. [CrossRef]

122. Liu, Y.; Gao, S.; Gao, C.; Zong, K.; Hu, Z.; Ling, W. Garnet-rich granulite xenoliths from the Hannuoba basalts, North China: Petrogenesis and implications for the Mesozoic crust-mantle interaction. *J. Earth Sci.* **2010**, *21*, 669–691. [CrossRef]

123. Moriguti, T.; Shibata, T.; Nakamura, E. Lithium, boron and lead isotope and trace element systematics of Quaternary basaltic volcanic rocks in northeastern Japan: Mineralogical controls on slab-derived fluid composition. *Chem. Geol.* **2004**, *212*, 81–100. [CrossRef]

124. Tomascak, P.B.; Widom, E.; Benton, L.D.; Goldstein, S.L.; Ryan, J.G. The control of lithium budgets in island arcs. *Earth Planet. Sci. Lett.* **2002**, *196*, 227–238. [CrossRef]

125. Grove, T.; Parman, S.; Bowring, S.; Price, R.; Baker, M. The role of an H$_2$O-rich fluid component in the generation of primitive basaltic andesites and andesites from the Mt. Shasta region, N California. *Contrib. Mineral. Petrol.* **2002**, *142*, 375–396. [CrossRef]

126. Handley, H.K.; Macpherson, C.G.; Davidson, J.P.; Berlo, K.; Lowry, D. Constraining fluid and sediment contributions to subduction-related magmatism in Indonesia: Ijen volcanic complex. *J. Pet.* **2007**, *48*, 1155–1183. [CrossRef]

127. Magna, T.; Wiechert, U.; Grove, T.L.; Halliday, A.N. Lithium isotope fractionation in the southern Cascadia subduction zone. *Earth Planet. Sci. Lett.* **2006**, *250*, 428–443. [CrossRef]

128. Decitre, S.; Deloule, E.; Reisberg, L.; James, R.; Agrinier, P.; Mével, C. Behavior of Li and its isotopes during serpentinization of oceanic peridotites. *Geochem. Geophys Geosyst* **2002**, *3*, 1–20. [CrossRef]

129. Lui-Heung, C.; Gieskes, J.M.; Chen-Feng, Y.; Edmond, J.M. Lithium isotope geochemistry of sediments and hydrothermal fluids of the Guaymas Basin, Gulf of California. *Geochim. Cosmochim. Acta* **1994**, *58*, 4443–4454. [CrossRef]

130. Tomascak, P.B.; Carlson, R.W.; Shirey, S.B. Accurate and precise determination of Li isotopic compositions by multi-collector sector ICP-MS. *Chem. Geol.* **1999**, *158*, 145–154. [CrossRef]

131. Chan, L.H.; Frey, F.A. Lithium isotope geochemistry of the Hawaiian plume: Results from the Hawaii Scientific Drilling Project and Koolau Volcano. *Geochem. Geophys. Geosyst.* **2003**, *4*, 8707. [CrossRef]

132. Ryan, J.G.; Langmuir, C.H. The systematics of lithium abundances in young volcanic rocks. *Geochim. Cosmochim. Acta* **1987**, *51*, 1727–1741. [CrossRef]

133. Bindeman, I.N.; Ponomareva, V.V.; Bailey, J.C.; Valley, J.W. Volcanic arc of Kamchatka: A province with high-δ^{18}O magma sources and large-scale ^{18}O/^{16}O depletion of the upper crust. *Geochim. Cosmochim. Acta* **2004**, *68*, 841–865. [CrossRef]

134. Bindeman, I. Oxygen isotopes in mantle and crustal magmas as revealed by single crystal analysis. *Rev. Miner. Geochem.* **2008**, *69*, 445–478. [CrossRef]

135. Auer, S.; Bindeman, I.; Wallace, P.; Ponomareva, V.; Portnyagin, M. The origin of hydrous, high-δ^{18}O voluminous volcanism: Diverse oxygen isotope values and high magmatic water contents within the volcanic record of Klyuchevskoy volcano, Kamchatka, Russia. *Contrib Miner. Pet.* **2009**, *157*, 209–230. [CrossRef]

136. Dorendorf, F.; Wiechert, U.; Wörner, G. Hydrated sub-arc mantle: A source for the Kluchevskoy volcano, Kamchatka/Russia. *Earth Planet. Sci. Lett.* **2000**, *175*, 69–86. [CrossRef]

137. Perkins, G.B.; Sharp, Z.D.; Selverstone, J. Oxygen isotope evidence for subduction and rift-related mantle metasomatism beneath the Colorado Plateau–Rio Grande rift transition. *Contrib. Miner. Pet.* **2006**, *151*, 633–650. [CrossRef]

138. Harmon, R.S.; Hoefs, J. Oxygen isotope heterogeneity of the mantle deduced from global 18 O systematics of basalts from different geotectonic settings. *Contrib Miner. Pet.* **1995**, *120*, 95–114. [CrossRef]

139. Wang, Z.; Eiler, J. Insights into the origin of low-δ^{18}O basaltic magmas in Hawaii revealed from in situ measurements of oxygen isotope compositions of olivines. *Earth Planet. Sci. Lett.* **2008**, *269*, 377–387. [CrossRef]

140. Wei, C.S.; Zheng, Y.F.; Zhao, Z.F.; Valley, J.W. Oxygen and neodymium isotope evidence for recycling of juvenile crust in northeast China. *Geology* **2002**, *30*, 375–378. [CrossRef]

141. Muehlenbachs, K. Alteration of the oceanic crust and the ^{18}O history of seawater. In *Stable Isotopes in High Temperature Geological Processes*; Valley, J.W., Taylor, H.P., O'Neil, J.R., Eds.; De Gruyter: Berlin, Germany, 1986; pp. 125–444.

142. Cooper, K.M.; Eiler, J.M.; Asimow, P.D.; Langmuir, C.H. Oxygen isotope evidence for the origin of enriched mantle beneath the mid Atlantic ridge. *Earth Planet. Sci. Lett.* **2004**, *220*, 297–316. [CrossRef]

143. Demény, A.; Vennemann, T.W.; Hegner, E.; Nagy, G.; Milton, J.A.; Embey-Isztin, A.; Homonnay, Z.; Dobosi, G. Trace element and C–O–Sr–Nd isotope evidence for subduction-related carbonate–silicate melts in mantle xenoliths (Pannonian Basin, Hungary). *Lithos* **2004**, *75*, 89–113. [CrossRef]

144. Eiler, J.M.; Farley, K.A.; Valley, J.W.; Hauri, E.; Craig, H.; Hart, S.R.; Stolper, E.M. Oxygen isotope variations in ocean island basalt phenocrysts. *Geochim. Cosmochim. Acta* **1997**, *61*, 2281–2293. [CrossRef]

145. Xia, Q.K.; Dallai, L.; Deloule, E. Oxygen and hydrogen isotope heterogeneity of clinopyroxene megacrysts from Nushan Volcano, SE China. *Chem. Geol.* **2004**, *209*, 137–151. [CrossRef]

146. Liu, P.P.; Teng, F.Z.; Dick, H.J.B.; Zhou, M.F.; Chung, S.L. Magnesium isotopic composition of the oceanic mantle and oceanic Mg cycling. *Geochim. Cosmochim. Acta* **2017**, *206*, 151–165. [CrossRef]

147. Handler, M.R.; Baker, J.A.; Schiller, M.; Bennett, V.C.; Yaxley, G.M. Magnesium stable isotope composition of Earth's upper mantle. *Earth Planet. Sci. Lett.* **2009**, *282*, 306–313. [CrossRef]

148. Ionov, D.A. Petrology of mantle wedge lithosphere: New data on supra-subduction zone peridotite xenoliths from the andesitic Avacha volcano, Kamchatka. *J. Pet.* **2010**, *51*, 327–361. [CrossRef]

149. Teng, F.Z.; Wadhwa, M.; Helz, R.T. Investigation of magnesium isotope fractionation during basalt differentiation: Implications for a chondritic composition of the terrestrial mantle. *Earth Planet. Sci. Lett.* **2007**, *261*, 84–92. [CrossRef]

150. Yang, W.; Teng, F.Z.; Zhang, H.F. Chondritic magnesium isotopic composition of the terrestrial mantle: A case study of peridotite xenoliths from the North China craton. *Earth Planet. Sci. Lett.* **2009**, *288*, 475–482. [CrossRef]

151. Wimpenny, J.; Yin, Q.Z.; Tollstrup, D.; Xie, L.W.; Sun, J. Using Mg isotope ratios to trace Cenozoic weathering changes: A case study from the Chinese Loess Plateau. *Chem. Geol.* **2014**, *376*, 31–43. [CrossRef]

152. Li, W.Y.; Teng, F.Z.; Xiao, Y.; Huang, J. High-temperature inter-mineral magnesium isotope fractionation in eclogite from the Dabie orogen, China. *Earth Planet. Sci. Lett.* **2011**, *304*, 224–230. [CrossRef]

153. Von Strandmann, P.A.E.P.; Elliott, T.; Marschall, H.R.; Coath, C.; Lai, Y.J.; Jeffcoate, A.B.; Ionov, D.A. Variations of Li and Mg isotope ratios in bulk chondrites and mantle xenoliths. *Geochim. Cosmochim. Acta* **2011**, *75*, 5247–5268. [CrossRef]

154. Hacker, B.R. H_2O subduction beyond arcs. *Geochem. Geophys. Geosyst.* **2008**, *9*, Q03001. [CrossRef]

155. Magni, V.; Faccenna, C.; van Hunen, J.; Funiciello, F. How collision triggers backarc extension: Insight into Mediterranean style of extension from 3-D numerical models. *Geology* **2014**, *42*, 511–514. [CrossRef]

156. Peacock, S.M.; Wang, K. Seismic consequences of warm versus cool subduction metamorphism: Examples from southwest and northeast Japan. *Science* **1999**, *286*, 937–939. [CrossRef]

157. Van Keken, P.E.; Hacker, B.R.; Syracuse, E.M.; Abers, G.A. Subduction factory: 4. Depth-dependent flux of H_2O from subducting slabs worldwide. *J. Geophys. Res.* **2011**, *116*, B01401. [CrossRef]

158. Zheng, Y.F.; Chen, R.X.; Xu, Z.; Zhang, S.B. The transport of water in subduction zones. *Sci. China Earth Sci.* **2016**, *59*, 651–682. [CrossRef]

159. Ko, Y.T.; Kuo, B.Y.; Wang, K.L.; Lin, S.C.; Hung, S.H. The southwestern edge of the Ryukyu subduction zone: A high Q mantle wedge. *Earth Planet. Sci. Lett.* **2012**, *335–336*, 145–153. [CrossRef]

Journal of
Marine Science and Engineering

Article

Trace Element Evidence of Subduction-Modified Mantle Material in South Mid-Atlantic Ridge 18–21°S Upper Mantle

Tianxiao Ji [1,2] and Zhigang Zeng [1,2,3,*]

1   Seafloor Hydrothermal Activity Laboratory, CAS Key Laboratory of Marine Geology and Environment, Institute of Oceanology, Chinese Academy of Sciences, Qingdao 266071, China
2   University of Chinese Academy of Sciences, Beijing 100049, China
3   Laboratory for Marine Mineral Resources, Qingdao National Laboratory for Marine Science and Technology, Qingdao 266071, China
*   Correspondence: zgzeng@ms.qdio.ac.cn

Abstract: Mid-ocean ridge basalts (MORBs), produced at mid-ocean ridge where the continents and subduction zones are distant, are the product of partial melting of the upper mantle and their chemical composition can provide information about the mantle itself. The geochemical characteristics of MORBs enable us to be more informed about the geological processes of the upper mantle below the mid-ocean ridge, and assist us in understanding mantle heterogeneity and geodynamic processes. In this paper, new data of major elements, trace elements, and Nd-Hf isotopes of south mid-Atlantic ridge (SMAR) 18–21°S MORBs are presented. TAS diagram shows that the samples belong to subalkaline basalt compositional field. Trace elements (e.g., $(La/Sm)_N = 0.49$–0.79) show that the samples are N-MORBs. However, the primitive mantle-normalized trace element patterns showed that the studied samples were clearly enriched in Rb, U, Pb, and other fluid-mobile elements. Meanwhile, the trace element ratios, such as Nb/U and Ce/Pb, are also significantly different from the typical N-MORB. Combined with the Nd-Hf isotopic composition, we propose that these anomalies are not related to continental crust material, delaminated subcontinental lithospheric mantle (SCLM), recycled sediments, direct supply of mantle plume, nor are they the result of subduction directly affecting the mantle source, but are caused by the incorporation of mantle material modified by subduction.

Keywords: south mid-Atlantic ridge (SMAR); mid-ocean ridge basalt (MORB); subduction; mantle heterogeneity

Citation: Ji, T.; Zeng, Z. Trace Element Evidence of Subduction-Modified Mantle Material in South Mid-Atlantic Ridge 18–21°S Upper Mantle. *J. Mar. Sci. Eng.* **2023**, *11*, 441. https://doi.org/10.3390/jmse11020441

Academic Editor: Sergio Bonomo

Received: 30 December 2022
Revised: 17 January 2023
Accepted: 17 January 2023
Published: 17 February 2023

1. Introduction

Mantle is heterogeneous. Isotopically, the mantle can be divided into depleted MORB mantle (DMM), enriched mantle 1 and 2 (EM1 and EM2), high 'μ' = $^{238}U/^{204}Pb$ (HIMU), and other mantle end members [1,2]. Different plumes have different characteristics of enriched mantle end-members. For example, St. Helena plume is the representative of HIMU mantle end-members, and Tristan-Gough plume represents the EM1 mantle end-members [3,4]. OIB is the product of the mantle plume, and its geochemical composition can indicate the origin of the mantle end-member. For example, HIMU mantle end-members are usually hypothesized to be associated with the recycled oceanic crust [3]. MORBs are derived from the DMM, and are produced at the mid-ocean ridge [5] which is distant from continents and subduction zones. Therefore, the composition of MORBs is less contaminated by the continental material and subduction associated process, and is generally considered to be relatively uniform. However, increasing evidence [6–12] is showing that the mantle source of MORB is not as uniform as initially hypothesized and it is even heterogeneous at different scales. Dupre and Allegre [6] found that the Pb and Sr isotopic compositions of MORBs from the Indian Ocean are more radioactive than MORBs from the other ocean. Mougel et al. [12] sampled the East Pacific Rise (EPR) segment at a high density within the length of only 15 km, and found that the MORB source region was highly heterogeneous even at this small scale. Subduction of oceanic plate, detachment of SCLM, and mantle plume are the reasons of upper mantle heterogeneity [4,12–22]. Studying which process could cause upper mantle heterogeneity can assist us in understanding the dynamics of the Earth.

The more incompatible the elements are, the more inclined they are to enter the melt phase during melting and crystallization. Elements with similar compatibility (such as Ce and Pb, Nb and U, etc.) have little relative change in their contents in rocks due to their similar geochemical behavior in magmatic processes. Therefore, the ratio of these element pairs in rocks is basically unchanged compared with their mantle source, and can be used to reflect the chemical composition characteristics of the source area [13]. Nb/U and Ce/Pb ratios of MORBs are relatively uniform [13,23,24], and significantly higher than those of the continental material and island arc volcanics [13]. Therefore, the anomalies in the ratios of Nb/U and Ce/Pb in MORB can be used as indicators of contamination caused by recycled sediments, delaminated subcontinental lithospheric mantle, continental crust material, and subduction of oceanic plate.

Yang et al. [21] proposed a back-arc basin basalt (BABB) filter based on Nb/U, Ce/Pb, Rb/Nb, and Ba/Nb ratios. Using this filter, it was found that BABB-like MORBs are widespread in the Atlantic, Indian, and Arctic oceans, but rare in the Pacific. This compositional distribution is explained by the subduction shield model around the Pacific Ocean. Circum-Pacific subducted slabs are mostly continuous for at least the last 180 million years and at least to transition zone depth, and would strictly limit the contribution of the slab flux to sub-Pacific mantle [21]. In this model, the mid-Atlantic (from ~33°N to ~34°S) is hardly affected by subduction due to the presence of a slab window (the subduction of the Chile ridge). However, the slab window does not exist since the Pacific subduction began. Was the mid-Atlantic mantle really not affected by subduction?

In this study, we present new whole-rock major element, trace element, and Nd-Hf isotope data of SMAR 18–21°S MORBs, combined with published geochemical data, to investigate whether the area was affected by subduction.

2. Geological Background and Samples

The South Atlantic is located between Africa and South America (Figure 1), with Olavtoppen Island at the southernmost end and Romanche Trench at the northernmost end, which gradually opened from south to north during the breakup of west Gondwana from ~134 Ma [25,26]. There are several hotspots in the South Atlantic including Ascension, Circe, St. Helena, Tristan, Discovery, and Shona (yellow circles in Figure 1), among which St. Helena is closest to the study area. The St. Helena hotspot is linked by the St. Helena seamount chain to West Africa, but there is no corresponding continent flood basalt [27] (Figure 1). Geochemically, St. Helena is a mantle plume with typical HIMU mantle end-member characteristics. Mantle plumes in the South Atlantic have extensive effects on different ridge segments of SMAR [28–30], and are associated with the breakup of Gondwana [31].

During the China Ocean Survey (COS) Expeditions No. 22 and 26 cruises of the R/V DAYANGYIHAO, 16 MORBs from 14 stations in this study were collected by TV-grab (TVG) at depths of 2075–3386 m from the axis of the SMAR 18–21°S, which is located at the extension line of St. Helena seamount chain and 750 km from St. Helena Island. The main rock types of this area include N-MORB and E-MORB, but the radioisotope compositions, especially the $^{206}Pb/^{204}Pb$ ratio, are more radioactive than those of typical MORBs, indicating that these basalts are clearly affected by enriched components [26,32,33]. At present, the St. Helena plume is still affecting the upper mantle beneath the study area through fluid channels [26].

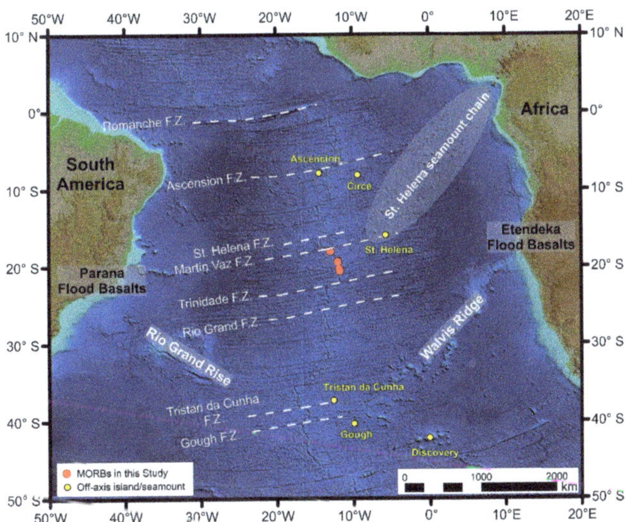

Figure 1. Geological background and the locations of study samples. Red circles represent new MORBs samples in this study and gray circles represent MORBs samples in previous studies. Off-axis islands and seamounts are shown by yellow circles. The cited island/seamount basalts data are downloaded from PetDB (http://www.earthchem.org/petdb; accessed on 31 May 2021).

3. Analytical Methods

3.1. Preparation

Prior to the geochemical analysis, samples were crushed into small chips less than 2 mm in size, where altered samples were removed under a binocular microscope. Then, the remaining rock fragments were placed in ultrapure water for ultrasonic cleaning with the water being changed every 15 min until it was clear after cleaning. The dried samples were placed in absolute ethanol for ultrasonic cleaning to remove the organic matter on the surface. After a second round of drying, the clean samples were handpicked again under a binocular microscope to avoid alteration. Finally, the clean rock fragments were ground to 200 mesh with an agate mortar for subsequent geochemical analysis.

3.2. Whole-Rock Major Element and Trace Element Analyses

Major element data for whole-rock samples were obtained by X-ray fluorescence (XRF) spectrometry on fused glass discs using a PANalytical AXIOS Minerals instrument at the Rock-Mineral Preparation and Analysis Lab, the Institute of Geology and Geophysics (IGG; Beijing, China), Chinese Academy of Sciences (CAS), following the procedures described similarly in [34]. Loss on ignition (LOI) was measured as the weight loss of the samples, which was obtained independently by igniting 0.5 g of dry sample aliquot in a porcelain crucible for 1.5 h at 1000 °C in a muffle furnace. Analysis results for the reference standard (GSR-3) for major elements are consistent with the recommended values within the analytical error, and detailed data are presented in Table S1.

Whole-rock trace element contents were analyzed using a PlasmaQuant-MS Elite ICP–MS instrument at the State Key Laboratory of Ore Deposit Geochemistry (SKLODG), Institute of Geochemistry, Chinese Academy of Sciences (IGCAS), China. The powdered samples (50 mg) were dissolved with a 1 mL HF + 1 mL HNO3 mixture in high-pressure Teflon bombs at ~185 °C for 35 h. Rh was used as an internal standard to monitor signal drift during counting. The analytical precision was generally better than 10%. Analysis results for three international reference standards (OU-6, AGV-2, and GBPG-1) for trace elements are consistent with the recommended values within the analytical error, and detailed data are presented in Table S1.

3.3. Nd-Hf Isotope Analyses

Nd isotope analysis was performed at the University of Science and Technology of China (USTC), Hefei, China, following the procedures described similarly in [35,36]. Whole-rock powders of approximately 100 mg were weighed and placed in 15 mL of Teflon stuffy tanks and dissolved in a mixture of 2–3 mL of purified HF solution and 8–10 drops of purified $HClO_4$ solution. Decomposition of refractory phases was ensured by heating the samples in a Teflon tank at 120 °C for approximately 7 days. After the samples were completely dissolved, the sample solutions were dried on a hot plate at 120 °C, and then heated to 150 °C to completely remove the HF and $HClO_4$. Next, 3 mL of purified 6 N HCl solution was added to the sample tanks two times to clear the inside of the tanks and then dried again. The sample residues were redissolved with 1 mL of purified 3 N HCl solution to prepare for chemical separation and purification. Light rare earth elements were isolated on quartz columns by conventional ion exchange chromatography with a 5-mL resin bed of Bio-Rad AG50W–X12, 200–400 mesh. Nd was separated from other rare earth elements on quartz columns using 1.7 mL of Teflon powder coated with HDEHP and di(2-ethylhexyl) orthophosphoric acid, as the cation exchange medium. All isotopic measurements were performed on a Finnigan MAT 262 mass spectrometer. Nd was loaded as phosphate on preconditioned Re filaments, and measurements were performed in a Re double filament configuration. The $^{143}Nd/^{144}Nd$ ratios were normalized to $^{146}Nd/^{144}Nd = 0.7219$.

The measurement accuracies of the $^{143}Nd/^{144}Nd$ ratios for the samples were better than 0.003%. Procedural blanks were <100 pg for Nd. The precision for all the measured isotopic ratios is given in 2σ uncertainty. Details of the analytical uncertainties are presented in Table S2. During the period of data collection, repeated measurements on the Jndi standard solutions gave average $^{143}Nd/^{144}Nd$ ratios of 0.512115 ±0.000006 (2σ, $n = 7$). The results of Nd isotopic analyses on the standard materials BCR-2 and AGV-2 (basalt powder) gave average $^{143}Nd/^{144}Nd$ ratios of 0.512635 ± 0.000009 (2σ, $n = 2$) and 0.512789 ± 0.000009 (2σ, $n = 2$).

Whole-rock Hf isotopic analysis was performed at Guizhou Tongwei Analytical Technology Co., Ltd., China. About 50–100 mg of rock powder samples were dissolved with a mixture of concentrate nitric acid and hydrofluoric acid in Teflon bombs at 185 °C in the oven for 3 days, and dried on a hot plate at 80 °C. Hf was initially washed from the column using a mixture of 0.2 N HBr + 0.5 N HNO_3 (collected) and was separated using a Bio-Rad AG50W–X8 cation exchange column (Bio-Rad Laboratories, Hercules, CA, USA). Hf was first washed using 1.5 N HCl (collected), then was collected on a column using HDEHP(di(2-ethylhexyl)-coated Teflon powder. The previously collected HFSE Hf was heated until dry and then redissolved in 3.0 N HCl. Finally, Hf was extracted from the column using 2.0 N HF and collected in a 10 mL PFA (preconditioned perfluoroalkoxy) beaker. The Hf-bearing elution was gently evaporated until dry and subsequently redissolved in 1.0 mL of 2 wt% HNO_3. These diluted solutions were introduced into a Nu Plasma HR MC-ICP-MS with a DSN-100 dissolution nebulizing system to determine the whole-rock Hf isotopic composition. Instrument bios and mass fractionation were corrected by normalization raw ratios to $^{179}Hf/^{177}Hf = 0.7325$. Seven measurements of W-2a and BHVO-2 yielded average ratios of $^{176}Hf/^{177}Hf = 0.282741 \pm 7$ (2σ, $n = 7$) and 0.283103 ± 6 (2σ, $n = 7$).

Analysis results for these international reference standards for Nd-Hf isotopes are consistent with the recommended values, which can be found in GeoReM (georem.mpch-mainz.gwdg.de), within the analytical error.

4. Results

4.1. Major and Trace Elements

All 16 samples were analyzed for whole-rock major and trace elements and yielded low LOI values (<0.5 wt%; Table S1), suggesting that the samples were fresh. The major oxide contents described below were recalculated on an LOI-free bias. The content of MgO is 7.37–9.56 wt%, and the average MgO content is 8.4 wt%, suggesting that the samples

 *biomedicines*

Review

Deciphering the Functions of Telomerase Reverse Transcriptase in Head and Neck Cancer

Tsung-Jang Yeh [1,2,3], Chi-Wen Luo [4,5], Jeng-Shiun Du [1,2,3], Chien-Tzu Huang [1,2], Min-Hung Wang [1,2], Tzer-Ming Chuang [1,3], Yuh-Ching Gau [1,2,3], Shih-Feng Cho [1,2,3], Yi-Chang Liu [1,2], Hui-Hua Hsiao [1,2], Li-Tzong Chen [3,6,7], Mei-Ren Pan [2,8,9], Hui-Ching Wang [1,2,3,*] and Sin-Hua Moi [2,*]

[1]   Division of Hematology & Oncology, Department of Internal Medicine, Kaohsiung Medical University Hospital, Kaohsiung Medical University, Kaohsiung 807, Taiwan
[2]   Graduate Institute of Clinical Medicine, College of Medicine, Kaohsiung Medical University, Kaohsiung 807, Taiwan
[3]   Center for Cancer Research, Kaohsiung Medical University, Kaohsiung 807, Taiwan
[4]   Department of Surgery, Kaohsiung Medical University Hospital, Kaohsiung 807, Taiwan
[5]   Department of Cosmetic Science and Institute of Cosmetic Science, Chia Nan University of Pharmacy and Science, Tainan 717, Taiwan
[6]   Division of Gastroenterology, Department of Internal Medicine, Kaohsiung Medical University Hospital, Center for Cancer Research, Kaohsiung Medical University, Kaohsiung 807, Taiwan
[7]   National Institute of Cancer Research, National Health Research Institutes, Tainan 704, Taiwan
[8]   Drug Development and Value Creation Research Center, Kaohsiung Medical University, Kaohsiung 807, Taiwan
[9]   Department of Medical Research, Kaohsiung Medical University Hospital, Kaohsiung 807, Taiwan
*   Correspondence: joellewang66@gmail.com (H.-C.W.); moi9009@gmail.com (S.-H.M.); Tel.: +886-(7)-3121101 (ext. 2512#418) (S.-H.M.)

Citation: Yeh, T.-J.; Luo, C.-W.; Du, J.-S.; Huang, C.-T.; Wang, M.-H.; Chuang, T.-M.; Gau, Y.-C.; Cho, S.-F.; Liu, Y.-C.; Hsiao, H.-H.; et al. Deciphering the Functions of Telomerase Reverse Transcriptase in Head and Neck Cancer. *Biomedicines* **2023**, *11*, 691. https://doi.org/10.3390/biomedicines11030691

Academic Editor: Vui King Vincent-Chong

Received: 9 February 2023
Revised: 19 February 2023
Accepted: 22 February 2023
Published: 24 February 2023

Abstract: Head and neck cancers (HNCs) are among the ten leading malignancies worldwide. Despite significant progress in all therapeutic modalities, predictive biomarkers, and targeted therapies for HNCs are limited and the survival rate is unsatisfactory. The importance of telomere maintenance via telomerase reactivation in carcinogenesis has been demonstrated in recent decades. Several mechanisms could activate telomerase reverse transcriptase (TERT), the most common of which is promoter alternation. Two major hotspot TERT promoter mutations (C228T and C250T) have been reported in different malignancies such as melanoma, genitourinary cancers, CNS tumors, hepatocellular carcinoma, thyroid cancers, sarcomas, and HNCs. The frequencies of TERT promoter mutations vary widely across tumors and is quite high in HNCs (11.9–64.7%). These mutations have been reported to be more enriched in oral cavity SCCs and HPV-negative tumors. The association between TERT promoter mutations and poor survival has also been demonstrated. Till now, several therapeutic strategies targeting telomerase have been developed although only a few drugs have been used in clinical trials. Here, we briefly review and summarize our current understanding and evidence of TERT promoter mutations in HNC patients.

Keywords: head and neck cancer; telomerase reverse transcriptase (TERT); promoter mutations; prognosis

1. Introduction

Head and neck cancers (HNCs) arising in the oral cavity, oropharynx, larynx, and hypopharynx were the seventh most common cancer worldwide in 2018 [1] and approximately seven to eight hundred thousand new cases are diagnosed each year worldwide [2]. The most common type, squamous cell carcinoma, is a highly lethal group of heterogeneous neoplasms often diagnosed at an advanced stage [3]. Tobacco and alcohol consumption are the main etiological factors [4]. Betel quid chewing and infection by oncogenic human papillomavirus (HPV) types 16 and 18 have emerged as important etiological factors for a subset of HNCs in the oral cavity and oropharynx, respectively [5–10]. HPV-positive malignancies represent 5–20% of all HNCs and 40–90% of those arising from the oropharynx [11].

The prevalence of HPV-driven HNCs has been dramatically increasing in developed countries, predominantly affecting middle-aged white men, non-smokers, non-drinkers, or mild-to-moderate drinkers with a higher socioeconomic status and better performance statuses than those with HPV-unrelated SCCs [11,12]. The treatment of HNC is generally multimodal, including surgery, chemotherapy, and radiotherapy, and differs according to disease stage, anatomical location, and surgical accessibility. However, despite significant progress in all therapeutic modalities, the 5-year overall survival (OS) rate of HNC patients remains unsatisfactory [13–15].

In the era of biomarker-driven personalized cancer therapy, several biomarkers have been proposed as prognostic and predictive factors in different cancers, such as KIT mutations in gastrointestinal stromal tumors, EGFR mutations in lung cancer, and HER2 overexpression in breast cancer [16]. However, unlike other cancer types, there are limited predictive biomarkers and targeted therapies for HNCs [4,16–18]. With the development of advanced technical approaches, genome, and exome analyses have provided a comprehensive view of genetic alterations in HNC and uncovered potential new therapeutic opportunities [17,19–24]. In addition to commonly mutated genes, such as *TP53, CDKN2A, CCND1, PIK3CA*, and *NOTCH1*, telomerase reverse transcriptase (*TERT*) promoter mutations have been detected in a significant proportion of HNC patients [3,4,12,13,15,25–29]. *TERT* is located on chromosome 5p15.33 in humans and is an integral and essential part of the telomerase holoenzyme, which plays a key role in cancer formation. Mostly, telomerase activity was increased by upregulation of *TERT* expression via several genetic and epigenetic alterations, and *TERT* promoter mutations are known as the most important [30]. However, the incidence of *TERT* promoter mutations varies in the head and neck subsites, and the association between *TERT* promoter mutations and outcomes is unclear. Therefore, in this review, we summarize our current understanding and evidence of *TERT* promoter mutations in HNC patients.

2. Telomeres and Telomerase in Normal Cells

Telomeres are the physical ends of eukaryotic linear chromosomes [31,32]. In human cells, telomeres are composed of TTAGGG nucleotide repeats with a $3'$ single-stranded overhang, and the variation ranges from 3 to 20 kilobase pairs [12,33]. They are bound by a six-member protein complex known as shelterin. Telomeres cover the coding DNA at the end to avoid loss of genetic information in linear DNA, act as a cap to prevent degradation by a nucleolytic attack, and prevent aberrant activation of a DNA damage response (DDR), which could lead to inappropriate processing of telomeres and as sites for double-strand break repair [31,34,35].

Telomerase, a specialized reverse transcriptase, is a large multi-subunit ribonucleoprotein complex that synthesizes telomeric DNA sequences and provides a molecular basis for unlimited proliferative potential [36]. Telomerase comprises two major components: the telomeric RNA component (also known as the telomerase RNA component: TERC or TR) and the telomerase reverse transcriptase (TERT), which is encoded by the TERT gene. The TERT is located in the human chromosome band 5p15.33, and the TERC is located at 3q26.3 [37]. The TERC serves as the template for telomere hexamer repeat additions onto the DNA, and the TERT is responsible for the reverse transcribing hexamer repeats onto the chromosomal ends [33,38–40]. TERT expression is silenced during development, unlike the TERC and other constitutively expressed telomerase components [35].

Telomerase is a key telomere length maintenance mechanism and is present in germline, hematopoietic, stem, and other rapidly renewing cells [41]. However, in most normal somatic cells, telomerase activity is extremely low or absent. Therefore, loss of telomeric repeats occurs at each round of DNA replication, after which the telomeres are reduced to a critical length [42]. Critical telomere attrition elicits a DDR that mediates cell cycle arrest and leads to replicative senescence or apoptosis via the p53 or Rb tumor suppressor pathways [35,43]. Telomere attrition acts as a barrier to replicative immortality, also called a "mitotic clock" that limits the cell cycle number and further triggers cellular senescence [33,35]. Although rare, in the absence of telomerase, some cells employ another DNA recombination mechanism, termed alternative lengthening of telomeres (ALT), which reverses telomere attrition to bypass senescence [44].

3. Telomere and Telomerase in Cancer Cells

Telomere length and telomerase activity are crucial for cellular immortalization, tumorigenesis, and cancer progression. Telomere maintenance via telomerase reactivation is a nearly universal hallmark of cancer cells [35,45,46]. The vast majority of cancers overcome replicative senescence by upregulating TERT expression and telomerase activity [35]. Telomerase activity is upregulated in 80–90% of malignancies, enabling unlimited replication of cancer cells, similar to embryonic and stem cells [13,15,47]. For the remaining 10–15% of cancers, upregulation of telomerase activity is achieved through the ALT pathway [48].

There are several ways to upregulate telomerase activity and activate the normally silent human TERT (hTERT) gene. The mechanisms of hTERT activation include chromosomal rearrangements (i.e., duplications, amplifications, insertions, interchromosomal changes, inverted orientations, or deletions), TERT promoter somatic mutations, epigenetic modifications (i.e., DNA methylation, or post-transcriptional regulation by microRNAs), transcriptional activators or repressors, TERT gene polymorphism and alternative splicing (i.e., pre-mRNA alternative splicing of the TERT gene) [30,49–51]. In a pan-cancer genomics study, Barthel et al. detected TERT expression in 73% of the 6835 total tumor samples, which were associated with TERT point mutations, rearrangements, DNA amplifications, and transcript fusions. Among the TERT-expressing samples, there were 31% TERT promoter mutations, 3% TERT amplifications, 3% TERT structural variants, 5% TERT promoter structural variants, and 53% TERT promoter methylation [51]. Some of these mechanisms may interact with each other and have a synergistic effect on TERT expression [30].

Besides the canonical role of telomerase in telomere maintenance, there are also some non-canonical functions (telomere length-independent mechanisms) in tumorigenesis, such as the regulation of metabolic mechanisms, epigenetic regulation, and modulation of chromatin, oxidative stress protection, RNA silencing, signal transduction pathways (Wnt and c-MYC signaling pathways), enhanced mitochondrial function, cell adhesion, and migration [30,52–58].

4. Telomerase Reverse Transcriptase (TERT) Promoter Mutations

Among the several mechanisms of hTERT activation, TERT somatic promoter mutations are the most common non-coding driver mutations in cancer [30,59] and occurred at a high frequency in over 50 cancer types [60]. They have been reported in two major hotspots (mainly C > T transitions), which are located at −124 and −146 base pairs upstream of the transcriptional start site on chromosome 5 and are designated as C228T and C250T, respectively [13,25,32,61,62]. A less frequent hTERT promoter mutation −57 base pairs upstream of the transcriptional start site with an A > C transition (at position 1,295,161 on chromosome 5) has been found to be a disease-segregating germline mutation in a melanoma-prone family [63]. Other less frequent, yet recurrent, mutations on chromosome 5 have also been discovered in cancers at the following positions: 1,295,228 C > A, 1,295,248–1,295,243 CC > TT, and 1,295,161 A > C [64].

TERT somatic promoter mutations are predominantly heterozygous and lead to the generation of an 11 bp sequence, CCCGGAAGGGG, which is similar to the E26 transformation-specific (ETS) factor binding motif [60,65]. Then, ETS binding factors, such as GA-binding protein (GABP), are recruited. This recruitment resulted in direct transcriptional activation of hTERT expression and promoted an epigenetic shift from a repressed to active chromatin conformation [35,65–68]. These promoter mutations were proven to be associated with higher levels of TERT mRNA, TERT protein, telomerase enzymatic activity, and telomere length in a study of 23 human urothelial cancer cell lines [69].

Two TERT promoter hotspot mutations, C228T and C250T are the most common; however, their frequencies vary widely across tumors from different sites (Table 1). These mutations occur most frequently in cancers with low rates of self-renewal [25] and are rare in pediatric and young adult cancers [3,70]. The highest frequencies of TERT promoter mutations have been reported in melanoma, bladder cancer, urothelial carcinoma, CNS tumors, hepatocellular carcinoma, thyroid cancer, basal cell carcinoma, and cutaneous squamous cell carcinoma. Due to the variety of sarcoma subtypes, the prevalence of TERT promoter mutations varies widely, and the highest TERT promoter mutation rate is reported in myxoid liposarcoma (79.1%) [25].

Table 1. Frequency spectrum of hTERT promoter mutations across different cancer types.

Cancer Type	Mutation Frequency (%)	Reference
Malignant melanoma	17.0–85.0	[61,63,71,72]
Genitourinary cancers		
Bladder cancer	59.0–85.0	[25,61,73–76]
Urothelial carcinomas	29.5–64.5	[77,78]
Kidney cancers	0	[61]
Prostate Cancer	0	[79]
CNS tumors		
Glioblastoma	54.0–84.0	[61,70,73,78,80]
Other gliomas (ependymoma, astrocytoma, mixed glioma, oligodendroglioma)	2.7–78.0	[25,64,70,78]
Medulloblastoma	33.3–65.0	[70,78]
Hepatocellular carcinoma	31.4–59.0	[25,78,81–84]
Thyroid cancer (papillary, follicular, poorly differentiated, and anaplastic carcinomas)	3.4–46.3	[61,85–87]
Gastrointestinal stromal tumor	0–3.8	[61,88]
Malignant pleural mesothelioma	11.3	[89]
Atypical fibroxanthomas	93.0	[90]
Sarcomas (chondrosarcoma, fibrosarcoma, myxofibrosarcoma, myxoid liposarcoma, osteosarcoma, pleomorphic dermal sarcomas)	4.3–79.1	[25,90,91]
Basal cell carcinoma of the skin	73.8	[92]
Squamous cell carcinoma of the skin	20.0–74.0	[25,92,93]
Squamous cell carcinoma of esophageal	1.6	[94]
Squamous cell carcinoma of penile	48.6	[95]
Squamous cell carcinoma of the head and neck	11.9–64.7	[3,4,13,15,25–29,32,93,96,97]
Squamous cell carcinoma of the cervix	0–21.4	[25,26,93,96]
Breast cancer, colorectal cancer, ovarian cancer, esophageal adenocarcinoma, acute myeloid leukemia, chronic lymphoid leukemia, pancreatic cancer, and testicular carcinoma	0–5.0	[61,78]

5. TERT Promoter Mutations in Head and Neck Squamous Cell Carcinoma

5.1. The Frequency of TERT Promoter Mutations

For HNCs, the frequency of TERT promoter mutations varied significantly among previous studies. These differences could be explained by the tumor subsite, sample size, methodological sensitivity, risk factors, and population ethnicity (Table 2).

were less affected by low-pressure crystallization. All samples are plotted in the basalt area in a total alkali vs. silica (TAS) plot, and are classified as low-K subalkaline basalts (Figure 2).

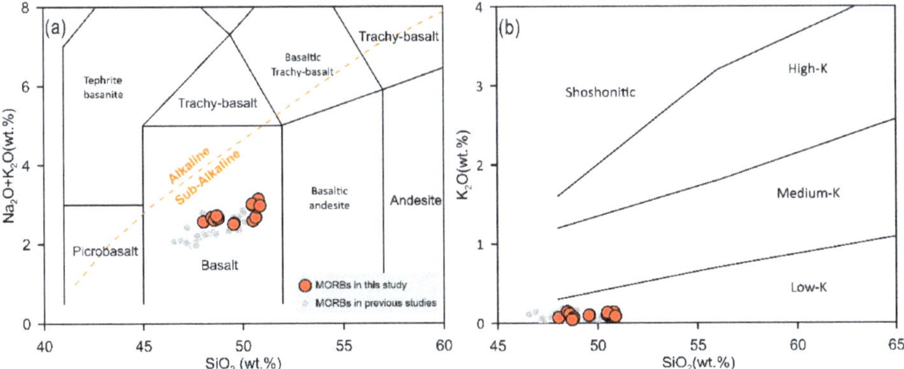

Figure 2. (**a**) Total alkali vs. SiO$_2$ diagram. The alkaline-subalkaline discrimination line is from [37]. (**b**) K$_2$O vs. SiO$_2$ diagram. The discrimination lines are from [38]. The red and gray circles represent the SMAR basalts in this study and in previous studies [26,32,39], respectively. Primitive mantle and N-MORB data are from [40].

The (La/Sm)$_N$ ("N" denotes normalization to chondrite) ratio of the samples ranges from 0.49 to 0.79, and the Zr/Nb ratio ranges from 17.1 to 79.1, which indicate that the samples belong to the typical N-MORB. Figure 3a shows the primitive mantle-normalized trace element patterns of the study samples. In addition to the highly incompatible elements expected to be heavily depleted overall, the most remarkable feature is that all samples have clear positive anomalies of U and Pb (U/U* = 1.05–3.52, Pb/Pb*= 1.01–2.62; U* = (Th$_N$ + Nb$_N$)/2, Pb*= (Ce$_N$ + Pr$_N$)/2; "N" denotes normalization to the primitive mantle), negative anomalies of Nb, and Rb enrichment in some samples. The chondrite-normalized REE patterns diagram (Figure 3b) shows that most samples are depleted in LREE, and some samples have a relatively flat REE pattern or even a slight depletion of HREE. Meanwhile, most samples have negative Eu anomalies, with the lowest Eu/Eu*(Eu* = (Sm$_N$ + Gd$_N$)/2; "N" denotes normalization to the primitive mantle) being 0.86.

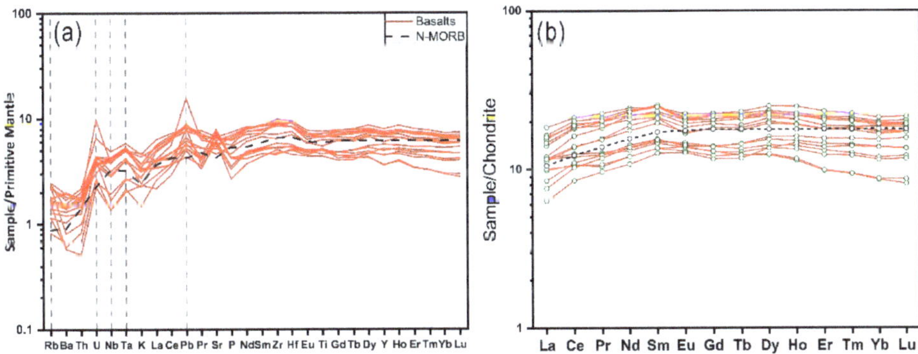

Figure 3. (**a**) Primitive mantle-normalized trace element patterns for the analyzed samples. (**b**) Chondrite-normalized trace element patterns for the analyzed samples. Primitive mantle, chondrite, and N-MORB data are from [40].

4.2. Nd-Hf Isotopic Compositions

All sixteen samples were analyzed for whole-rock Nd isotopic composition and eight samples were analyzed for whole-rock Hf isotopic composition. The isotope data are shown in Table S2. $^{143}Nd/^{144}Nd$ ratio ranges from 0.513030 to 0.513176, and $^{176}Hf/^{177}Hf$ ratio ranges from 0.282960 to 0.283188, which are in the range of previous studies [26,33], and confirm their geochemically depleted character. εNd-εHf plot (Figure 4) shows that all samples are plotted below the mantle array and were positively correlated.

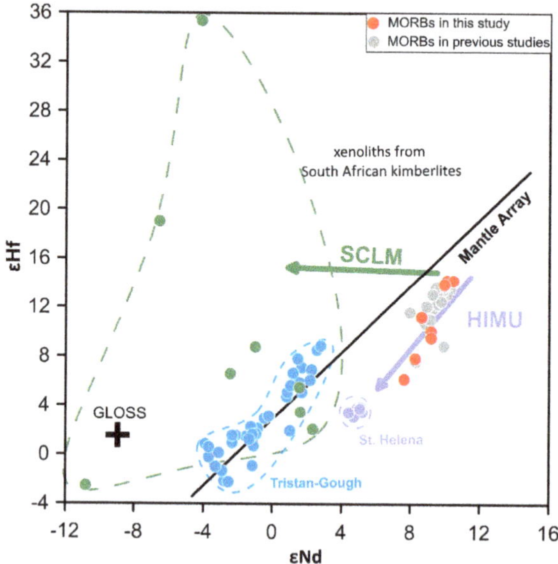

Figure 4. Plots of εNd vs. εHf. Red dots represent new MORBs in this study and gray dots represent SMAR 18–21°S MORBs in [26,32,39]. Black cross represents global subducting sediment (GLOSS), the composition of which is from [41,42]. The purple area represents the basalts of St. Helena seamount chain, and represents HIMU end-member. The blue area represents the basalts of Walvis ridge, and represents EM1 component. The green area represents the low-temperature garnet-peridotite xenoliths from Cretaceous South African kimberlites, and represents SCLM. The basalts from the St. Helena and the Tristan-Gough plumes are from PetDB (http://www.earthchem.org/petdb), the xenoliths from Cretaceous South African kimberlites are from [43]. Nd-Hf mantle array is from [44].

5. Discussion

5.1. Abnormal Trace Element Ratios

Ce/Pb and Nb/U ratios are relatively uniform in MORBs worldwide [13,24] at 25 ± 5 and 47 ± 10 (black dashed lines in Figure 5; [45]), respectively. However, Ce/Pb and Nb/U ratios in our samples vary widely (Figure 5), with an average value of 18 and 29, respectively. Figure 5 shows that regardless of the degree of depletion, Ce/Pb and Nb/U ratios of our samples are lower than the corresponding values (25 for Ce/Pb, 47 for Nb/U), with the lowest values of 9 and 13, respectively, which corresponds to the clear enrichment of U and Pb in the samples. In general, Ce/Pb ratio is positively correlated with Nb/U ratio (correlation coefficient is 0.63).

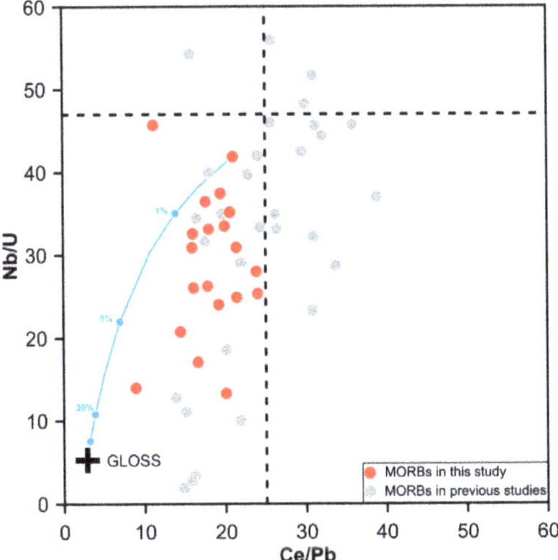

Figure 5. Plots of Ce/Pb vs. Nb/U. Red dots represent new MORBs in this study and gray dots represent SMAR 18–21°S MORBs in [26,32,39]. Black cross represents global subducting sediment (GLOSS), the composition of which is from [41]. Two dashed lines intersect at Nb/U = 47 and Ce/Pb = 25, which are from [13]. The blue line is the mixing line between GLOSS and SMAR-61. The details of end-members and mixing results are shown in Table S3.

The ratio of fluid-mobile element to fluid-immobile element, such as Ba/Nb (3.4–4.8), Ba/Th (59.7–94.6), Rb/La (0.22–0.71), Rb/Nb (0.35–1.04) ratios, was higher than the typical N-MORB (2.7 for Ba/Nb, 52.5 for Ba/Th, 0.22 for Rb/La, 0.24 for Rb/Nb; [40]). The plot of Rb/Nb vs. Ba/Nb (Figure 6a) shows that the Rb/Nb ratio is positively correlated with the Ba/Nb ratio (correlation coefficient is 0.62), except for the Rb-enriched MORBs. In the plot of Rb/La vs. Ba/Th (Figure 6b), Rb/La ratio of our samples is significantly positively correlated with Ba/Th ratio (correlation coefficient is 0.80).

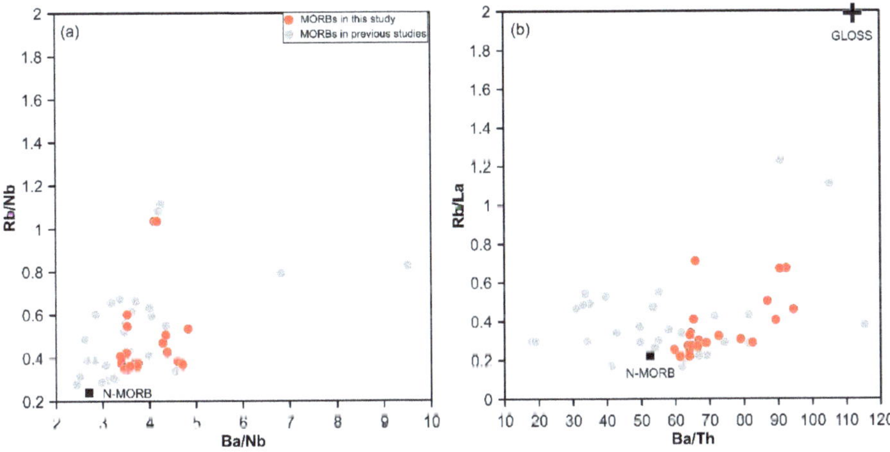

Figure 6. Plots of (**a**) Ba/NB vs. Rb/Nb. (**b**) Ba/Th-Rb/La. Red dots represent new MORBs in this study and gray dots represent SMAR 18–21°S MORBs in [26,32,39]. Black square represents the typical N-MORB, the composition of which is from [40]. Black cross represents global subducting sediment (GLOSS), the composition of which is from [41].

Combined with the concentration of fluid-mobile elements, such as U and Pb, in the samples, subduction-related fluid process is a potential explanation for the abnormal trace elements ratios. In addition, continental crust material [46], recycled sediments [41], subduction modified mantle [47–50] have high Ba/Nb, Ba/Th, Rb/La, and Rb/Nb ratios, and higher U and Pb contents. Their injection into the mantle can also be the cause of this compositional anomaly. In addition to the above reasons, Le Roux et al. [51] proposed that the direct supply of mantle plume material was one of the reasons for the changes in some trace element ratios in MORBs.

5.2. The origin of Trace Element Anomalies
5.2.1. Continental Crust and SCLM

The continental crust materials have similar trace element characteristics as the samples in this study, such as low Ce/Pb and Nb/U ratios, high U and Pb contents and Ba/Nb, Ba/Th, Rb/La, and Rb/Nb ratios [46]. Furthermore, delaminated lower continental crust (LCC) has been found in the mantle beneath East Pacific Rise (EPR), Southeast Indian ridge (SEIR), and mid-Atlantic ridge (MAR) [10,12,15,16,52]. Therefore, the continental crust material is a possible origin of the trace element anomalies. In addition, the Atlantic Ocean was formed by the breakup of the west Gondwana [25]; therefore, the addition of SCLM should also be considered.

The addition of continental crust can significantly reduce the Nb/U ratio in the upper mantle, thus obtaining the trace element compositions consistent with the observed phenomenon. However, the εNd value of LCC is significantly lower than the MORB source [17]. Therefore, the addition of continental crust also changes the Nd isotopic composition of the MORB source, and thus MORB presents a mixed curve in the Ce/Pb–εNd plot. Figure 7 shows that the Nb/U ratio of MORBs in this study changes dramatically (from 45.7 to 13.3) within a limited range of εNd values. This indicates that there is no identifiable continental crust material in the upper mantle beneath SMAR 18–21°S.

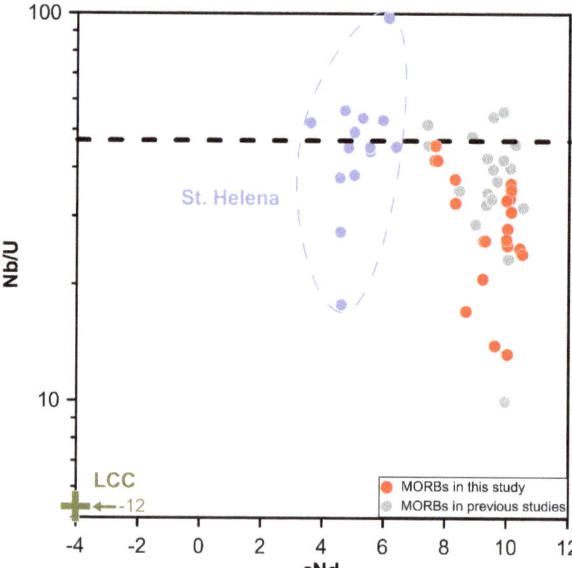

Figure 7. Plots of εNd vs. Nb/U. Red dots represent new MORBs in this study and gray dots represent SMAR 18–21°S MORBs in [26,32]. Black square represents the typical N-MORB, the composition of which is from [40]. Olive cross represents mean isotopic composition of mafic granulite which is from [17] and references therein.

Hanan et al. [16] proposed that the occurrence of LCC material in a MORB source would decouple the Nd-Hf isotopes of MORBs, and make MORB plots point to the position above the Nd-Hf mantle array in the εNd-εHf plot. As shown in Figure 4, the samples have good correlation and all fall under the Nd-Hf mantle array. This is consistent with our conclusion that there is no identifiable continental crust material in the upper mantle of the study area.

For the reason of up to 3 Ga of radiogenic ingrowth following multiple episodes of mantle metasomatism, underplating, and melt extraction events, subcontinental lithosphere is isotopically heterogeneous [51]. However, SCLM beneath south Africa shows a strong subduction zone geochemical signature [53–55], with a relative depletion in Nb. Therefore, SCLM seems to be another possible origin of the trace element anomalies. Garnet and clinopyroxene are the main major host minerals for Nd and Hf in the mantle with concentrations mostly in the range of 0.01–15 (up to ~50) ppm Nd and 0.001–2 (up to ~8) ppm Hf in clinopyroxene, and 0.01–5 (up to ~20) ppm Nd and 0.02–1.5 (up to ~2.5) ppm Hf in garnet. Hf isotopic compositions of garnet and clinopyroxene are generally more radiogenic for their enrichment in Lu [43]. Therefore, the injection of SCLM will lead to a higher εHf value at a given εNd value. Whole-rock Nd-Hf isotope data for low-temperature garnet-peridotite xenoliths from Cretaceous South African kimberlites were plotted in Figure 4. The result shows that SCLM was distant from the trend formed by the SMAR 18–21°S MORB samples; therefore, there was no clear influence of subcontinental lithospheric mantle in the mantle beneath SMAR 18–21°S. In addition, Hanan et al. [16] ruled out the influence of SCLM since the amount of SCLM required to cause a clear abnormal ratio of trace elements was significantly large.

5.2.2. Direct Supply of Mantle Plume Material

Ridge-plume interaction is an important cause of mantle heterogeneity. If the mantle plume exists near the mid-ocean ridge, the mantle plume material will affect the mantle beneath the mid-ocean ridge, increasing the degree of isotopic heterogeneity in the MORBs source that was relatively uniform [11,56,57]. The study area is located in the southwest extension direction of the St. Helena seamount chain, only about 750 km away from the seamount chain. Previous studies [26,28–30,32,33] proved that the mantle source was affected by the St. Helena plume through Sr-Nd-Pb isotope data and geophysical evidence. The purple area in Figure 4 represents the isotopic composition region of the basalt produced by the St. Helena plume. Figure 4 shows that the Nd and Hf isotopes of the MORB samples are well correlated, and point to the St. Helena plume area, which is consistent with previous conclusions.

Le Roux et al. [51] proposed that some anomalies of trace elements in MORBs were caused by the direct supply of adjacent plume materials. If the anomalies of trace elements in this study are caused by the direct supply of plume materials, the basalts produced by the mantle plume should generally have low Ce/Pb and Nb/U ratios. However, Figure 7 shows that only few St. Helena basalts have a low Nb/U ratio, while most have a constant Nb/U ratio typical of oceanic basalts, with an average Nb/U ratio of 44, which is consistent with previous studies [13,23,24]. The injection of plume materials cannot reduce the Nb/U ratio of source mantle. Moreover, suppose that the direct supply of plume materials really is the origin of the trace element anomalies, the abnormal trace element ratios should be related to the isotopes. Namely, the more radioactive the isotopic composition of MORB is, the more clear the abnormal ratio of the trace elements will be. However, the correlation between the ratio of trace elements and isotopic composition is not clear (Figure 8). Therefore, the direct supply of St. Helena material does not result in the abnormal trace element ratios.

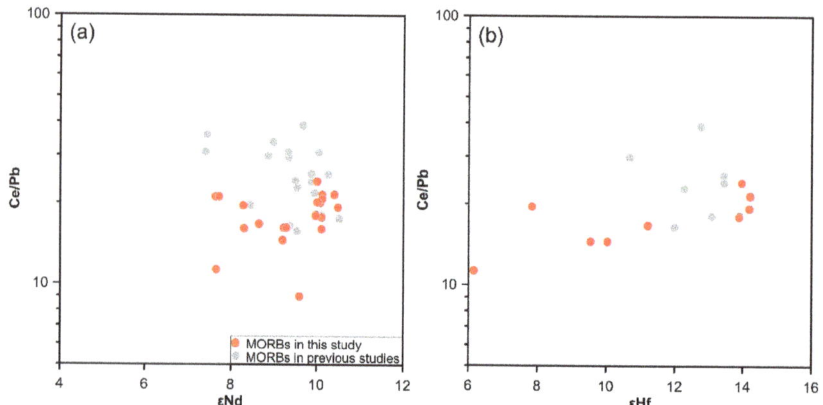

Figure 8. Plots of (**a**) εNd (**b**) εHf vs. Nb/U. Red dots represent new MORBs in this study and gray dots represent SMAR 18–21°S MORBs in [26,32].

5.2.3. Recycled Sediments

Sediments can be subducted into the mantle and thus affect the geochemical characteristics of the mantle [58,59]. The composition of subducted sediments is highly variable [41,60], but in general, recycled sediments have identifiable trace element and isotope characteristics. Therefore, recycled sediments are commonly used to explain the isotope and trace element system in MORB [42,51,52,61,62].

The plot of the GLOSS in Figures 4, 5 and 6b is located in the extension direction of the MORB sample array, and seems to indicate that it contributes to the source. However, after quantitative modeling (Figure 5), it was found that at least 10–20% of the sediment in the source was required to produce the observed trace element ratio anomalies. Even in BABB, such as the Manus Basin, the proportion of sediment is only <5% [63]. Clearly, the presence of this large proportion of sediments in the mantle beneath the mid-ocean ridge, which is distant from the subduction zone, is unreasonable. Moreover, due to the high ratio of ^{87}Sr/^{86}Sr in the sediments [41], this proportion of the sediments would cause the Sr isotopic composition of the samples to deviate significantly from the MORB range. Previous data [26,32,33] showed that the Sr isotopic composition (0.702398–0.702996, except for SA4A in [33]) in SMAR 18–21°S MORBs was not significantly higher than the normal MORB range. Therefore, we ruled out the sediment as an option. In addition, the Sr/Nd ratio of the sample (10.6–27.7, average 17.6) did not support the presence of recycled sediment (~12, [41]) in the source.

5.2.4. Subduction

The subduction signals of MORB or OIB are often interpreted as the result of the processes through which subduction directly affects the source mantle [21] or the existence of subduction modified mantle [17,61,64] in the source, and the Ba/Nb ratio is one of the most common indicators of subduction signals [65]. In our study, although the Ba/Nb ratio of our samples did not reach the standard (Ba/Nb > 6) of the BABB filter proposed by Yang et al. [21], it is still significantly higher than the typical N-MORB (Ba/Nb ~2.7; [40]). Considering that the study area is affected by the St. Helena plume, the addition of plume material will weaken the subduction signal in the study area. Therefore, we cannot rule out the possibility that the mantle in the study area is affected by subduction only by the fact that the Ba/Nb ratio is not sufficiently enough. In addition, Nb negative anomalies are important indicators of subduction, which are commonly seen in rocks produced in back-arc basins, island arcs, and other subduction-associated tectonic environments [47,63].

Furthermore, Ba/Th and Th/Nb ratios can be used to indicate the influence of subduction fluid and melt, respectively [48,51,65]. Ba/Nb vs. Ba/Th, Ba/Nb vs. Th/Nb plots

(Figure 9) show that the variation of Ba/Nb ratio in the samples was mainly controlled by the variation of Ba/Th ratio. Moreover, Ba/Th vs. Th/Nb plot (Figure 10) shows that if the study area is indeed affected by subduction, it should be mainly affected by the subduction fluid. Most of the subduction fluids are released shallowly in the subduction zone [66]; therefore, we can rule out the direct effects of the subduction zone. The spatially and temporally closest subduction zone which subducts/subducted toward the present SMAR 18–21°S is the eastern Pacific subduction zone, but the subduction zone and SMAR 18–21°S were separated by the South American continent, even before the south Atlantic opened. As the Atlantic opens, the distance becomes increasingly larger. For shallow subduction fluids to be directly affected in this distance, it would require an unreasonably small subducted angle and an extremely long subducted distance. Therefore, the hypothesis that the subduction of the eastern Pacific directly affects the source mantle is ruled out. The ancient subduction zone associated with the Rheic Ocean was located on the northern side of Africa and South America [67,68]. Similar to the eastern Pacific subduction zone, subduction fluids are difficult to be affected in the study area due to the distance. Therefore, only the process of subduction modified mantle transmitted to the study area through the mantle convection can make sense.

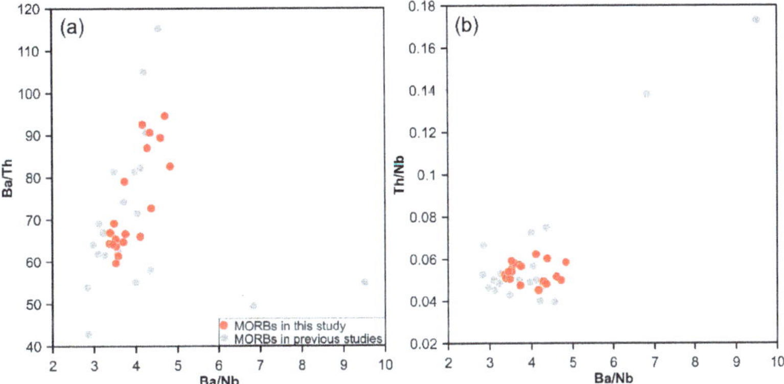

Figure 9. Plots of Ba/Nb vs. (**a**) Ba/Th (**b**) Th/Nb. Red dots represent new MORBs in this study and gray dots represent SMAR 18–21°S MORBs in [26,32].

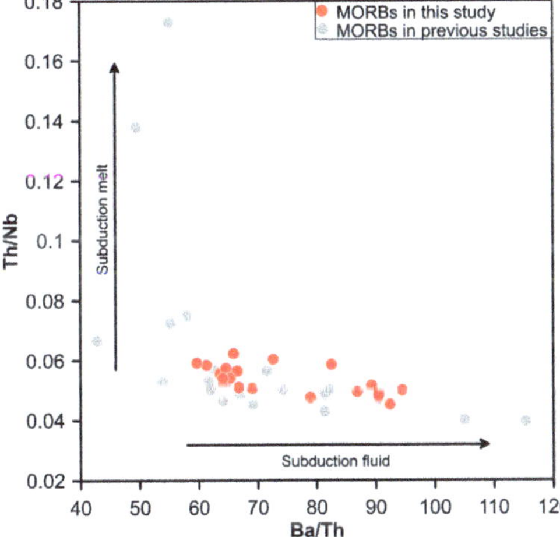

Figure 10. Plots of Ba/Th vs. Th/Nb. Red dots represent new MORBs in this study and gray dots represent SMAR 18–21°S MORBs in [26,32].

6. Conclusions

In this study, 16 basalt samples from 14 sites on the axis of SMAR 18–21°S were collected and analyzed for the composition of major element, trace element, and Nd-Hf isotope. The results are shown as follows:

1. These samples belong to the low-K subalkaline basalt, which is classified into N-MORB according to the $(La/Sm)_N$ ratio, and Nd-Hf isotope composition shows the characteristics of depletion. However, compared with typical N-MORB samples, the fluid-mobile elements, such as U, Pb, and Rb are enriched. Meanwhile, Ce/Pb and Nb/U ratios are higher than the constant value, while Ba/Nb, Ba/Th, Rb/La, Rb/Nb, and other trace element ratios are significantly higher than the typical N-MORB. In addition, our samples all have negative anomalies of Nb.
2. Plot of εNd-εHf shows that the mantle beneath SMAR 18–21°S is affected by the St. Helena plume. However, the anomalies of trace element in the samples are not caused by the direct supply of plume material. In addition, we exclude the possibility of continental crust material, subcontinental lithospheric mantle, and recycled sediments. Furthermore, we propose that the anomalies of trace element are related to subduction.

The influence of subduction on the samples is mainly caused by the subduction fluid. Based on the geographical location, it seems unreasonable that the subduction directly affects the mantle beneath SMAR 18–21°S. It is more likely that the mantle modified by the subduction fluid is transmitted to the mantle beneath SMAR 18–21°S through mantle convection.

Supplementary Materials: The following supporting information can be downloaded at: https://www.mdpi.com/article/10.3390/jmse11020441/s1, Table S1: Whole rock major and trace elements compositions; Table S12: Whole Rock Nd-Hf isotopic compositions; Table S3: The mixing calculations of ratios of trace element between GLOSS and SMAR-61.

Author Contributions: Conceptualization, T.J. and Z.Z.; methodology, T.J.; software, T.J.; validation, T.J.; formal analysis, T.J.; investigation, T.J. and Z.Z.; resources, Z.Z.; data curation, T.J.; writing—original draft preparation, T.J.; writing—review and editing, T.J. and Z.Z.; visualization, T.J.; supervision, Z.Z.; project administration, Z.Z.; funding acquisition, Z.Z. All authors have read and agreed to the published version of the manuscript.

Funding: This research received no external funding or This research was funded by the National Natural Science Foundation of China (Grant No. 91958213), Strategic Priority Research Program (B) of the Chinese Academy of Sciences (Grant No. XDB42020402), National Basic Research Program of China (Grant No. 2013CB429700), and the Taishan Scholars Program.

Informed Consent Statement: Not applicable.

Data Availability Statement: All the data are given in the Supporting Information.

Acknowledgments: We are grateful for the valuable comments and suggestions from the anonymous reviewers and editors. This work was supported by the National Natural Science Foundation of China (Grant No. 91958213), Strategic Priority Research Program (B) of the Chinese Academy of Sciences (Grant No. XDB42020402), National Basic Research Program of China (Grant No. 2013CB429700), and the Taishan Scholars Program.

Conflicts of Interest: The authors declare no conflict of interest.

References

1. Zindler, A.; Hart, S.R. Chemical Geodynamics. *Ann. Rev. Earth Planet. Sci.* **1986**, *14*, 493–571. [CrossRef]
2. Hart, S.R. Heterogeneous Mantle Domains: Signatures, Genesis and Mixing Chronologies. *Earth Planet. Sci. Lett.* **1988**, *90*, 273–296. [CrossRef]
3. Willbold, M.; Stracke, A. Trace Element Composition of Mantle End-members: Implications for Recycling of Oceanic and Upper and Lower Continental Crust. *Geochem. Geophys. Geosyst.* **2006**, *7*, 2005GC001005. [CrossRef]
4. White, W.M. Isotopes, DUPAL, LLSVPs, and Anekantavada. *Chem. Geol.* **2015**, *419*, 10–28. [CrossRef]
5. Workman, R.K.; Hart, S.R. Major and Trace Element Composition of the Depleted MORB Mantle (DMM). *Earth Planet. Sci. Lett.* **2005**, *231*, 53–72. [CrossRef]

6. Dupré, B.; Allègre, C.J. Pb–Sr Isotope Variation in Indian Ocean Basalts and Mixing Phenomena. *Nature* **1983**, *303*, 142–146. [CrossRef]
7. Hart, S.R. A Large-Scale Isotope Anomaly in the Southern Hemisphere Mantle. *Nature* **1984**, *309*, 753–757. [CrossRef]
8. Pearce, J.A.; Kempton, P.D.; Nowell, G.M.; Noble, S.R. Hf–Nd Element and Isotope Perspective on the Nature and Provenance of Mantle and Subduction Components in Western Pacific Arc–Basin Systems. *J. Petrol.* **1999**, *40*, 33. [CrossRef]
9. Niu, Y.; Bideau, D.; Hékinian, R.; Batiza, R. Mantle Compositional Control on the Extent of Mantle Melting, Crust Production, Gravity Anomaly, Ridge Morphology, and Ridge Segmentation: A Case Study at the Mid-Atlantic Ridge 33–35° N. *Earth Planet. Sci. Lett.* **2001**, *186*, 383–399. [CrossRef]
10. Meyzen, C.M.; Ludden, J.N.; Humler, E.; Luais, B.; Toplis, M.J.; Mével, C.; Storey, M. New Insights into the Origin and Distribution of the DUPAL Isotope Anomaly in the Indian Ocean Mantle from MORB of the Southwest Indian Ridge: INDIAN OCEAN DUPAL ISOTOPE ANOMALY. *Geochem. Geophys. Geosyst.* **2005**, *6*, Q11K11. [CrossRef]
11. Gale, A.; Laubier, M.; Escrig, S.; Langmuir, C.H. Constraints on Melting Processes and Plume-Ridge Interaction from Comprehensive Study of the FAMOUS and North Famous Segments, Mid-Atlantic Ridge. *Earth Planet. Sci. Lett.* **2013**, *365*, 209–220. [CrossRef]
12. Mougel, B.; Agranier, A.; Hemond, C.; Gente, P. A Highly Unradiogenic Lead Isotopic Signature Revealed by Volcanic Rocks from the East Pacific Rise. *Nat. Commun.* **2014**, *5*, 4474. [CrossRef] [PubMed]
13. Hofmann, A.W. Mantle Geochemistry: The Message from Oceanic Volcanism. *Nature* **1997**, *385*, 219–229. [CrossRef]
14. Kamenetsky, V.S.; Maas, R.; Sushchevskaya, N.M.; Norman, M.D.; Cartwright, I.; Peyve, A.A. Remnants of Gondwanan Continental Lithosphere in Oceanic Upper Mantle: Evidence from the South Atlantic Ridge. *Geology* **2001**, *29*, 243–246. [CrossRef]
15. Escrig, S.; Capmas, F.; Dupré, B.; Allègre, C.J. Osmium Isotopic Constraints on the Nature of the DUPAL Anomaly from Indian Mid-Ocean-Ridge Basalts. *Nature* **2004**, *431*, 59–63. [CrossRef]
16. Hanan, B.B.; Blichert-Toft, J.; Pyle, D.G.; Christie, D.M. Contrasting Origins of the Upper Mantle Revealed by Hafnium and Lead Isotopes from the Southeast Indian Ridge. *Nature* **2004**, *432*, 91–94. [CrossRef]
17. Janney, P.E.; Le Roex, A.P.; Carlson, R.W. Hafnium Isotope and Trace Element Constraints on the Nature of Mantle Heterogeneity beneath the Central Southwest Indian Ridge (13° E to 47° E). *J. Petrol.* **2005**, *46*, 2427–2464. [CrossRef]
18. Stracke, A.; Hofmann, A.W.; Hart, S.R. FOZO, HIMU, and the Rest of the Mantle Zoo: THE MANTLE ZOO. *Geochem. Geophys. Geosyst.* **2005**, *6*, Q05007. [CrossRef]
19. Stracke, A. Earth's Heterogeneous Mantle: A Product of Convection-Driven Interaction between Crust and Mantle. *Chem. Geol.* **2012**, *330–331*, 274–299. [CrossRef]
20. Zhang, G.-L.; Luo, Q.; Zhao, J.; Jackson, M.G.; Guo, L.-S.; Zhong, L.-F. Geochemical Nature of Sub-Ridge Mantle and Opening Dynamics of the South China Sea. *Earth Planet. Sci. Lett.* **2018**, *489*, 145–155. [CrossRef]
21. Yang, A.Y.; Langmuir, C.H.; Cai, Y.; Michael, P.; Goldstein, S.L.; Chen, Z. A Subduction Influence on Ocean Ridge Basalts Outside the Pacific Subduction Shield. *Nat. Commun.* **2021**, *12*, 4757. [CrossRef] [PubMed]
22. Fang, W.; Dai, L.-Q.; Zheng, Y.-F.; Zhao, Z.-F. Molybdenum Isotopes in Mafic Igneous Rocks Record Slab-Mantle Interactions from Subarc to Postarc Depths. *Geology* **2022**, *51*, 3–7. [CrossRef]
23. Halliday, A.N.; Lee, D.-C.; Tommasini, S.; Davies, G.R.; Paslick, C.R.; Godfrey Fitton, J.; James, D.E. Incompatible Trace Elements in OIB and MORB and Source Enrichment in the Sub-Oceanic Mantle. *Earth Planet. Sci. Lett.* **1995**, *133*, 379–395. [CrossRef]
24. Hofmann, A.W.; Class, C.; Goldstein, S.L. Size and Composition of the MORB+OIB Mantle Reservoir. *Geochem. Geophys. Geosyst.* **2022**, *23*, e2022GC010339. [CrossRef]
25. Ito, G.; van Keken, P.E. Hot Spots and Melting Anomalies. In *Treatise on Geophysics*; Elsevier: Amsterdam, The Netherlands, 2007; pp. 371–435, ISBN 978-0-444-52748-6.
26. Zhang, H.; Yan, Q.; Li, C.; Liu, X.; Yang, Y.; Wang, C.; Hua, Q.; Zhu, Z.; Zhang, H.; Zhao, R. Tracing Material Contributions from Saint Helena Plume to the South Mid-Atlantic Ridge System. *Earth Planet. Sci. Lett.* **2021**, *572*, 117130. [CrossRef]
27. Wilson, M. Magmatism and Continental Rifting during the Opening of the South Atlantic Ocean: A Consequence of Lower Cretaceous Super-Plume Activity? *Geol. Soc. Lond. Spec. Publ.* **1992**, *68*, 241–255. [CrossRef]
28. Schilling, J.-G. Upper Mantle Heterogeneities and Dynamics. *Nature* **1985**, *314*, 62–67. [CrossRef]
29. Hanan, B.B.; Kingsley, R.H.; Schilling, J.-G. Pb Isotope Evidence in the South Atlantic for Migrating Ridge—Hotspot Interactions. *Nature* **1986**, *322*, 137–144. [CrossRef]
30. Fontignie, D.; Schilling, J.-G. Mantle Heterogeneities beneath the South Atlantic: A Nd-Sr-Pb Isotope Study along the Mid-Atlantic Ridge (3° S–46° S). *Earth Planet. Sci. Lett.* **1996**, *142*, 209–221. [CrossRef]
31. Storey, B.C. The Role of Mantle Plumes in Continental Breakup: Case Histories from Gondwanaland. *Nature* **1995**, *377*, 301–308. [CrossRef]
32. Zhang, H.; Shi, X.; Li, C.; Yan, Q.; Yang, Y.; Zhu, Z.; Zhang, H.; Wang, S.; Guan, Y.; Zhao, R. Petrology and Geochemistry of South Mid-Atlantic Ridge (19° S) Lava Flows: Implications for Magmatic Processes and Possible Plume-Ridge Interactions. *Geosci. Front.* **2020**, *11*, 1953–1973. [CrossRef]
33. Zhong, Y.; Zhang, X.; Sun, Z.; Liu, J.; Li, W.; Ma, Y.; Liu, W.; Xia, B.; Guan, Y. Sr–Nd–Pb–Hf Isotopic Constraints on the Mantle Heterogeneities beneath the South Mid-Atlantic Ridge at 18–21° S. *Minerals* **2020**, *10*, 1010. [CrossRef]
34. Xue, D.-S.; Su, B.-X.; Zhang, D.-P.; Liu, Y.-H.; Guo, J.-J.; Guo, Q.; Sun, J.-F.; Zhang, S.-Y. Quantitative Verification of 1:100 Diluted Fused Glass Beads for X-Ray Fluorescence Analysis of Geological Specimens. *J. Anal. At. Spectrom.* **2020**, *35*, 2826–2833. [CrossRef]

35. Chen, F.; Hegner, E.; Todt, W. Zircon Ages and Nd Isotopic and Chemical Compositions of Orthogneisses from the Black Forest, Germany: Evidence for a Cambrian Magmatic Arc. *Int. J. Earth Sci.* **2000**, *88*, 791–802. [CrossRef]

36. Chen, F.; Li, X.-H.; Wang, X.-L.; Li, Q.-L.; Siebel, W. Zircon Age and Nd–Hf Isotopic Composition of the Yunnan Tethyan Belt, Southwestern China. *Int. J. Earth Sci.* **2007**, *96*, 1179–1194. [CrossRef]

37. Irvine, T.N.; Baragar, W.R.A. A Guide to the Chemical Classification of the Common Volcanic Rocks. *Can. J. Earth Sci.* **1971**, *8*, 523–548. [CrossRef]

38. Rickwood, P.C. Boundary Lines within Petrologic Diagrams Which Use Oxides of Major and Minor Elements. *Lithos* **1989**, *22*, 247–263. [CrossRef]

39. Zhong, Y.; Liu, W.; Sun, Z.; Yakymchuk, C.; Ren, K.; Liu, J.; Li, W.; Ma, Y.; Xia, B. Geochemistry and Mineralogy of Basalts from the South Mid-Atlantic Ridge (18.0°–20.6° S): Evidence of a Heterogeneous Mantle Source. *Minerals* **2019**, *9*, 659. [CrossRef]

40. Sun, S.-S.; McDonough, W.F. Chemical and Isotopic Systematics of Oceanic Basalts: Implications for Mantle Composition and Processes. *Geol. Soc. Lond. Spec. Publ.* **1989**, *42*, 313–345. [CrossRef]

41. Plank, T.; Langmuir, C.H. Tracing Trace Elements from Sediment Input to Volcanic Output at Subduction Zones. *Nature* **1993**, *362*, 739–743. [CrossRef]

42. Chauvel, C.; Lewin, E.; Carpentier, M.; Arndt, N.T.; Marini, J.-C. Role of Recycled Oceanic Basalt and Sediment in Generating the Hf–Nd Mantle Array. *Nat. Geosci.* **2008**, *1*, 64–67. [CrossRef]

43. Bedini, R. Isotopic Constraints on the Cooling of the Continental Lithosphere. *Earth Planet. Sci. Lett.* **2004**, *223*, 99–111. [CrossRef]

44. Vervoort, J.D.; Blichert-Toft, J. Evolution of the Depleted Mantle: Hf Isotope Evidence from Juvenile Rocks through Time. *Geochim. Cosmochim. Acta* **1999**, *63*, 533–556. [CrossRef]

45. Hofmann, A.W.; Jochum, K.P.; Seufert, M.; White, W.M. Nb and Pb in Oceanic Basalts: New Constraints on Mantle Evolution. *Earth Planet. Sci. Lett.* **1986**, *79*, 33–45. [CrossRef]

46. Rudnick, R.L.; Gao, S. Composition of the Continental Crust. In *Treatise on Geochemistry*; Elsevier: Amsterdam, The Netherlands, 2014; pp. 1–51, ISBN 978-0-08-098300-4.

47. Park, S.-H.; Lee, S.-M.; Kamenov, G.D.; Kwon, S.-T.; Lee, K.-Y. Tracing the Origin of Subduction Components beneath the South East Rift in the Manus Basin, Papua New Guinea. *Chem. Geol.* **2010**, *269*, 339–349. [CrossRef]

48. Nielsen, S.G.; Shu, Y.; Auro, M.; Yogodzinski, G.; Shinjo, R.; Plank, T.; Kay, S.M.; Horner, T.J. Barium Isotope Systematics of Subduction Zones. *Geochim. Cosmochim. Acta* **2020**, *275*, 1–18. [CrossRef]

49. Richter, M.; Nebel, O.; Maas, R.; Mather, B.; Nebel-Jacobsen, Y.; Capitanio, F.A.; Dick, H.J.B.; Cawood, P.A. An Early Cretaceous Subduction-Modified Mantle underneath the Ultraslow Spreading Gakkel Ridge, Arctic Ocean. *Sci. Adv.* **2020**, *6*, eabb4340. [CrossRef]

50. Villalobos-Orchard, J.; Freymuth, H.; O'Driscoll, B.; Elliott, T.; Williams, H.; Casalini, M.; Willbold, M. Molybdenum Isotope Ratios in Izu Arc Basalts: The Control of Subduction Zone Fluids on Compositional Variations in Arc Volcanic Systems. *Geochim. Cosmochim. Acta* **2020**, *288*, 68–82. [CrossRef]

51. le Roux, P.J.; le Roex, A.P.; Schilling, J.-G.; Shimizu, N.; Perkins, W.W.; Pearce, N.J.G. Mantle Heterogeneity beneath the Southern Mid-Atlantic Ridge: Trace Element Evidence for Contamination of Ambient Asthenospheric Mantle. *Earth Planet. Sci. Lett.* **2002**, *203*, 479–498. [CrossRef]

52. Hanan, B.B.; Blichert-Toft, J.; Hemond, C.; Sayit, K.; Agranier, A.; Graham, D.W.; Albarède, F. Pb and Hf Isotope Variations along the Southeast Indian Ridge and the Dynamic Distribution of MORB Source Domains in the Upper Mantle. *Earth Planet. Sci. Lett.* **2013**, *375*, 196–208. [CrossRef]

53. Duncan, A.R. The Karoo Igneous Province—A Problem Area for Inferring Tectonic Setting from Basalt Geochemistry. *J. Volcanol. Geotherm. Res.* **1987**, *32*, 13–34. [CrossRef]

54. Turner, S.; Hawkesworth, C.; Gallagher, K.; Stewart, K.; Peate, D.; Mantovani, M. Mantle Plumes, Flood Basalts, and Thermal Models for Melt Generation beneath Continents: Assessment of a Conductive Heating Model and Application to the Paraná. *J. Geophys. Res.* **1996**, *101*, 11503–11518. [CrossRef]

55. Peate, D.W.; Hawkesworth, C.J.; Mantovani, M.M.S.; Rogers, N.W.; Turner, S.P. Petrogenesis and Stratigraphy of the High-Ti/Y Urubici Magma Type in the Parana Flood Basalt Province and Implications for the Nature of 'Dupal'-Type Mantle in the South Atlantic Region. *J. Petrol.* **1999**, *40*, 451–473. [CrossRef]

56. Zhang, Y.-S.; Tanimoto, T. Ridges, Hotspots and Their Interaction as Observed in Seismic Velocity Maps. *Nature* **1992**, *355*, 45–49. [CrossRef]

57. Wei, X.; Zhang, G.-L.; Castillo, P.R.; Shi, X.-F.; Yan, Q.-S.; Guan, Y.-L. New Geochemical and Sr-Nd-Pb Isotope Evidence for FOZO and Azores Plume Components in the Sources of DSDP Holes 559 and 561 MORBs. *Chem. Geol.* **2020**, *557*, 119858. [CrossRef]

58. Tatsumi, Y. The Subduction Factory: How It Operates in the Evolving Earth. *Gsa Today* **2005**, *15*, 4. [CrossRef]

59. Sun, W.; Teng, F.-Z.; Niu, Y.-L.; Tatsumi, Y.; Yang, X.-Y.; Ling, M.-X. The Subduction Factory: Geochemical Perspectives. *Geochim. Cosmochim. Acta* **2014**, *143*, 1–7. [CrossRef]

60. Lin, P.-N. Trace Element and Isotopic Characteristics of Western Pacific Pelagic Sediments: Implications for the Petrogenesis of Mariana Arc Magmas. *Geochim. Cosmochim. Acta* **1992**, *56*, 1641–1654. [CrossRef]

61. Rehkamper, M.; Hofmann, A.W. Recycled Ocean Crust and Sediment in Indian Ocean MORB. *Earth Planet. Sci. Lett.* **1997**, *147*, 93–106. [CrossRef]

62. Ahmad, Q.; Wille, M.; Rosca, C.; Labidi, J.; Schmid, T.; Mezger, K.; König, S. Molybdenum Isotopes in Plume-Influenced MORBs Reveal Recycling of Ancient Anoxic Sediments. *Geochem. Persp. Let.* **2022**, *23*, 43–48. [CrossRef]
63. Wang, X.; Huang, P.; Huang, H.; Hu, N. Geochemical and Isotopic Compositions of East Rift Lavas from the Manus Basin: Implications for the Origin of Subduction Components. *Geol. J.* **2020**, *55*, 7429–7442. [CrossRef]
64. Yu, H.-M.; Nan, X.-Y.; Wu, F.; Widom, E.; Li, W.-Y.; Kuentz, D.; Huang, F. Barium Isotope Evidence of a Fluid-Metasomatized Mantle Component in the Source of Azores OIB. *Chem. Geol.* **2022**, *610*, 121097. [CrossRef]
65. Leat, P.T.; Livermore, R.A.; Millar, I.L.; Pearce, J.A. Magma Supply in Back-Arc Spreading Centre Segment E2, East Scotia Ridge. *J. Petrol.* **2000**, *41*, 845–866. [CrossRef]
66. Grove, T.L.; Till, C.B.; Krawczynski, M.J. The Role of H_2O in Subduction Zone Magmatism. *Annu. Rev. Earth Planet. Sci.* **2012**, *40*, 413–439. [CrossRef]
67. Murphy, J.B.; Keppie, J.D.; Nance, R.D.; Dostal, J. Comparative Evolution of the Iapetus and Rheic Oceans: A North America Perspective. *Gondwana Res.* **2010**, *17*, 482–499. [CrossRef]
68. Nance, R.D.; Gutiérrez-Alonso, G.; Keppie, J.D.; Linnemann, U.; Murphy, J.B.; Quesada, C.; Strachan, R.A.; Woodcock, N.H. A Brief History of the Rheic Ocean. *Geosci. Front.* **2012**, *3*, 125–135. [CrossRef]

Journal of
*Marine Science
and Engineering*

Article

Detrital Zircon Provenance in the Sediments in the Southern Okinawa Trough

Bowen Zhu [1,2,3] and Zhigang Zeng [1,2,4,*]

1   CAS Key Laboratory of Marine Geology and Environment, Institute of Oceanology, Chinese Academy of Sciences, Qingdao 266071, China; zhubowen@qdio.ac.cn
2   University of Chinese Academy of Sciences, Beijing 100049, China
3   Center for Ocean Mega-Science, Chinese Academy of Sciences, Qingdao 266071, China
4   Laboratory for Marine Mineral Resources, Qingdao Pilot National Laboratory for Marine Science and Technology, Qingdao 266061, China
*   Correspondence: zgzeng@ms.qdio.ac.cn

Abstract: The provenance of sediments in the Southern Okinawa Trough since the late Holocene has been a controversial scientific issue during the past 20 years. Previous studies based on isotope proxies generally indicated Taiwanese rivers as the primary source in the Southern Okinawa Trough since the late Holocene. Based on the zircon U-Pb geochronology, this study identified how sediments from the Yangtze River/East China Sea shelf had contributed significantly to the Southern Okinawa Trough in the past 624 a BP. Notably, this study found two Paleoarchean zircon grains, which indicated they originated from older orogenic belts. These data shed new light on the provenance of sediments, and a partial supply from the mainland of China cannot be excluded.

Keywords: detrital zircon; sediment provenance; Okinawa Trough

1. Introduction

As a momentous "sink" in the East Asia continental margin, the Okinawa Trough is a crucial area to study continental-oceanic interactions and "source-to-sink" processes. Sediments in the Okinawa Trough record the evolutionary history of sea level, ocean circulation, East Asian monsoon, and human activities since the late Pleistocene [1,2]. During the Holocene, terrigenous sediments from various sources were deposited in the East China Sea (ECS) shelf and Okinawa Trough, such as large rivers on the Chinese mainland (the Yangtze River and Yellow River), the "mountain stream type" small and medium rivers on Taiwan Island, and small rivers on the Korean Peninsula and the Ryukyu Island Arc [2–7]. Owing to the sudden change in the Kuroshio Current to reach its present position approximately 7.1 ka [1], sediments in the Southern Okinawa Trough (SOT) record continuous climate and ocean signals, which provides geological evidence for tracing the sedimentary response of the evolution of Kuroshio Current. Simultaneously, the rapid deposition rate in the SOT indicates high sediment supply [8–11]. Therefore, defining the provenance of sediments in the SOT is essential to reveal the evolution of sedimentary environment.

In the last two decades, there have been several research projects conducted on provenance of sediments in the SOT. However, these have not yet led to decisive results. Due to the different research indicators, previous studies have obtained inconsistent conclusions on the provenance of the SOT since the Holocene. For example, records of Sr Nd isotopic compositions indicate that the provenances of ODP-1202B and H4-S3 were mainly derived from Taiwanese rivers during the past 3.0 ka BP [5,10]. However, a study based on the Sr-Pb isotopic composition of the sediments of RC14-91 core argued that loess and the sediments from the Yangtze River account for 40% of the total [12]. Moreover, the compositions of the total organic carbon (TOC) and total nitrogen (TN), hydrocarbons, long-chain n-alkanes, and fatty acids in surface sediments from the SOT are different from Taiwan

Citation: Zhu, B.; Zeng, Z. Detrital Zircon Provenance in the Sediments in the Southern Okinawa Trough. *J. Mar. Sci. Eng.* **2022**, *10*, 142. https://doi.org/10.3390/jmse10020142

Academic Editor: Antoni Calafat

Received: 2 December 2021
Accepted: 18 January 2022
Published: 21 January 2022

Publisher's Note: MDPI stays neutral with regard to jurisdictional claims in published maps and institutional affiliations.

river sediments [13,14], indicating that there may be other potential sources in the SOT. Notably, under the ocean circulation system of the Western Pacific since the Holocene, it seems improbable for the sediments from the Yangtze River or the ECS to enter the SOT. Therefore, it is necessary to use more accurate provenance fingerprints to determine whether the Yangtze River or the East China Sea are potential sources for the sediments of the SOT.

The detrital zircon geochronology, which has been used in the ECS [6], the South China Sea [15,16], and the Yellow Sea [17], has become one method used to trace the provenance of sediments in the Chinese marginal seas. In this study, we also reconsidered previous hypothesis, and report detrital zircon U-Pb age distributions from H4-S2, aiming to identify the sediment provenance in the SOT since 624 a BP.

2. Regional Setting

The ECS has a wide continental shelf with complex terrain and hydrological conditions, so sediments are commonly affected by two sources: large and medium rivers in the southeastern China mainland (such as the Yangtze River, Oujiang, and Minjiang) and small and medium rivers (such as Zhuoshui river) in Taiwan [18]. The sediment discharge of the Yangtze River has reached 680 Mt/a historically [19], but with the construction of the Three Gorges Dam, this amount has reduced to 100 Mt/a [20]. At the same time, Minjiang and Oujiang, along the southeastern coast of the Chinese mainland, also have significant sediment loads, contributing 17–20 Mt of sediments per year [21]. On the other side of the Taiwan Strait, Zhuoshui river, the longest river in Taiwan, carries 54 Mt of sediments per year into the ECS [22]. As an essential provenance area of SOT, Lanyang river in northeastern Taiwan can transport 6–9 Mt of sediments to the SOT every year [23–25]. The above rivers are the potential provenance areas of sediments in this study.

Ocean currents affect the transport of sediments in the ECS. For example, sediments from the Yangtze River, under the action of the Zhejiang-Fujian Coastal Current (ZFCC) and Changjiang Diluted Water, affect the composition of sediments in most areas of the East China Sea [26]. Meanwhile, most of the sediments from the Changjiang are confined to the East China Sea shelf by the Taiwan Warm Current, forming the mud wedge, and are difficult to transport to the area east of 123° E [26,27].

The area around SOT has developed complex topography characterized by the ECS slope, the North Mien-Hua Canyon, the Mien-Hua Canyon, Keelung Valley, and I-Lan Ridge (Figure 1). The hydrologic environment in this region is influenced by ocean large-scale and mesoscale ocean dynamic processes, including the Kuroshio Current, internal tides, upwelling, and cyclonic eddies (Figure 1b) [28–32].

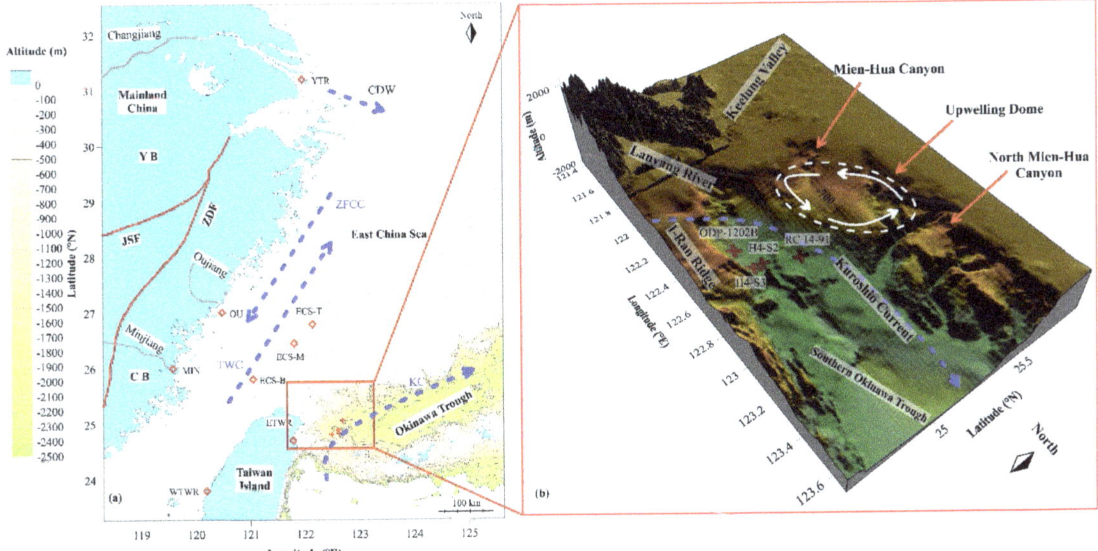

Figure 1. (**a**) Currents, cores, and potential provenance areas distribution in the East China Sea (modified from references [5,8,26,33–35]), and (**b**) the location of cores in the Southern Okinawa Trough. YTR, the Yangtze River (Changjiang) mouth; OU, Oiujiang; MIN, Minjiang; ECS-T, ECS-M, and ECS-B, East China Sea shelf; WTWR, Zhuoshui river mouth; ETWR, Lanyang river mouth. CDW, Changjiang Diluted Water; ZFCC, the Zhejiang-Fujian Coastal Current; TWC, the Taiwan Warm Current; KC, Kuroshio Current. Y B, the Yangtze Block; C B, the Cathaysia Block; JSF, Jiang-Shao Fault; ZDF, Zhenghe-Dapu Fault. The location of upwelling dome, valley, ridge, and canyon modified from reference [8]. The pathways of currents and upwelling dome in this figure are not a representation of the actual location. The potential provenance areas are represented by red diamonds, with specific latitude and longitude from the references [6,33,36,37]. Cores in the Southern Okinawa Trough are represented by red crosses, with specific locations from the references [8,10,12]. Topographic data comes from https://www.gebco.net/ (accessed on 2 December 2021).

3. Materials and Methods

3.1. Samples and Age Model

H4-S2, a 477 cm core on the ocean floor at 1505 m in depth, located in the SOT (Figure 1b). This study focused on the detrital zircon from five layers in H4-S2 (26–66 cm, 112–152 cm, 192–232 cm, 332–372 cm, and 402–442 cm) (Figure 2). The depositional age at 477 cm is 624 a BP (Figure 2).

3.2. Methods

3.2.1. Detrital Zircon U-Pb Age and Th/U Analysis

We collected five samples from H4-S2 (Figure 2) to investigate the provenance of the detrital zircons. The details of the samples are provided in Table S2. Since samples are rare, only 400 g of sediment for each sample were processed. Zircon grains were extracted from the sediments by using conventional heavy liquid and magnetic separation procedures.

For detrital zircon U-Pb age analysis samples, more than 1000 grains were selected. A subset of 400 grains was randomly selected under a binocular microscope, fixed, transferred to an epoxy mount, and polished to expose the midsection. Cathodoluminescence (CL) images were used to obtain the internal structure of grains and aid in selecting dating points. The above work was completed at the Institute of Resources of the Chinese Academy of Geological Sciences.

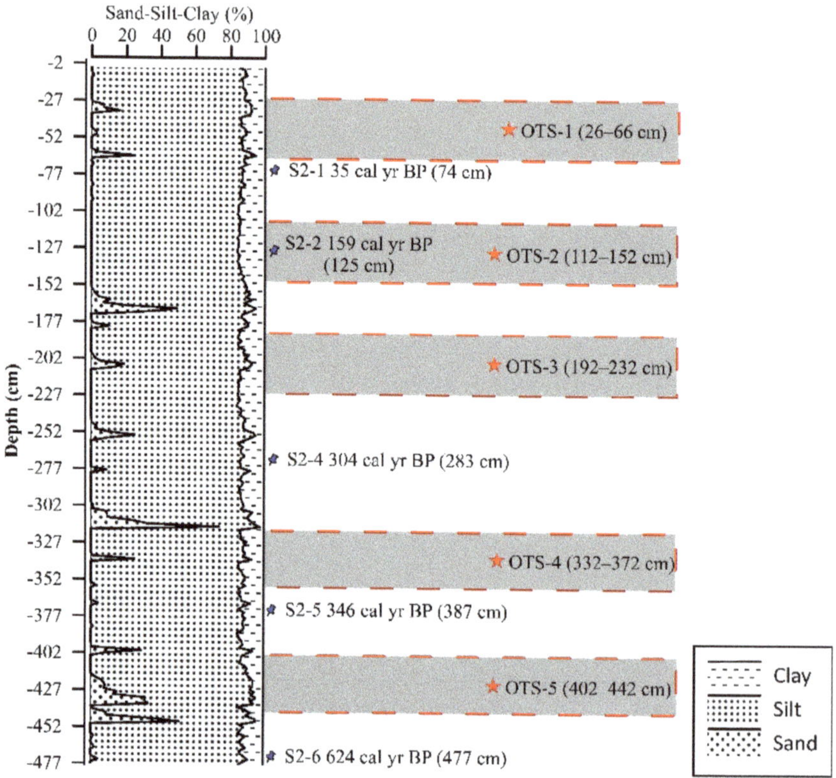

Figure 2. The sedimentological column of the H4-S2. The level of detrital zircon sample is marked by red stars and gray rectangles with red boundaries, and its depth placed on the right. The grain size and dating data of H4-S2 are sourced from reference [38].

Zircon U-Pb dating and Th/U analysis were conducted using the LA-ICP-MS instrument in the Mineral and Fluid Inclusion Microanalysis Laboratory of the Institute of Geology, Chinese Academy of Geological Sciences, Beijing. The NWR 193UC laser ablation system (Elemental Scientific Lasers, Omaha, NE, USA) is equipped with a Coherent Excistar 200 excimer laser and a Two Volume 2 ablation cell. The ablation system was coupled to an Agilent 7900 ICP-MS (Agilent, Santa Clara, CA, USA). An external energy meter was used to ensure that the input laser fluence value matched the actual energy of the sample well before analysis. The zircon mounts were cleaned ultrasonically in ultrapure water. Before analysis, the mounts were cleaned again using AR-grade methanol, and each spot was preablated for five shots (~0.3 μm in depth) to remove potential surface contamination. The analyses were performed using a 25 μm diameter spot size at 5 Hz, 2 J/cm^2 laser fluence.

The Iolite software package was used for data reduction. Zircon 91500 was used as the primary standard, and GJ-1 and Plešovice were used as secondary standard. The 91500 standard was analyzed twice, and both GJ-1 and Plešovice were analyzed once every 10–12 analyses for the sample. Typically, 35–40 s of the sample signals was acquired after 20 s of gas background measurement. The exponential function was used to calibrate the downhole fractionation. NIST 610 and 91Zr were used to calibrate the trace element concentrations as external reference materials and internal standard elements, respectively. The measured ages of the reference materials in this batch are listed as follows: 91500 (1061.5 ± 3.2 Ma, 2σ), GJ-1 (604 ± 6 Ma, 2σ), and Plešovice (340 ± 4 Ma, 2σ), which agreed with the nominal values well within uncertainty.

ICPMSDataCal9.2 was used to process the experimental data obtained through the above method [39]. For detrital zircon with a ^{207}Pb/^{206}Pb age less than 1400 Ma, ^{206}Pb/^{238}U ages with a degree of concordance greater than 90% were selected, and for detrital zircons with a ^{207}Pb/^{206}Pb age greater than 1400 Ma, ^{207}Pb/^{206}Pb ages greater than 90% were selected [40–42].

3.2.2. Visual Analysis of the Detrital Zircon Age Distribution

The detrital zircon U-Pb age distribution was visualized as kernel density estimation (KDE) plots with an adaptive bandwidth of 30 Ma, and multidimensional scaling (MDS) map by using an R package for statistical provenance analysis [43]. The above method can transform the differences in age distribution into the shape of the curve and the distance in two-dimensional space. The more similar the distribution is, the more similar the shape will be and the smaller the distance will be.

4. Results

4.1. Detrital Zircon Grain-Size and U-Pb Age Distribution of the Southern Okinawa Trough over the Past 700 Years

Most zircon grains are igneous in origin according to oscillatory zoning (Figure 3a) and Th/U > 0.1 (Figure 3b) [33,44,45]. Thus, 75–93 valid ages (427 in total) were obtained for each sample (Table S1).

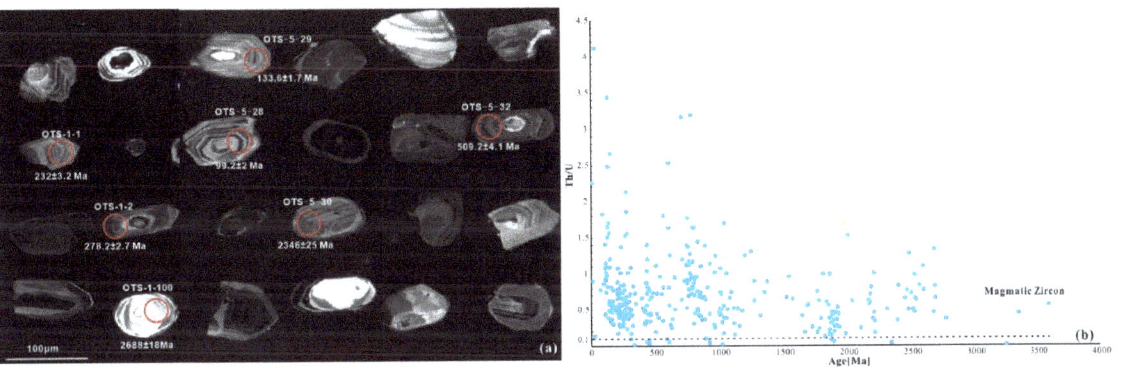

Figure 3. Cathodoluminescence images and U-Th ratios of detrital zircon grains from sediments in H4-S2. (**a**) Cathodoluminescence images; (**b**) U-Th ratios. Red circles indicate the dating points.

Since the plot of 25 µm was selected, the size of the zircon grains was relatively fine. The equivalent spherical diameter (ESD), which is the cube root of the product of lengths of the three axes [46], was distributed from 36.21–89.18 µm and mainly distributed from 36.21–62.92 µm. The mean ESD of analyzed zircon grains in OTS-1 to OTS-5 are 50.83, 48.89, 52.47, 53.07, and 61.08 µm, respectively (Table S2).

All samples show seven primary groups: 200–100 Ma, 300–200 Ma, 500–400 Ma, 900–700 Ma, 1.1–1.0 Ga, 2.0–1.8 Ga, and 2.7–2.5 Ga (Figure 4). However, there are some differences between each sample; for example, the main peaks in OTS-1 and OTS-5 appear in the 200–100 Ma group, while peaks in OTS-2 and OTS-4 lie from 300 to 200 Ma, and the peak of OTS-3 is older than others and falls in the 900–700 Ma group. Moreover, OTS-4 has more grains distributed in the range of 1.2–1.0 Ga, while other samples did not show a similar situation. Furthermore, Cenozoic grains all appear in five samples, including OTS-1-92 (28.53 ± 0.86 Ma), OTS-2-94 (29.12 ± 0.3 Ma), OTS-3-83 (58.6 ± 1.6 Ma), OTS-4-83 (9.73 ± 0.43 Ma), and OTS-5-65 (13.75 ± 0.64 Ma) (Table S1). Paleoarchean grains are only found at OTS-2-31 (3597 ± 16 Ma), and OTS-4-51 (3282 ± 12 Ma) (Table S1).

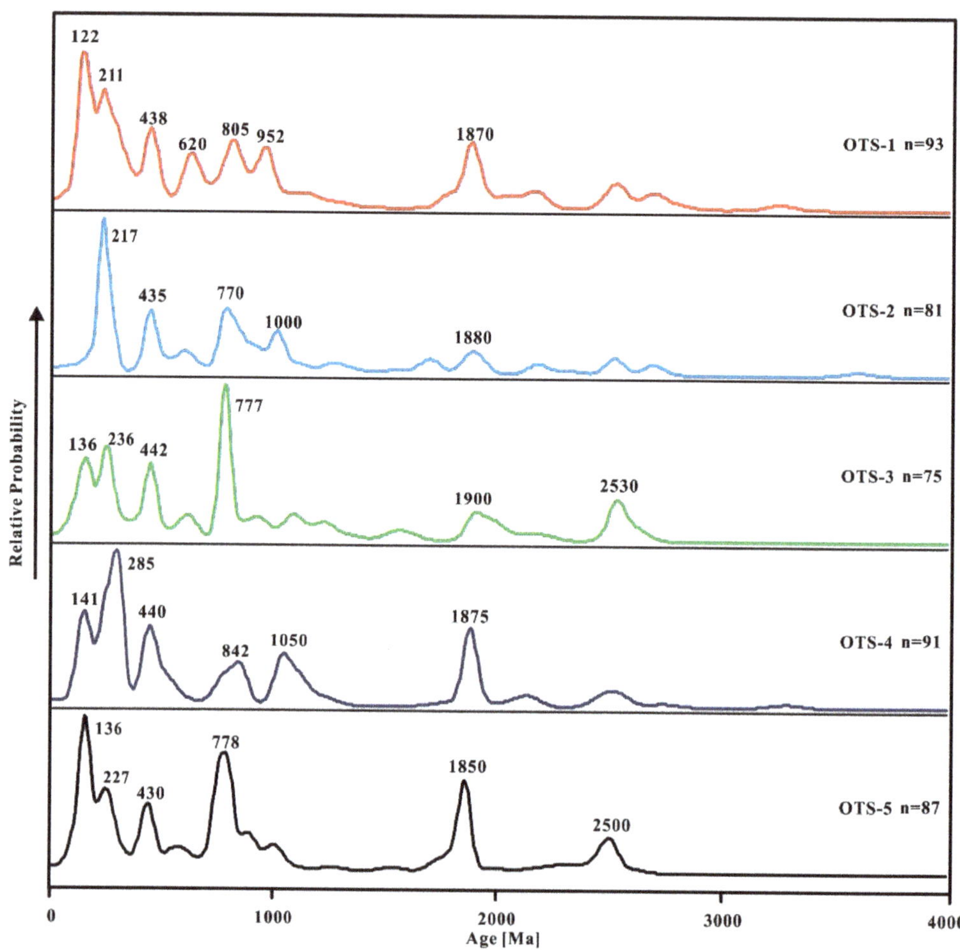

Figure 4. Kernel density estimation (KDE) plots for detrital zircon U-Pb ages of OTS-1 to OTS-5 in H4-S2. Please refer to Table S1 for detailed data. The results of OTS-1 and OTS-2 are reprocessing of the previous data [47]. The values of major age peaks are marked in this figure.

4.2. Detrital Zircon U-Pb Age Distribution from the Potential Provenance of the Southern Okinawa Trough

Recent studies have shown that detrital zircon from sediments in modern rivers may well reproduce the expected age distribution of the zircon-bearing bedrock source area [33,48]. To better define the provenance of detrital zircon grains in the southern Okinawa Trough, we consider the East China Sea shelf, Yangtze River, Lanyang river, Zhuoshui river, Minjiang, and Oujiang as potential provenance areas (Figure 1a).

The U-Pb age distribution of the detrital zircon in the potential provenance areas is mainly characterized by the following characteristics (Figure 5): (1) The detrital zircon in the sediments of the Lanyang river and Zhuoshui river is represented by seven main age groups: 200–100 Ma, 300–200 Ma, 550–360 Ma, 850–700 Ma, 1.1–0.9 Ga, 2.0–1.8 Ga, and 2.6–2.4 Ga. The age distribution of the Zhuoshui river is relatively simple, with a higher proportion of Phanerozoic zircons, but the Zhuoshui river has a more complex age composition and more Precambrian zircons [33]. (2) The detrital zircon of Changjiang-derived sediments shows five major age groups: 300–100 Ma, 600–400 Ma, 900–700 Ma, 2.0–1.7 Ga, and 2.6–2.2 Ga [36,49]. The distribution characteristics of Changjiang-derived and Huanghe-derived sediments are completely different, with four main age groups: 500–200 Ma, 1000–800 Ma, 2.0–1.5 Ga, and 2.8–2.2 Ga [50]. In addition, the age peak of the Huanghe-derived sediments appeared in the Paleoproterozoic, which is also an important signal to distinguish the two provenance areas (Changjiang and Huanghe). (3) The age distribution of the detrital zircon in Oujiang is simple, distributed in the ranges of 200–60 Ma and 2.4–1.8 Ga; (4) Minjiang is mainly distributed in the range of 1.0 Ga–60 Ma, and the remaining particles were distributed in the range of 3.3–1.2 Ga [36,37,51]. The age ranges of Oujiang and Minjiang are obviously younger than those of Changjiang and Huanghe, which also provides a basis for distinguishing different provenance areas for this study. (5) Due to the different locations, the age distribution of the East China Sea shelf is different. ECS-B has four main age groups: 600–60 Ma, 1.2 Ga–700 Ma, 1.9–1.6 Ga, and 2.7–2.4 Ga; ECS-M has two main age groups: 500–3 Ma and 1000–600 Ma; and ECS-T has three main age groups: 1.2 Ga–100 Ma, 2.1–1.8 Ga, and 2.6–2.4 Ga [6].

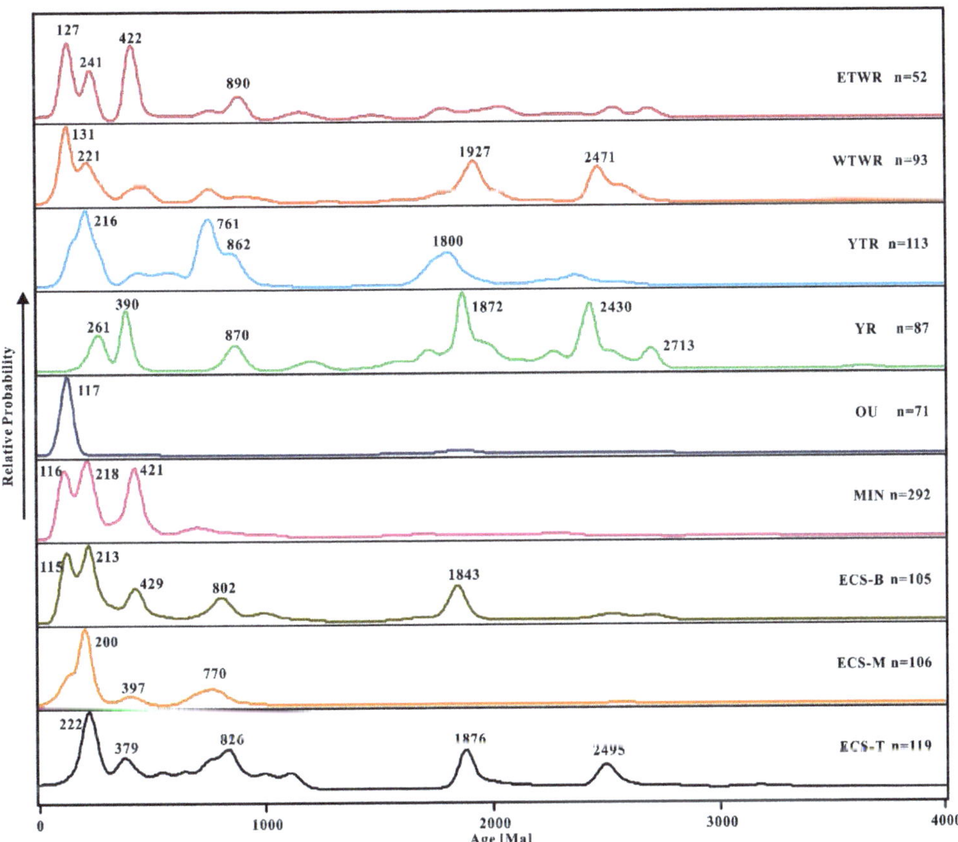

Figure 5. Kernel density estimation (KDE) plots for detrital zircon U-Pb ages of Lanyang river mouth [33], Zhuoshui river mouth [33], the Yangtze River mouth [36], Yellow River mouth [50], Oujiang mouth [34], Minjiang mouth [37], and the East China Sea shelf [6]. Please refer to Table S1 for detailed data. The values of major age peaks are marked in this figure.

5. Discussion

5.1. Detrital Zircon Provenance of the Southern Okinawa Trough in the Past 700 Years

A recent study suggested that the detrital zircon may not be representative of the provenance of bulk sediments [52]. Therefore, the provenance identification results of our study may only indicate the provenance of detrital zircons.

The 700–900 Ma age component, especially the peak at ~761 Ma, is the diagnostic characteristic of the Yangtze River [36]. This identification sign of the Yangtze River is of great significance to distinguish other potential provenance areas of SOT, since there is no significant peak at ~761 Ma in other provenance areas except the Yangtze River (Figure 5). This component and peak prevail in OTS-3 and OTS-5 (Figure 4), indicating that sediments from the Yangtze River and southern Okinawa Trough are of the same origin. In addition to the 700–900 Ma age group, 100–200 Ma, 200–300 Ma, and 400–500 Ma are also the primary age components in H4-S2 (Figure 4). These groups cannot be used to accurately determine their sources since they coexist in the Lanyang river, Zhuoshui river, Minjiang, and the East China Sea shelf (Figure 5). The same feature is also present in the 1.8–2.0 Ga and 2.5–2.7 Ga age groups. However, it is simpler due to the fact that only the Zhuoshui river, the Yangtze River, and the East China Sea shelf provide for these groups. Although the age distribution characteristics of the Yellow River and Oujiang are listed in this study (Figure 5), they are not potential provenance areas for the following reasons: (1) Oujiang-derived sediments only contain Mesozoic grains, which are not similar to any H4-S2 samples; (2) the modern Yellow River estuary is so far away from the southern Okinawa Trough that there is no evidence of sediments migrating to this region; and (3) sediments from the Yellow River are characterized by high concentrations of Archean zircon, which are completely different from H4-S2 (Figures 4 and 5). The MDS maps more clearly show the source-sink relationship between H4-S2 and provenance areas (Figure 6). OTS-2, OTS-3, and OTS-5 were close to YTR and ECS-T. OTS-1 and OTS-4 were closer to ETWR (Figure 6). The results of MDS indicated that Lanyang river may be the source of detrital zircon grains in OTS-1 and OTS-4. Grains in OTS-2, OTS-3 and OTS-5 may be derived from the East China Sea shelf and Yangtze River. All of the provenance analyses indicated that the detrital zircon, which was found in H4-S2 in the past 624 a BP, originated from a mixture of sediment supplies from Taiwanese rivers, the East China Sea shelf, and the Yangtze River. This understanding is consistent with previous provenance identification results of geochemistry and mineralogy in this region [8,12,53,54]. Thus, it is believed that detrital zircons in the SOT have consistently recorded sediment supplies from the Yangtze River and the East China Sea shelf since 624 a BP.

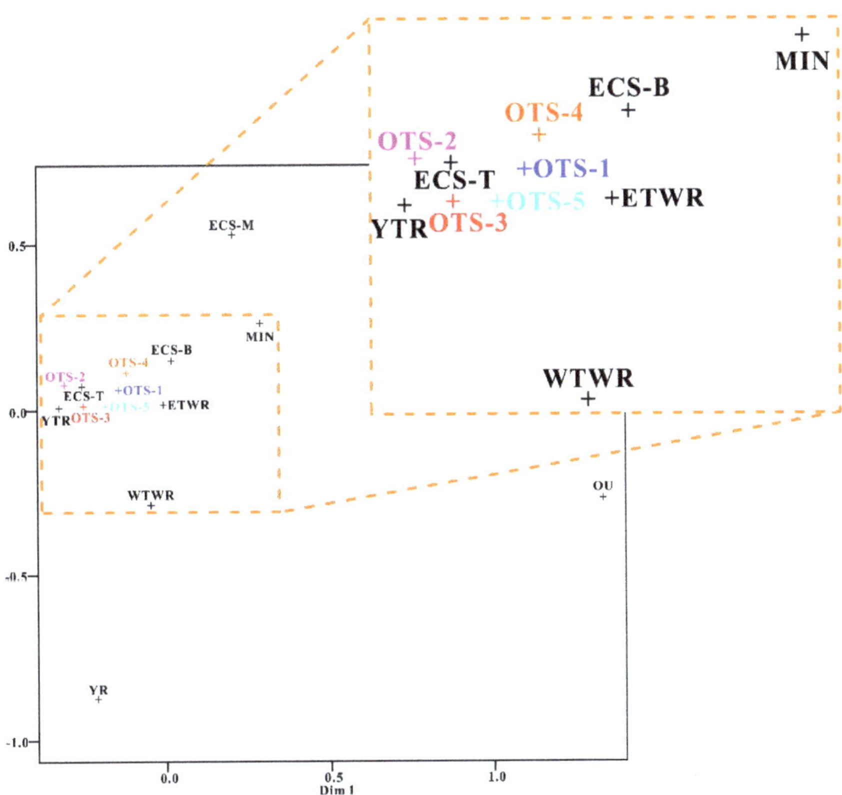

Figure 6. Multidimensional scaling (MDS) plot for detrital zircon U-Pb ages of H4-S2 and potential provenance areas. The orange dotted line shows an enlargement of part of this figure.

Most grains analyzed in this study have a size of 36.21–89.18 μm and are mainly concentrated in the silty fraction, which is common in H4-S2 (Figure 2). In previous studies, the grain-size effect on the detrital zircon age distributions was mentioned [55]. For the case of this study, although the median grain size distributions of detrital zircon in OTS-1, OTS-2, OTS-3, and OTS-4 are relatively finer than OTS-5, no specific correlation of zircon age and grain size was noted (Figure 7). There was also no correlation between the grain size and zircon age in the same sample, meaning that older zircon grains were not necessarily finer (Figure 7).

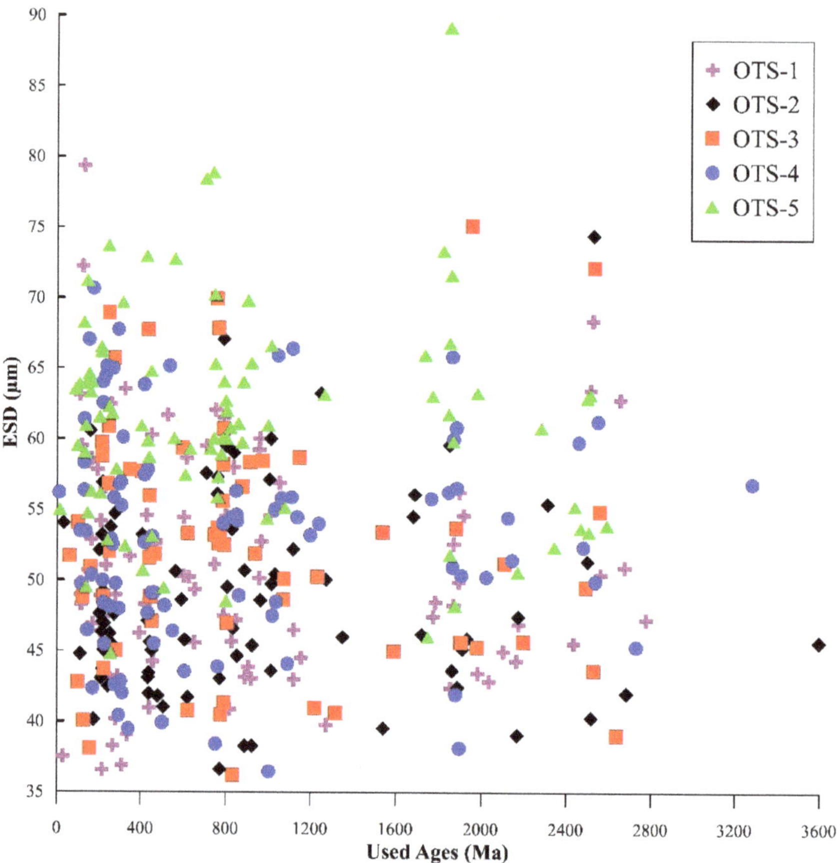

Figure 7. Distribution plots of detrital zircon U-Pb ages and ESD which be used in this study. Please refer to Table S2 for detailed data.

5.2. Paleoarchean Detrital Zircon: Singal or Noise?

Although there are only two Paleoarchean detrital zircon grains (OTS-2-31: 3597 ± 16 Ma and OTS-4-51: 3282 ± 12 Ma), they may be important provenance indicators since the provenance areas around the southern Okinawa Trough could not provide such old grains. In order to find the source of these two grains, we collected the reported Paleoarchean grains from the south China (the Yangtze Block and Cathaysia Block). Detailed locations, data, and references are listed in Table S3. As shown in Figure 8, Paleoarchean zircon grains appear to be distributed in various regions throughout the south China. Grains of similar age exist in the sediments of the middle and upper reaches of the Minjiang river [37], which should be direct evidence that Paleoarchean grains from the south China could through the Minjiang river to the East China Sea. Another evidence is that Paleoarchean detrital zircon has been found in the main stream of the Yangtze River (Hukou, Shigu and Panzhihua) and its tributaries (Hanjiang and Ganjiang) [36,49].

Therefore, the two Paleoarchean grains in this study may not be noise, but weak signal to indicate provenance. Although sporadic detrital zircon data from the East China Sea do not contain such old zircons [6], ancient grains from the Yangtze River, Yangtze Block, and Cathaysia Block may all be sources of OTS-2-31 and OTS-4-51.

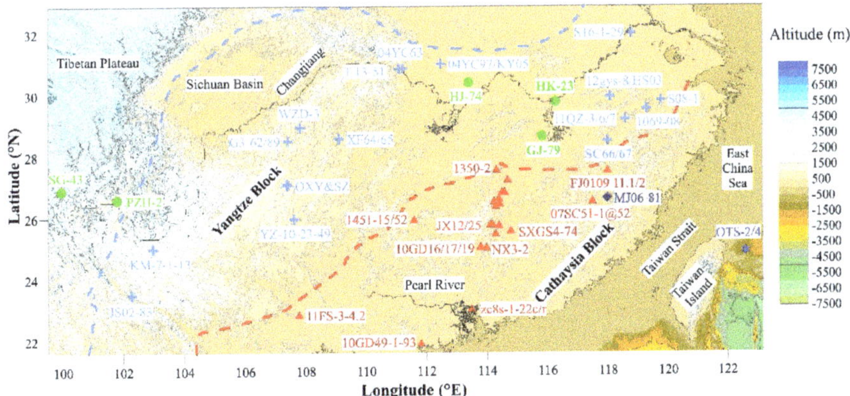

Figure 8. Distribution of Paleoarchean detrital zircon grains in the south China and East China Sea. Grains in the Yangtze Block, Cathaysia Block, Yangtze River Basin, and Minjiang are marked by light blue cross, red triangle, green circle, and dark blue diamond, respectively. The location of the Yangtze Block and the Cathaysia Block modified from references [33–35]. Please refer to Table S3 and references therein for detailed data of ages and location. The light blue and red dotted lines roughly define the position of the Yangtze Block and the Cathaysia Block, respectively. Topographic data comes from https://www.gebco.net (accessed on 1 December 2021).

5.3. Why the Yangtze River Zircons Appear in the Souerhtern Okinawa Trough?

Previous studies on the sediment sources of ECS indicated that ECS shelf received a large number of materials from the Yangtze River, and the contribution rate of the Yangtze River could reach 72% and even higher [6,56]. Similarly, physical oceanography and heavy mineral studies have shown that the Yangtze River sediments can be transported southward to the East China Sea Shelf via the ZFCC and transported further east by the puncture front [26,57,58]. Moreover, the ECS shelf is characterized by the extensive development of residual deposits [59]. With increasing offshore distance, the grain size of surface sediments in the ECS shelf is gradually enhanced [26], and the content of stable minerals increased significantly [60], indicating a large amount of Pleistocene residual sand accumulation on the outer shelf and slope of the East China Sea. Hence, the sediments from the ancient and modern Yangtze River may be jointly preserved in the outer shelf and slope of the ECS and become the source for sediments in the SOT.

Previous studies have suggested that mass wasting and density currents caused by occasional events, such as earthquakes, typhoons, and heavy floods, in the hinterland resulted in the transport of sediments on the slope to the seafloor [23,61]. In addition, the widely developed submarine canyon at the junction of the East China Sea shelf and Okinawa Trough is a natural channel for sediment transport [62]. Particles from the shelf are transported by grain flows through submarine canyons into the Okinawa Trough [63]. Many coarse-grained layers in H4-S2 record the extensive development of turbidity current events in the past 624 a BP [38]. Therefore, the detrital zircons in H4-S2 are derived from the sediments of the modern Yangtze River and the residual deposits on the ECS. ZFCC may be an important mechanism for the transport of sediments from the Yangtze River to the ECS shelf. Gravity flow may be the transport mechanism of sediment from the ECS to the SOT. This could also explain the presence of Paleoarchean detrital zircons in H4-S2. Since this study only reported zircon data from five layers, it is hard to verify whether this transport process is continuous, especially with gravity flow.

6. Conclusions

Based on detrital zircon U-Pb geochronology provenance identification, our study provides direct evidence that sediments from the Southern Okinawa Trough recorded sediments from Taiwan rivers, the East China Sea shelf, and the Yangtze River in the last 700 years. At the same time, two Paleoarchean detrital zircon grains probably originated from the Cathaysia Block and Yangtze Block in the south China and were transported into the East China Sea by the Yangtze River or Minjiang river. Detrital zircon grains in this study are mostly in the silty fraction and may only indicate the provenance of the silt in the sediments or only the provenance of the detrital zircon. Detrital zircons from the modern Yangtze River were transported by the Min-Zhejiang coastal current, and settled on the East China Sea shelf. They and residual sediments in the East China Sea shelf were transported by gravity flow through submarine canyons and eventually deposited in the Southern Okinawa Trough.

Detrital zircon U-Pb geochronology is a powerful tool for distinguishing the sediments from Taiwan rivers, the East China Sea shelf, and the Yangtze River. A deep understanding of the quantitative provenance analysis and marine dynamics evolution in the Southern Okinawa Trough may be improved by the availability of longer time-scale cores and more grains of detrital zircon.

Supplementary Materials: The following are available online at https://www.mdpi.com/article/10.3390/jmse10020142/s1. Table S1: Isotopic data and U-Pb ages of detrital zircon in H4-S2 and potential provenance areas; Table S2: ESD compositions of detrital zircon grains in H4-S2; Table S3: Locations, ages, and references of Paleoarchean detrital zircon grains in the south China.

Author Contributions: Methodology, B.Z.; software, B.Z.; validation, B.Z.; investigation, B.Z.; resources, Z.Z.; data curation, B.Z.; writing—original draft preparation, B.Z.; writing—review and editing, B.Z.; visualization, B.Z.; supervision, Z.Z.; project administration, Z.Z.; funding acquisition, Z.Z. All authors have read and agreed to the published version of the manuscript.

Funding: This work was funded by the National Natural Science Foundation of China (Grant No. 91958213), the International Partnership Program of Chinese Academy of Sciences (Grant No. 133137KYSB20170003), the Special Fund for the Taishan Scholar Project of the Shandong Province (Grant No. ts201511061), the National Special Fund for the 13th Five Year Plan of COMRA (Grant No. DYI35-G2–01–02), and the National Basic Research Program of China (Grant No. 2013CB429700), the Natural Science Foundation of Shandong Province (Grant No. ZR2019QD016).

Institutional Review Board Statement: Not applicable.

Informed Consent Statement: Not applicable.

Data Availability Statement: All data and references involved in this study are listed in the Tables S1–S3.

Conflicts of Interest: The authors declare no conflict of interest.

References

1. Li, G.; Li, P.; Liu, Y.; Qiao, L.; Ma, Y.; Xu, J.; Yang, Z. Sedimentary system response to the global sea level change in the East China Seas since the last glacial maximum. *Earth-Sci. Rev.* **2014**, *139*, 390–405. [CrossRef]
2. Li, Q.; Zhang, Q.; Li, G.; Liu, Q.; Chen, M.-T.; Xu, J.; Li, J. A new perspective for the sediment provenance evolution of the middle Okinawa Trough since the last deglaciation based on integrated methods. *Earth Planet. Sci. Lett.* **2019**, *528*, 115839. [CrossRef]
3. Dou, Y.; Yang, S.; Liu, Z.; Clift, P.D.; Yu, H.; Berne, S.; Shi, X. Clay mineral evolution in the central Okinawa Trough since 28 ka: Implications for sediment provenance and paleoenvironmental change. *Palaeogeogr. Palaeoclimatol. Palaeoecol.* **2010**, *288*, 108–117. [CrossRef]
4. Dou, Y.; Yang, S.; Liu, Z.; Shi, X.; Li, J.; Yu, H.; Berne, S. Sr-Nd isotopic constraints on terrigenous sediment provenances and Kuroshio Current variability in the Okinawa Trough during the late Quaternary. *Palaeogeogr. Palaeoclimatol. Palaeoecol.* **2012**, *365*, 38–47. [CrossRef]
5. Dou, Y.; Yang, S.; Shi, X.; Clift, P.D.; Liu, S.; Liu, J.; Li, C.; Bi, L.; Zhao, Y. Provenance weathering and erosion records in southern Okinawa Trough sediments since 28 ka: Geochemical and Sr-Nd-Pb isotopic evidences. *Chem. Geol.* **2016**, *425*, 93–109. [CrossRef]
6. Huang, X.; Song, J.; Yue, W.; Wang, Z.; Mei, X.; Li, Y.; Li, F.; Lian, E.; Yang, S. Detrital zircon U-Pb ages in the east China seas: Implications for provenance analysis and sediment budgeting. *Minerals* **2020**, *10*, 398. [CrossRef]
7. Xu, Z.; Lim, D.; Li, T.; Kim, S.; Jung, H.; Wan, S.; Chang, F.; Cai, M. REEs and Sr-Nd isotope variations in a 20 ky-sediment core from the middle Okinawa Trough, East China Sea: An in-depth provenance analysis of siliciclastic components. *Mar. Geol.* **2019**, *415*, 105970. [CrossRef]

8.	Diekmann, B.; Hofmann, J.; Henrich, R.; Fuetterer, D.K.; Roehl, U.; Wei, K.-Y. Detrital sediment supply in the southern Okinawa trough and its relation to sea-level and Kuroshio dynamics during the late Quaternary. *Mar. Geol.* **2008**, *255*, 83–95. [CrossRef]

9.	Feng, X.; Wu, Y.; Yang, B.; Shan, X.; Liu, J. Records of hyperpycnal flow deposits in the southwestern okinawa trough and their paleoclimatic response since 1.3 ka. *Acta Sedimentol. Sin.* **2021**, *39*, 739–750. (In Chinese with English abstract)

10.	Hu, S.; Zeng, Z.; Fang, X.; Yin, X.; Chen, Z.; Li, X.; Zhu, B.; Qi, H. Increasing terrigenous sediment supply from Taiwan to the southern Okinawa Trough over the last 3000 years evidenced by Sr-Nd isotopes and geochemistry. *Sediment. Geol.* **2020**, *406*, 105725. [CrossRef]

11.	Li, C.; Jiang, B.; Li, A.; Li, T.; Jiang, F. Sedimentation rates and provenance analysis in the southwestern Okinawa Trough since the mid-Holocene. *Chin. Sci. Bull.* **2009**, *54*, 1234–1242. [CrossRef]

12.	Bentahila, Y.; Ben Othman, D.; Luck, J.-M. Strontium, lead and zinc isotopes in marine cores as tracers of sedimentary provenance: A case study around Taiwan orogen. *Chem. Geol.* **2008**, *248*, 62–82. [CrossRef]

13.	Jeng, W.L.; Huh, C.A. A comparison of sedimentary aliphatic hydrocarbon distribution between the southern Okinawa Trough and a nearby river with high sediment discharge. *Estuar. Coast. Shelf Sci.* **2006**, *66*, 217–224. [CrossRef]

14.	Kao, S.J.; Lin, F.J.; Liu, K.K. Organic carbon and nitrogen contents and their isotopic compositions in surficial sediments from the East China Sea shelf and the southern Okinawa Trough. *Deep. Sea Res. Part II Top. Stud. Oceanogr.* **2003**, *50*, 1203–1217. [CrossRef]

15.	Shao, L.; Cao, L.; Pang, X.; Jiang, T.; Qiao, P.; Zhao, M. Detrital zircon provenance of the Paleogene syn-rift sediments in the northern South China Sea. *Geochem. Geophys. Geosyst.* **2016**, *17*, 255–269. [CrossRef]

16.	Zhong, L.; Li, G.; Yan, W.; Xia, B.; Feng, Y.; Miao, L.; Zhao, J. Using zircon U-Pb ages to constrain the provenance and transport of heavy minerals within the northwestern shelf of the South China Sea. *J. Asian Earth Sci.* **2017**, *134*, 176–190. [CrossRef]

17.	Choi, T.; Lee, Y.I.; Orihashi, Y.; Yi, H.-I. The provenance of the southeastern Yellow Sea sediments constrained by detrital zircon U-Pb age. *Mar. Geol.* **2013**, *337*, 182–194. [CrossRef]

18.	Liu, X.; Li, A.; Dong, J.; Lu, J.; Huang, J.; Wan, S. Provenance discrimination of sediments in the Zhejiang-Fujian mud belt, East China Sea. Implications for the development of the mud depocenter. *J. Asian Earth Sci.* **2018**, *151*, 1–15. [CrossRef]

19.	Oguri, K.; Matsumoto, E.; Yamada, M.; Saito, Y.; Iseki, K. Sediment accumulation rates and budgets of depositing particles of the East China Sea. *Deep. Sea Res. Part II Top. Stud. Oceanogr.* **2003**, *50*, 513–528. [CrossRef]

20.	Yang, S.Y.; Jung, H.S.; Lim, D.I.; Li, C.X. A review on the provenance discrimination of sediments in the Yellow Sea. *Earth-Sci. Rev.* **2003**, *63*, 93–120. [CrossRef]

21.	Jia, J.; Gao, J.; Cai, T.; Li, Y.; Yang, Y.; Wang, Y.P.; Xia, X.; Li, J.; Wang, A.; Gao, S. Sediment accumulation and retention of the Changjiang (Yangtze River) subaqueous delta and its distal muds over the last century. *Mar. Geol.* **2018**, *401*, 2–16. [CrossRef]

22.	Dadson, S.J.; Hovius, N.; Chen, H.G.; Dade, W.B.; Hsieh, M.L.; Willett, S.D.; Hu, J.C.; Horng, M.J.; Chen, M.C.; Stark, C.P.; et al. Links between erosion, runoff variability and seismicity in the Taiwan orogen. *Nature* **2003**, *426*, 648–651. [CrossRef] [PubMed]

23.	Hsu, S.C.; Lin, F.J.; Jeng, W.L.; Chung, Y.C.; Shaw, L.M.; Hung, K.W. Observed sediment fluxes in the southwesternmost Okinawa Trough enhanced by episodic events: Flood runoff from Taiwan rivers and large earthquakes. *Deep. Sea Res. Part I Oceanogr. Res. Pap.* **2004**, *51*, 979–997. [CrossRef]

24.	Jeng, W.L.; Lin, S.; Kao, S.J. Distribution of terrigenous lipids in marine sediments off northeastern Taiwan. *Deep. Sea Res. Part II Top. Stud. Oceanogr.* **2003**, *50*, 1179–1201. [CrossRef]

25.	Kao, S.J.; Liu, K.K. Estimating the suspended sediment load by using the historical hydrometric record from the Lanyang-Hsi watershed. *Terr. Atmos. Ocean. Sci.* **2001**, *12*, 401–414. [CrossRef]

26.	Zhang, K.; Li, A.; Huang, P.; Lu, J.; Liu, X.; Zhang, J. Sedimentary responses to the cross-shelf transport of terrigenous material on the East China Sea continental shelf. *Sediment. Geol.* **2019**, *384*, 50–59. [CrossRef]

27.	Liu, J.P.; Xu, K.H.; Li, A.C.; Milliman, J.D.; Velozzi, D.M.; Xiao, S.B.; Yang, Z.S. Flux and fate of Yangtze river sediment delivered to the East China Sea. *Geomorphology* **2007**, *85*, 208–224. [CrossRef]

28.	Andres, M.; Wimbush, M.; Park, J.H.; Chang, K.I.; Lim, B.H.; Watts, D.R.; Ichikawa, H.; Teague, W.J. Observations of Kuroshio flow variations in the East China Sea. *J. Geophys. Res. Oceans* **2008**, *113*, C05013. [CrossRef]

29.	Chang, H.; Xu, Z.; Yin, B.; Hou, Y.; Liu, Y.; Li, D.; Wang, Y.; Cao, S.; Liu, A.K. Generation and propagation of M_2 internal tides modulated by the Kuroshio northeast of Taiwan. *J. Geophys. Res. Oceans* **2019**, *124*, 2728–2749. [CrossRef]

30.	Hsu, S.C.; Lin, F.J.; Jeng, W.L.; Tang, T.Y. The effect of a cyclonic eddy on the distribution of lithogenic particles in the southern East China Sea. *J. Mar. Res.* **1998**, *56*, 813–832. [CrossRef]

31	Nakamura, H.; Nishina, A.; Ichikawa, H.; Nonaka, M.; Sasaki, H. Deep countercurrent beneath the Kuroshio in the Okinawa Trough. *J. Geophys. Res. Oceans* **2008**, *113*, C06030. [CrossRef]

32.	Yu, H.S.; Chow, J. Cenozoic basins in northern Taiwan and tectonic implications for the development of the eastern Asian continental margin. *Palaeogeogr. Palaeoclimatol. Palaeoecol.* **1997**, *131*, 133–144. [CrossRef]

33.	Deng, K.; Yang, S.; Li, C.; Su, N.; Bi, L.; Chang, Y.-P.; Chang, S.-C. Detrital zircon geochronology of river sands from Taiwan: Implications for sedimentary provenance of Taiwan and its source link with the east China mainland. *Earth-Sci. Rev.* **2017**, *164*, 31–47. [CrossRef]

34.	Xu, X.; O'Reilly, S.Y.; Griffin, W.L.; Wang, X.; Pearson, N.J.; He, Z. The crust of Cathaysia: Age, assembly and reworking of two terranes. *Precambrian Res.* **2007**, *158*, 51–78. [CrossRef]

35.	Yang, J.; Gao, S.; Chen, C.; Tang, Y.; Yuan, H.; Gong, H.; Xie, S.; Wang, J. Episodic crustal growth of North China as revealed by U-Pb age and Hf isotopes of detrital zircons from modern rivers. *Geochim. Cosmochim. Acta* **2009**, *73*, 2660–2673. [CrossRef]

36. He, M.; Zheng, H.; Clift, P.D. Zircon U-Pb geochronology and Hf isotope data from the Yangtze River sands: Implications for major magmatic events and crustal evolution in Central China. *Chem. Geol.* **2013**, *360*, 186–203. [CrossRef]

37. Xu, Y.; Wang, C.Y.; Zhao, T. Using detrital zircons from river sands to constrain major tectono-thermal events of the Cathaysia Block, SE China. *J. Asian Earth Sci.* **2016**, *124*, 1–13. [CrossRef]

38. Yang, Y.M. *Study on the Characteristics of Turbidite Sediments Hosted Sulfides Deposit from the Southern Okinawa Trough*; University of Chinese Academy of Sciences Qingdao: Qingdao, China, 2021.

39. Liu, Y.; Hu, Z.; Zong, K.; Gao, C.; Gao, S.; Xu, J.; Chen, H. Reappraisement and refinement of zircon U-Pb isotope and trace element analyses by LA-ICP-MS. *Chin. Sci. Bull.* **2010**, *55*, 1535–1546. [CrossRef]

40. Fornelli, A.; Gallicchio, S.; Micheletti, F.; Langone, A. Preliminary U-Pb detrital zircon ages from Tufiti di Tusa formation (Lucanian Apennines, Southern Italy): Evidence of rupelian volcaniclastic supply. *Minerals* **2020**, *10*, 786. [CrossRef]

41. Fornelli, A.; Gallicchio, S.; Micheletti, F.; Langone, A. U-Pb detrital zircon ages from Gorgoglione Flysch sandstones in Southern Apennines (Italy) as provenance indicators. *Geol. Mag.* **2021**, *158*, 859–874. [CrossRef]

42. Spencer, C.J.; Kirkland, C.L.; Taylor, R.J.M. Strategies towards statistically robust interpretations of in situ U-Pb zircon geochronology. *Geosci. Front.* **2016**, *7*, 581–589. [CrossRef]

43. Vermeesch, P.; Resentini, A.; Garzanti, E. An R package for statistical provenance analysis. *Sediment. Geol.* **2016**, *336*, 14–25. [CrossRef]

44. Corfu, F.; Hanchar, J.M.; Hoskin, P.W.O.; Kinny, P. Atlas of zircon textures. *Rev. Mineral. Geochem.* **2003**, *53*, 469–500. [CrossRef]

45. Xie, J.; Yang, S.; Ding, Z. Methods and application of using detrital zircons to trace the provenance of loess. *Sci. China Earth Sci.* **2012**, *55*, 1837–1846. [CrossRef]

46. Malusà, M.G.; Garzanti, E. The sedimentology of detrital thermochronology. In *Fission-Track Thermochronology and Its Application to Geology*; Malusà, M.G., Fitzgerald, P.G., Eds.; Springer International Publishing: Cham, Switzerland, 2019; pp. 123–143.

47. Zhu, B.; Zeng, Z. Characteristic of Detrital Zircon U-Pb Geochronology of detrital zircon in the southern Okinawa Trough and its implication for sediment provenance. *Mar. Geol. Front.* **2021**, *38*. (In Chinese with English abstract)

48. Wang, W.; Bidgoli, T.; Yang, X.; Ye, J. Source-to-sink links between East Asia and Taiwan from Detrital Zircon geochronology of the Oligocene Huagang formation in the East China Sea Shelf Basin. *Geochem. Geophys. Geosyst.* **2018**, *19*, 3673–3688. [CrossRef]

49. He, M.; Zheng, H.; Bookhagen, B.; Clift, P.D. Controls on erosion intensity in the Yangtze River basin tracked by U-Pb detrital zircon dating. *Earth-Sci. Rev.* **2014**, *136*, 121–140. [CrossRef]

50. Zheng, P.; Li, D.; Chen, Y.; Hou, K.; Liu, C. Zircon U-Pb Ages of Clastic Sediment from the Outfall of the Yellow River and Their Geological Significance. *Geoscience* **2013**, *27*, 79–90. (In Chinese with English abstract)

51. Xu, Y.; Sun, Q.; Yi, L.; Yin, X.; Wang, A.; Li, Y.; Chen, J. Detrital zircons U-Pb age and Hf isotope from the western side of the Taiwan strait: Implications for sediment provenance and crustal evolution of the northeast Cathaysia Block. *Terr. Atmos. Ocean. Sci.* **2014**, *25*, 505–535. [CrossRef]

52. Li, Y.; Huang, X.; Nguyen Thi, H.; Lian, E.; Yang, S. Disentangle the sediment mixing from geochemical proxies and detrital zircon geochronology. *Mar. Geol.* **2021**, *440*, 106572. [CrossRef]

53. Hu, S.; Zeng, Z.; Fang, X.; Zhu, B.; Li, X.; Chen, Z. Rare earth element geochemistry of sediments from the southern Okinawa Trough since 3 ka: Implications for river-sea processes and sediment source. *Open Geosci.* **2019**, *11*, 929–947. [CrossRef]

54. Zhu, B.; Zeng, Z. Heavy mineral compositions of sediments in the southern Okinawa Trough and their provenance-tracing implication. *Minerals* **2021**, *11*, 1191. [CrossRef]

55. Yang, S.; Zhang, F.; Wang, Z. Grain size distribution and age population of detrital zircons from the Changjiang (Yangtze) River system, China. *Chem. Geol.* **2012**, *296*, 26–38. [CrossRef]

56. Qiao, S.; Shi, X.; Wang, G.; Zhou, L.; Hu, B.; Hu, L.; Yang, G.; Liu, Y.; Yao, Z.; Liu, S. Sediment accumulation and budget in the Bohai Sea, Yellow Sea and East China Sea. *Mar. Geol.* **2017**, *390*, 270–281. [CrossRef]

57. He, L.; Li, Y.; Zhou, H.; Yuan, D. Variability of cross-shelf penetrating fronts in the East China Sea. *Deep. Sea Res. Part II Top. Stud. Oceanogr.* **2010**, *57*, 1820–1826. [CrossRef]

58. Yuan, D.L.; Qiao, F.L.; Su, J. Cross-shelf penetrating fronts off the southeast coast of China observed by MODIS. *Geophys. Res. Lett.* **2005**, *32*, L19603. [CrossRef]

59. Qin, Y.S. *Geology of the East China Sea*; Science Press: Beijing, China, 1987. (In Chinese with English abstract)

60. Lu, K.; Qin, Y.; Wang, Z.; Huang, L.; Li, G. Heavy mineral provinces of the surface sediments in central-southern East China Sea and impilcations for provenance. *Mar. Geol. Lett.* **2019**, *35*, 20–26. (In Chinese with English abstract)

61. Ujiie, H.; Nakamura, T.; Miyamoto, Y.; Park, J.O.; Hyun, S.; Oyakawa, T. Holocene turbidite cores from the southern Ryukyu Trench slope: Suggestions of periodic earthquakes. *J. Geol. Soc. Jpn.* **1997**, *103*, 590–603. [CrossRef]

62. Sibuet, J.C.; Deffontaines, B.; Hsu, S.K.; Thareau, N.; Le Formal, J.P.; Liu, C.S.; Party, A.C.T. Okinawa trough backarc basin: Early tectonic and magmatic evolution. *J. Geophys. Res. Solid Earth* **1998**, *103*, 30245–30267. [CrossRef]

63. Liu, B.; Li, X.; Zhao, Y.; Zheng, Y.; Wu, J. Derbis transport on the western continental slope of the Okinaea Trough: Slumping and gravity flowing. *Oceanol. Limnol. Sin.* **2005**, *36*, 1–9. (In Chinese with English abstract)

Journal of
Marine Science and Engineering

Viewpoint

A Novel Device for the In Situ Enrichment of Gold from Submarine Venting Fluids

Zhigang Zeng [1,2,3,*], Xuebo Yin [1,4], Xiaoyuan Wang [1,4], Yu Yan [1,3] and Xueying Zhang [1]

[1] Seafloor Hydrothermal Activity Laboratory, CAS Key Laboratory of Marine Geology and Environment, Institute of Oceanology, Chinese Academy of Sciences, Qingdao 266071, China; xbyin@qdio.ac.cn (X.Y.); wangxiaoyuan@qdio.ac.cn (X.W.); yanyu@qdio.ac.cn (Y.Y.); xyzhang@qdio.ac.cn (X.Z.)

[2] Laboratory for Marine Mineral Resources, Qingdao National Laboratory for Marine Science and Technology, Qingdao 266071, China

[3] College of Marine Science, University of Chinese Academy of Sciences, Beijing 100049, China

[4] Center for Ocean Mega-Science, Chinese Academy of Sciences, 7 Nanhai Road, Qingdao 266071, China

* Correspondence: zgzeng@qdio.ac.cn; Tel.: +86-532-82898525

Abstract: Gold and other metals (Cu, Zn, Ag, etc.) are enriched in vent fluids, approximately 3–5 orders of magnitude higher than those in seawater, and this leads to the formation of sulfide enrichment in Cu, Zn, Au, and Ag deposited on the mid-ocean ridge and island arcs, as well as in back-arc basins. We developed a device that can extract the elements such as Cu, Zn, Au, and Ag from the vent fluids before the formation of the hydrothermal plume, sulfide deposit, and metalliferous sediment at the seafloor over a long period, which is beneficial to collecting hydrothermal resources effectively and avoiding the damage of ecological environments caused by mining the polymetallic sulfide resources. The application of this device will have significance for the development and utilization of seafloor hydrothermal resources, the sustainable development and implementation of the blue economy, and the construction of the marine ecological civilization in the future.

Keywords: metal element extraction; vent fluids; hydrothermal resources; seafloor mining

Citation: Zeng, Z.; Yin, X.; Wang, X.; Yan, Y.; Zhang, X. A Novel Device for the In Situ Enrichment of Gold from Submarine Venting Fluids. *J. Mar. Sci. Eng.* **2022**, *10*, 724. https://doi.org/10.3390/jmse10060724

Academic Editors: Gary Wilson and Markes E. Johnson

Received: 22 March 2022
Accepted: 20 May 2022
Published: 25 May 2022

Publisher's Note: MDPI stays neutral with regard to jurisdictional claims in published maps and institutional affiliations.

1. Introduction

Since the first observations of submarine hydrothermal discharge at the Galápagos Rift in 1977 [1], more than 700 confirmed submarine hydrothermal venting sites have been identified (http://www.interridge.org, accessed on 21 March 2020). Although the stability of hydrothermal activities is linked to the nature and location of the underlying heat source, as well as the subsurface fluid flow pathway, some show rapid variations, whereas others exhibit stable fluid chemistries over many years, such as the Trans-Atlantic Geotraverse (TAG) hydrothermal field on the Mid-Atlantic Ridge and South Cleft at the Juan de Fuca Ridge [2]; therefore, the release of vent fluids at the seafloor will persistently influence the elemental contribution to the ocean. Owing to the important role of hydrothermal fluid flow in transferring mass from the crust and mantle into the oceans, determining the magnitude of their flux has been the overriding question that many scientists have tried to assess [2–9]. It has been estimated that the fluid flux is ~375 × 10^{16} g/year at the ridge axis if 20% high-temperature (350 °C) and 80% low-temperature (5 °C) hydrothermal fluids are expelled from the seafloor. However, the fluid flux through hydrothermal plumes is ~11,000 × 10^{16} g/year if 20% of the fluids circulating at high temperature through young ocean crust are entrained into hydrothermal plumes [9]. The low-temperature fluid fluxes at ridge flanks could be greater, 2000–54,000 × 10^{16} g/year [8]. This combined flux is greater than the global riverine water flux of ~4000 × 10^{16} g/year and is sufficient for circulating the mass of the entire ocean (1.37 × 10^{21} kg) through the ocean crust in less than 1 million years [2,9,10].

The metals of the vent fluids are thought to be the result of the interaction of fluid with the ocean crust at high temperature [11]. In some cases, particularly in back- and island-arc settings, magmatic degassing can also add metals to the circulating fluids [2]. Therefore, in vent fluids, some metals are enriched relative to seawater, e.g., the concentration of Cu, Zn, Au, and Ag in vent fluids reach 162 µmol/kg, ~3100 µmol/kg, 250 pmol/kg, and 230 nmol/kg, respectively [11–14], which is approximately 3–5 orders of magnitude higher than those in seawater (Cu 0.004 µmol/kg, Zn 0.006 µmol/kg, Au 0.15 pmol/kg, and Ag 0.025 nmol/kg) [15,16]. This leads to the formation of sulfide enrichment in Cu (0.3–24.9 wt%), Zn (0.1–31.4 wt%), Au (0.1–88.9 ppm), and Ag (7–2305 ppm) deposited on the mid-ocean ridges and island-arcs, as well as in the back-arc basins [10,17]. Recently, it has been proposed that the base and precious metals can be obtained through cultivating seafloor sulfide deposits at artificial seafloor hydrothermal vents with further exploration for commercial mining [18]; however, the artificial hydrothermal vents created by boreholes could impact the ecological environment.

Therefore, if we are able to intercept the large amount of vent fluids and enrich Cu, Zn, Au, and Ag in the fluids artificially before the formation of the hydrothermal plume, sulfide deposit, and metalliferous sediment at the seafloor over a long period, the hydrothermal resources will be obtained effectively, and the damage to the ecological environment caused by mining the polymetallic sulfide resources can be avoided; thus, we developed a device that can enrich the elements such as Cu, Zn, Au, and Ag from the vent fluids. The application of this device will have significance for the development and utilization of seafloor hydrothermal resources and the preservation of a green and healthy ocean in the future.

2. Metal Element Extraction Device

In this proposal, for the enrichment and recovery of the base and precious metals from a vent fluid site, a metal element enrichment device (MEED) based on the "KEXUE" vessel is developed. A remote operated vehicle (ROV) is used to carry the MEED and place it near the seafloor hydrothermal vent sites. When the MEED is operated, the hydrothermal vent fluid is pumped into the device, and the particles are filtered into a tank. Gold and silver are extracted in the tank loaded with anion resin, and iron, copper, and zinc are extracted in the tank loaded with cation resin. When fully loaded the MEED is stopped and the device recovered. In the laboratory on the ship, the metal elements chelated by the anion/cation resin core equipped in the MEED will be quantitatively eluted. The design of the MEED is as follows:

The MEED includes a fluid recovery faucet (1); a telescopic pipe which is convenient for placement above the hydrothermal vent (2); a pipeline switch valve (3); a T-handle operated by a ROV manipulator (4); a rotatable joint, which controls the direction of the telescopic pipe (5); a vent fluid inlet pipe (6); a gravity settling tank for large particles in the vent fluid (7); a vent fluid centrifugal settling tank (8); a fine filter tank for filtering fine particles in the vent fluid (9) an ultrafiltration core for filtering soluble solid particles in the vent fluid (10); vent fluid sample tank (11); an anion adsorption tank (12) installed with an anion resin core (13); the adsorption of the anion resin is strengthened through a negative electrode column (14); a cation adsorption tank (15) installed with a cation resin core (16); a positive electrode column (17) strengthens the adsorption of cations such as iron, copper and zinc; a deep-water pump (18); a waste liquid outlet pipe discharging the waste liquid (19); a telescopic pipe (20), which discharges the waste liquid, and a protective shell (21) protecting the MEED from damage (see Figure 1). This design has been granted a patent (number 2021/05673) authorized by the republic of South Africa [19].

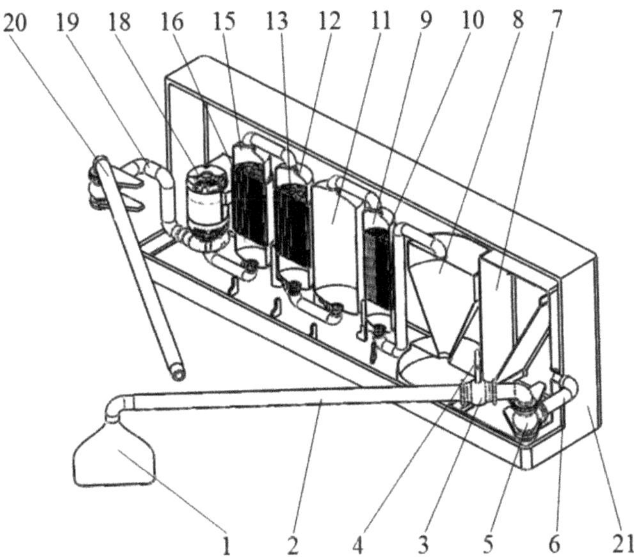

Figure 1. Design of the MEED; 1, fluid recovery faucet; 2, telescopic pipe; 3, pipeline switch valve; 4, T-handle; 5, rotatable joint; 6, vent fluid inlet pipe; 7, gravity settling tank; 8, vent fluid centrifugal settling tank; 9, fine filter tank; 10, ultrafiltration core; 11, hydrothermal fluid sample tank; 12, anion adsorption tank; 13, anion resin core; 14, negative electrode column; 15, cation adsorption tank; 16, cation resin core; 17, positive electrode column; 18, deep water pump; 19, outlet pipe; 20, telescopic pipe; and 21, protective shell.

The MEED and its fixing and recovery devices are carried by the ROV to the seafloor near the hydrothermal vent. The fluid recovery faucet (1) of the MEED is placed directly above the active hydrothermal vent, the length and angles of the suction telescopic pipe (2) and discharge telescopic pipe (20) are adjusted, and the discharge outlet of the telescopic pipe (20) is placed away from the vent to reduce the interference of the discharge liquid on the sampling. Then, the ROV manipulator is used to control the T-handle (4) to open the switch valve (3). Next, the power of the deep-water pump (18) is opened, and the negative (14) and the positive electrode columns (17) are connected. The vent fluid enters the gravity settling tank (7) through the suction faucet (1), the suction telescopic pipe (2), the switch valve (3), the rotatable joint (5), and the inlet pipe (6). Owing to the small diameter of the inlet pipe (6) and the large section at the entrance of the gravity settling tank (7), the flow rate of the fluid is greatly reduced after entering; thus, it is convenient for the settlement of large particles in the fluid, and the fluid impacts the inclined surface in the gravity settling tank (7), allowing for the large particles to accumulate gradually. The preliminarily separated fluid enters the centrifugal settling tank (8) and the supernatant liquid enter the fine filter tank (9) through the pipeline from the outlet. After the liquid flows through the ultrafiltration core (10), all particles have been removed at the centrifugal settling tank (8) and the fine filter tank (9). The liquid passes through the ultrafiltration core (10) and enters the tank (11). Then, the liquid flows from the fluid tank (11) into the anion adsorption tank (12). The Au, Pt, etc., anionic complexes are adsorbed by the anion resin core (13) under the combination of the anion resin core (13) and the negative electrode column (14). Then, the liquid enters into the cation adsorption tank (15), and the cations such as Cu^{2+}, Zn^{2+}, etc., are adsorbed by the cation resin core (16) under the combination of the cation resin core (16) and the positive electrode column (17). The waste liquid is discharged into the sea through the output pipe (19) and the discharge telescopic pipe (20) by the deep-water pump (18). After a set period, the pump will be stopped and the recovery apparatus reacted. When the ship sends the recovery signal, the recovery device and the connected MEED rise to the sea surface by automatically releasing the anchor system. The location is transmitted to the ship through the positioning signal transmission device. After receiving the signal, the ship salvages the MEED (see Figure 2).

Figure 2. Sketch of the MEED in operation.

3. Conclusions

We can combine the geophysical prospecting, geochemical exploration, and biological exploration techniques with drilling to explore the seafloor polymetallic sulfide and hydrothermal vent. Based on understanding the resource potential of seafloor polymetallic sulfide deposit by using the four exploring technique systems, employing the MEED can concentrate metal elements from vent fluids without destroying the hydrothermal ecological environment. The MEED is a potential tool for the sustainable development and implementation of a green and healthy ocean and the construction of marine ecological civilization.

Author Contributions: Z.Z.—Conceptualization, methodology, funding acquisition, writing and editing; X.Y.—methodology, writing and illustration; X.W.—writing and editing; Y.Y.—illustration; X.Z.—Conceptualization. All authors have read and agreed to the published version of the manuscript.

Funding: This research was funded by the NSFC Major Research Plan on West-Pacific Earth System Multispheric Interactions (grant number 91958213), the Special Fund for the Taishan Scholar Program of Shandong Province (grant number ts201511061), and the National Key Basic Research Program of China (grant number 2013CB429700).

Institutional Review Board Statement: Not applicable.

Informed Consent Statement: Not applicable.

Data Availability Statement: Data can be obtained from the corresponding author.

Conflicts of Interest: The authors declare no conflict of interest.

References

1. Edmond, J.M.; Measures, C.; McDuff, R.E.; Chan, L.H.; Collier, R.; Grant, B.; Corliss, J.B. Ridge crest hydrothermal activity and the balances of the major and minor elements in the ocean: The Galapagos data. *Earth Planet. Sci. Lett.* **1979**, *46*, 1–18. [CrossRef]
2. Humphris, S.E.; Klein, F. Progress in deciphering the controls on the geochemistry of fluids in seafloor hydrothermal systems. *Annu. Rev. Mar. Sci.* **2018**, *10*, 315–343. [CrossRef] [PubMed]
3. Wolery, T.J.; Sleep, N.H. Hydrothermal circulation and geochemical flux at mid-ocean ridge. *J. Geol.* **1976**, *84*, 249–275. [CrossRef]
4. Von Damm, K.L.; Edmond, J.M.; Grant, B.; Measures, C.I. Chemistry of submarine hydrothermal solutions at 21 °N, East Pacific Rise. *Geochim. Cosmochim. Acta* **1985**, *49*, 2197–2220. [CrossRef]

5. Elderfield, H.; Schultz, A. Mid-ocean ridge hydrothermal fluxes and the chemical composition of the ocean. *Annu. Rev. Earth Planet. Sci. Lett.* **1996**, *24*, 191–224. [CrossRef]
6. Davis, A.C.; Bickle, M.J.; Teagle, D.A.H. Imbalance in the oceanic strontium budget. *Earth Planet. Sci. Lett.* **2003**, *211*, 173–187. [CrossRef]
7. Zhai, S.K.; Wang, X.T.; Yu, Z.H. Heat and mass flux estimation of modern seafloor hydrothermal activity. *Acta Oceanol. Sin.* **2006**, *25*, 43–51.
8. Nielsen, S.G.; Rehkämper, M.; Teagle, D.A.; Butterfield, D.A.; Alt, J.C.; Halliday, A.N. Hydrothermal fluid fluxes calculated from the isotopic mass balance of thallium in the ocean crust. *Earth Planet. Sci. Lett.* **2006**, *251*, 120–133. [CrossRef]
9. German, C.R.; Seyfried, W.E. 8.7 Hydrothermal Processes. In *Treatise on Geochemistry*, 2nd; Holland, H.D., Turekian, K.K.B., Eds.; Elsevier: Oxford, UK, 2014; pp. 191–233.
10. Hannington, M.D.; de Ronde, C.E.J.; Petersen, S. *Sea-Floor Tectonics and Submarine Hydrothermal Systems*; Society of Economic Geologists: Littleton, CO, USA, 2005; pp. 111–141.
11. Von Damm, K.L. Seafloor hydrothermal activity: Black smoker chemistry and chimneys. *Annu. Rev. Earth Planet. Sci. Lett.* **1990**, *18*, 173–204. [CrossRef]
12. Douville, E.; Charlou, J.L.; Oelkers, E.H.; Bienvenu, P.; Colon, C.J.; Donval, J.P.; Appriou, P. The rainbow vent fluids (36°14′ N, MAR): The influence of ultramafic rocks and phase separation on trace metal content in Mid-Atlantic Ridge hydrothermal fluids. *Chem. Geol.* **2002**, *184*, 37–48. [CrossRef]
13. Fouquet, Y.; Stackelberg, U.V.; Charlou, J.L.; Donval, J.P.; Erzinger, J.; Foucher, J.P.; Herzig, P.; Mühe, R.; Soakai, S.; Wiedicke, M.; et al. Hydrothermal activity and metallogenesis in the Lau back-arc basin. *Nature* **1991**, *349*, 778–781. [CrossRef]
14. Kenison Falkner, K.; Edmond, J.M Gold in seawater. *Earth Planet. Sci. Lett.* **1990**, *98*, 208–221. [CrossRef]
15. Seyfried, W.E., Jr.; Seewald, J.S.; Berndt, M.E.; Kang, D.; Foustoukos, D.I. Chemistry of hydrothermal vent fluids from the Main Endeavour Field, northern Juan de Fuca Ridge: Geochemical controls in the aftermath of June 1999 seismic events. *J. Geophys. Res.* **2003**, *108*, 2429. [CrossRef]
16. Large, R.R.; Gregory, D.D.; Steadman, J.A.; Tomkins, A.G.; Lounejeva, E.; Danyushevsky, L.V.; Halpin, J.A.; Maslennikov, V.; Sack, P.J.; Mukherjee, I.; et al. Gold in the oceans through time. *Earth Planet. Sci. Lett.* **2015**, *428*, 139–150. [CrossRef]
17. Wohlgemuth-Ueberwasser, C.C.; Viljoen, F.; Petersen, S.; Vorster, C. Distribution and solubility limits of trace elements in hydrothermal black smoker sulfides: An in-situ LA-ICP-MS study. *Geochim. Cosmochim. Acta* **2015**, *159*, 16–41. [CrossRef]
18. Nozaki, T.; Ishibashi, J.I.; Shimada, K.; Nagase, T.; Takaya, Y.; Kato, Y.; Kawagucci, S.; Watsuji, T.; Shibuya, T.; Yamada, R.; et al. Rapid growth of mineral deposits at artificial seafloor hydrothermal vents. *Sci. Reps.* **2016**, *6*, 22163. [CrossRef] [PubMed]
19. Zhigang, Z.; Xuebo, Y.; Haiyan, Q. Device for Enriching Useful Elements in Fluid at Hydrothermal Vent. South. Africa Patent No. 2021/05673, 29 September 2021.

Journal of
*Marine Science
and Engineering*

Article

Simultaneous Determination of Fluorine and Chlorine in Marine and Stream Sediment by Ion Chromatography Combined with Alkaline Digestion in a Bomb

Yuhua Gao [1], Xiaoyuan Wang [2,3,4,*], Xianwen Fang [5], Xuebo Yin [2,4], Lu Chen [1], Jianling Bi [1], Yao Ma [6,7,8] and Shuai Chen [2,3,4]

[1] Shandong Institute of Geophysical & Geochemical Exploration, Jinan 250013, China; wtyykcsgyh@163.com (Y.G.); chery1110029@163.com (L.C.); bijianling0305@163.com (J.B.)
[2] Key Laboratory of Marine Geology and Environment, Institute of Oceanology, Chinese Academy of Sciences, Qingdao 266071, China; xbyin@qdio.ac.cn (X.Y.); Chenshuai@qdio.ac.cn (S.C.)
[3] Laboratory for Marine Mineral Resources, Qingdao National Laboratory for Marine Science and Technology, Qingdao 266071, China
[4] Center for Ocean-Mega Science, Chinese Academy of Sciences, Qingdao 266071, China
[5] Qingdao Ecological Environment Monitoring Center, Qingdao 266003, China; zcfxw@126.com
[6] College of Marine Resources and Environment, Hebei Normal University of Science and Technology, Qinhuangdao 066600, China; maoyaox@126.com
[7] Hebei Key Laboratory of Ocean Dynamics, Resources and Environments, Qinhuangdao 066600, China
[8] Qinhuangdao Key Laboratory of Marine Habitat and Resources, Qinhuangdao 066600, China
* Correspondence: wangxiaoyuan@qdio.ac.cn; Tel.: +86-532-82898541

Abstract: Fluorine and chlorine are important tracers for geochemical and environmental studies. In this study, a rapid alkaline digestion (NaOH) method for the simultaneous determination of fluorine and chlorine in marine and stream sediment reference samples using ion chromatography is developed. The proposed method suppresses the volatilization loss of fluorine and chlorine and decreases the matrix effects. The results are in good agreement with fluorine ~100%, chlorine ranging from 90 to 95% of the expected concentrations. The detection limits of this method were 0.05 μg/g for fluorine and 0.10 μg/g for chlorine. This method is simple, economical, precise and accurate, which shows great potential for the rapid simultaneous determination of fluorine and chlorine in large batches of geological and environmental samples commonly analyzed for environmental geochemistry studies.

Keywords: fluorine and chlorine; marine and stream sediment; ion chromatography; alkaline digestion; high pressure bomb

Citation: Gao, Y.; Wang, X.; Fang, X.; Yin, X.; Chen, L.; Bi, J.; Ma, Y.; Chen, S. Simultaneous Determination of Fluorine and Chlorine in Marine and Stream Sediment by Ion Chromatography Combined with Alkaline Digestion in a Bomb. *J. Mar. Sci. Eng.* **2022**, *10*, 93. https://doi.org/10.3390/jmse10010093

Academic Editor: Nathalie Fagel

Received: 7 December 2021
Accepted: 4 January 2022
Published: 11 January 2022

Publisher's Note: MDPI stays neutral with regard to jurisdictional claims in published maps and institutional affiliations.

1. Introduction

Fluorine and chlorine are of great interest in geological and environmental studies due to their special, highly mobile and volatile properties [1,2]. Fluorine is a minor constituent in a wide range of sedimentary minerals including phosphorites, phosphates, carbonates, silicates and clay minerals [3–6]. Chlorine is the dominant ligand that enables metal transport in the majority of hydrothermal solutions [6–8]. Thus, the content of fluorine, chlorine and ratios of element/Cl in sediment can be used as tracers for chemical evolution of fluids and water/rock interactions in low temperature sediment alteration [9,10] and high temperature hydrothermal systems [11–14], element recycling during subduction-related sediment melting [4,15], and early diagenesis of sediment [3]. Therefore, recent studies have focused on the precise determination of fluorine and chlorine in sediment. Several analytical techniques have been applied to the determination of fluorine and chlorine: by specific ion selective electrode [16–18], instrumental neutron activation analysis (INAA) [16], radiochemical neutron activation analysis (RNAA) [19–22], prompt gamma neutron activation analysis (PGNA) [23] or X-ray fluorescence spectrometry [24–27]. However, an ion selective electrode requires a rather complex preparation stage and the difficulties of other techniques are the requirement of special instruments and/or time-consuming processes [28]. Moreover, the detection limits of the determination by INAA or X-ray Fluorescence Spectrometry (XRFS) are generally too high, relative to usual abundances [29]. In contrast, ion chromatography, a compact and inexpensive instrument, is commonly used in many laboratories and is the most suitable method for sensitive and simultaneous determination of fluorine and chlorine [5,6,28–35]. However, it is difficult to quantitatively extract fluorine and chlorine from geological materials for ion chromatography analysis. So far, only a few methods have been used to extract fluorine and chlorine from geological samples, including pyrohydrolysis [5,6,29,31,32], alkaline fusion [5,17,28,30], microwave digestion [36], combustion [35] and NH_4HF_2 digestion with subsequent ammonium dilution [37]. However, the pyrohydrolysis method is not suitable for the analysis of a large batch of samples [37]; alkali fusion requires a high flux-to-sample ratio which results in high blank levels, total dissolved solids (TDS) content and matrix effects [5,28], microwave digestion has poor recoveries caused by incomplete digestion of sediment samples containing zircon or other refractory minerals [38], and the NH_4HF_2 digestion method cannot extract fluorine from geological materials [37]. Recently, a high-pressure digestion technique has been generally applied [8,36,39–41]. However, there are no reports about the simultaneous determination of fluorine and chlorine in sediment using this technique.

In this paper, a rapid alkaline digestion method for the simultaneous determination of fluorine and chlorine in marine and stream sediment reference samples using the high-pressure digestion bomb with a double inner arc seal design is described. The effects of the digestion parameters on the recoveries of fluorine and chlorine in sediment reference samples are described in detail. A small amount of the sample was digested, and the fluorine and chlorine were extracted completely. This method is practical and simple and can deal with a large number of samples simultaneously.

2. Experimental

2.1. Instrumentation

Experiments were carried out using ion chromatography (DX600, Dionex, CA, USA) at the Laboratory of Spartacus Testing Center, equipped with an anion exchange column in the suppression mode. Analytical conditions are reported in Table 1. Generally, the F- peak appeared at about 3.8 min after sample injection. Then, the Cl- peak appeared at about 4.7 min after the injection (Figure 1). One measurement cycle could be completed in ~10 min.

Table 1. Instrumental operating parameters used for ion chromatography analysis.

	Operating Parameters
Volume of Sample Injection Loop	25 μL
Column	IonPac AS14
Column Size	4 mm × 250 mm
Eluent	3.5 mmol/L Na_2CO_3 + 1 mmol/L $NaHCO_3$
Detector	Suppressed conductivity detector
Flow Rate	1.2 mL/min

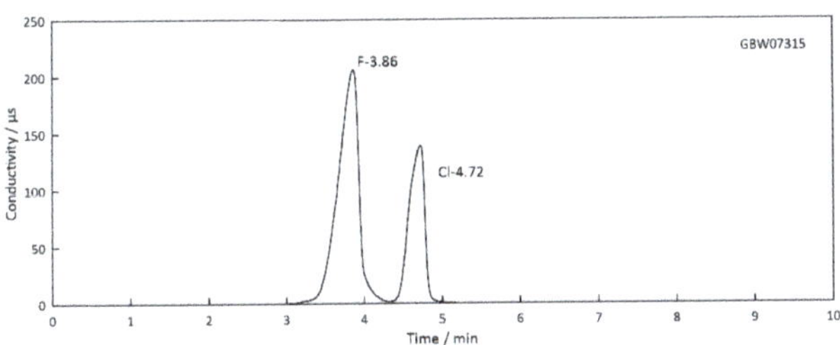

Figure 1. Chromatogram of marine sediment GBW07315.

2.2. Reagents and Certified Reference Materials

Alkaline digestion solution was prepared by diluting NaOH (AR, Aladdin, Shanghai, China) with pure water (18.2 MΩ·cm grade). Eluent solution was prepared just before analysis by diluting Na_2CO_3 (AR, Sinopharm, Shanghai, China) and $NaHCO_3$ (AR, Sinopharm, Shanghai, China) reagents with pure water. The calibration solutions were prepared by dilution from 1000 mg/L fluorine and chlorine standard solutions (National Research Center for Reference Materials, Beijing, China) with pure water. Three domestic reference materials GBW07315 (marine sediment from the CC area in the east pacific basin), GSD-9 (stream sediment from the Yangtze River) and GSD-10 (stream sediment from the catchment basin in Yishan, Guangxi Province) were used as reference samples. All these samples were in powder form with size less than 75 µm as originally prepared. Reference GBW07315 was used to optimize the alkaline digestion temperature and time.

2.3. Laboratory Ware

A screw-top PTFE-lined, corrosive-resistant digestion bomb with a volume of ~15 mL was used for this research (Figure 2). This bomb has a double inner arc seal design, the inner tank has an oval cross-section, the upper part of the inner tank plugs into the top of the lower part. The inner tank was pre-cleaned with 10% HNO_3 and heated to boiling for about 12 h at 120 °C, then rinsed with pure water.

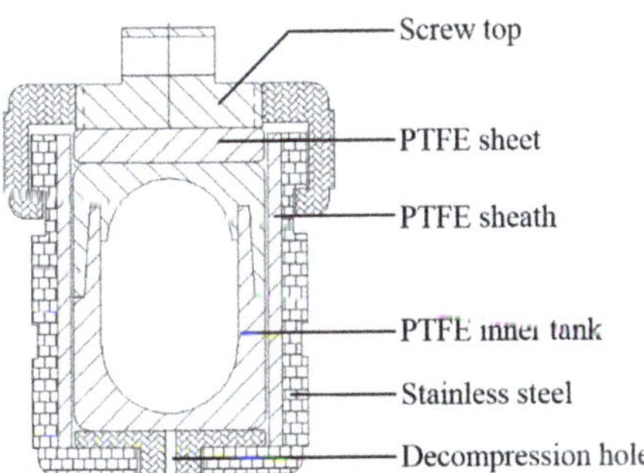

Figure 2. Sketch of the corrosive-resistant digestion bomb (Reproduced with permissions from Ref. [39], Copyright © 2018 Atomic Spectroscopy).

2.4. Sample Preparation

A total of 40 mg aliquot of sample powder was accurately weighed into the PTFE bombs and NaOH was added. The sealed bombs were then placed in an electric oven at 240 °C for 12 h. After cooling, 6 mL pure water was added, then the bombs were heated again in the electric oven at 180 °C for 12 h. After cooling again, the sample solution was transferred to a 15 mL centrifuge tube and diluted to 10 mL with pure water. After centrifuging for 8 min at 3000 rpm, 2 mL sample of the supernatant was transferred to a new polyethylene tube. The supernatant was measured by ion chromatography.

2.5. Calibration Curves and Limit of Detection

Fluoride and chloride calibration curves in the range of 1–10 µg/g and 1–200 µg/g were prepared respectively by dilution from 1000 mg/L standard solutions (National Research Center for Reference Materials, Beijing, China) with pure water (Figure 3). The correlation coefficient (R^2) of the linear calibration is 0.9999 for fluorine and 0.9997 for chlorine, respectively. The method's limit of detection (LOD, three times the standard deviation of the 6% NaOH blank solution for seven preparation blanks assuming a dilution factor of 250) for F and Cl were 0.05 µg/g and 0.10 µg/g, respectively.

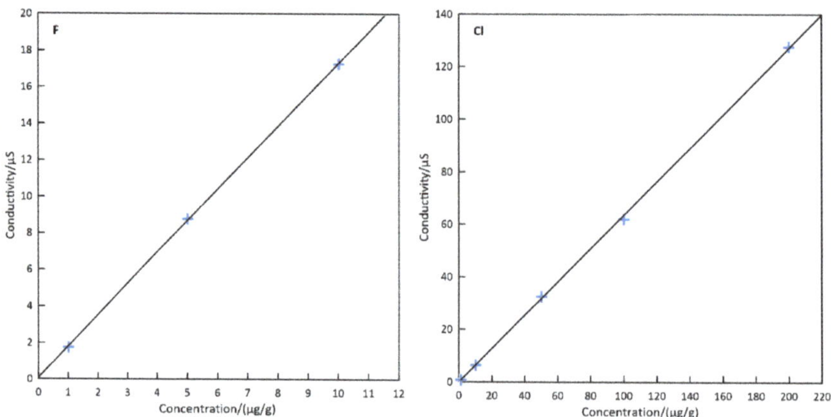

Figure 3. Fluoride and chloride calibration curves performed by ion chromatography.

3. Results

The measured concentrations of chlorine and fluorine were compared to the reference values provided for the standard material, GW07315, GSD-9 and GSD-10-1 (Table 2). Generally, the results are in good agreement with F- being ~100%, Cl ranging from 90–95% of the expected concentrations. The relative standard deviation between replicates was <6% and <10% for fluorine and chlorine, respectively.

Table 2. Results for GBW07315, GSD-9-1 and GSD-10-1, and Comparison with Certified Data.

	GWB07315		GSD-9-1		GSD-10-1	
	F (µg/g)	Cl (µg/g)	F (µg/g)	Cl (µg/g)	F (µg/g)	Cl (µg/g)
1	1055	33,037	467	49	149	41
2	1176	34,452	503	43	140	48
3	1099	34,992	482	46	151	45
4	1152	33,982	518	56	161	49
5	1058	36,987	503	45	144	49
6	1034	31,972	525	44	157	45
7	1156	33,650	490	48	142	40
Measured Average Value	1104	34,153	498	47	149	45
Reference Value	1100	36,000 ± 3000	494 ± 39	52 ± 11	149 ± 38	50
Relative Standard Deviation	5.2%	4.6%	4.0%	9.5%	5.2%	8.0%
Accuracy	100%	95%	101%	90%	100%	90%

4. Discussion

4.1. Effect of the Amount of NaOH

Although acid digestion has been commonly used for the decomposition of geological samples [38], mineral acids should be avoided to prevent losses of volatile halogens [37]. Alkaline fusion with NaOH can quantitatively extract fluorine [17] and chlorine [20–22] from geological materials, and the high-pressure digestion bomb requires a small amount of reagent, therefore NaOH was used as the digestion reagent. Figure 4 shows the agreements of F and Cl as a function of added NaOH amount for the digestion of 40 mg of GBW07315.

The agreements are expressed as ratios of the measured values relative to their reference values. It can be seen that F was completely recovered with 0.75 mL 6% NaOH, and the agreement for Cl was good with both 0.75 mL 6% NaOH and 0.50 mL 6% NaOH. Thus, the adopted optimum amount is 0.75 mL 6% NaOH.

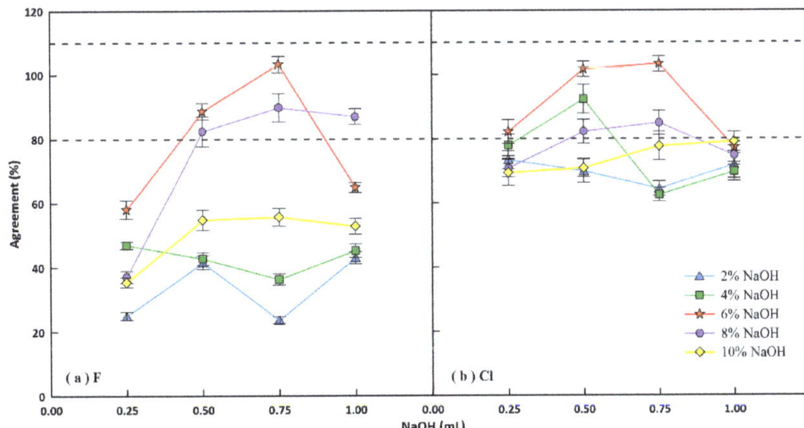

Figure 4. Agreement of (**a**) fluorine and (**b**) chlorine as a function of the amount of NaOH. The dotted lines delimit recoveries between 80% and 110%. The agreement is the ratio of the measured value relative to reference values. Error bars represent relative standard deviation ($n = 3$).

4.2. Effect of Digestion Temperature

Figure 5 illustrates the variation of the agreements for F and Cl in GBW07315 at different digestion temperatures (150–260 °C) with 0.75 mL 6% NaOH. The observation demonstrates that the digestion temperature is the critical factor for the complete recovery of F and Cl. The recoveries of both F and Cl increased from 150 °C to 240 °C and they were completely recovered at 240 °C and 260 °C. Therefore 240 °C was used as the optimum for further extractions.

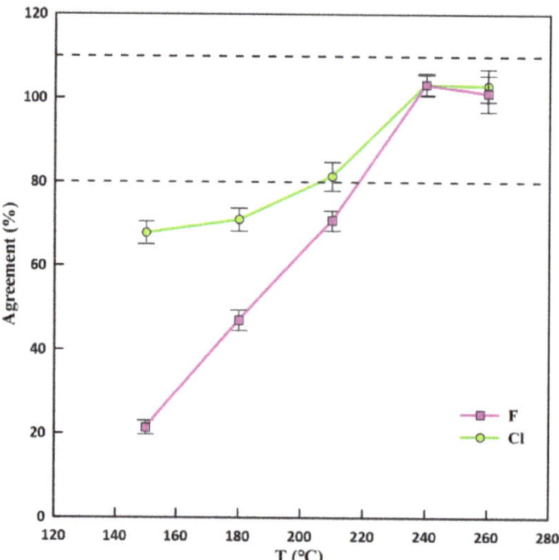

Figure 5. Agreement of fluorine and chlorine as a function of the digestion temperature with 0.75 mL 6% NaOH. The dotted lines delimit recoveries between 80% and 110%. The agreement is the ratio of the measured value relative to reference values. Error bars represent relative standard deviation (*n* = 3).

4.3. Method Efficiency

The recovery of the pyrohydrolysis method is not stable [8], probably due to the loss of the halogens or the incomplete extraction during the pyrohydrolysis process [21,22]. In contrast, the digestion bomb used in this study has a double inner arc seal design, allowing the evaporating material from the top to the bottom without clinging onto the inner tank wall, which provides an effective circulation and suppresses the volatilization loss of fluorine and chlorine. The present method consumes a small amount of alkaline (0.75 mL 6% NaOH can extract fluorine and chlorine from 40 mg sample completely) and decreases the matrix effects in comparison to the alkaline fusion method, which requires a large amount of flux and the matrix separation step leading to a high procedural blank [5,28,30]. The NH_4HF_2 digestion method has been proposed recently, which suppresses the volatilization loss of chlorine effectively owing to the formed ammonium salts with high boiling points, however, this method cannot determine fluorine and chlorine simultaneously [37]. In addition, compared to the complicated pyrohydrolysis, alkaline fusion and combustion methods [5,6,17,28–32,35], the rapid alkaline digestion method is simple and can deal with a large batch of samples.

5. Conclusions

Our results show that NaOH digestion in a high-pressure bomb with double inner arc seal design can be used for the quantitative extraction of fluorine and chlorine in sediments. The proposed method is economical and requires a small amount of the sample and reagent. This effective and simple method has no contamination problems, with good accuracy, and shows great potential for the determination of fluorine and chlorine in large batches of geological and environmental samples.

Author Contributions: Y.G.—Conceptualization, Data curation, formal analysis, writing and editing; X.W.—Conceptualization, funding acquisition, writing and editing; X.F.—formal analysis; X.Y.—Data curation, methodology, validation; L.C.—Data curation; J.B.—formal analysis; Y.M.—funding acquisition; S.C.—funding acquisition. All authors have read and agreed to the published version of the manuscript.

Funding: This research was funded by the Strategic Priority Research Program of the Chinese Academy of Sciences (grant number XDB42020402), the National Natural Science Foundation of China, grant number 91958213); the Natural Science Foundation of Shandong Province (grant number ZR2020MD068; ZR2020QD069), Natural Science Foundation of Hebei Province (grant number D2021407001).

Institutional Review Board Statement: Not applicable.

Informed Consent Statement: Not applicable.

Data Availability Statement: Data can be obtained from the corresponding author.

Acknowledgments: The authors are most grateful for the detailed comments and suggestions provided by the reviewers.

Conflicts of Interest: The authors declare no conflict of interest.

References

1. Bonifacie, M.; Jendrzejewski, N.; Agrinier, P.; Coleman, M.; Pineau, F.; Javoy, M. Pyrohydrolysis-IRMS determination of silicate chlorine stable isotope compositions. Application to oceanic crust and meteorite samples. *Chem. Geol.* **2007**, *242*, 187–201. [CrossRef]
2. Baasner, A.; Schmidt, B.C.; Webb, S.L. The effect of chlorine, fluorine and water on the viscosity of aluminosilicate melts. *Chem. Geol.* **2013**, *357*, 134–149. [CrossRef]
3. Rude, P.D.; Aller, R.C. Fluorine mobility during early diagenesis of carbonate sediment: An indicator of mineral transfor-mations. *Geochim. Cosmochim. Acta* **1991**, *55*, 2491–2509. [CrossRef]
4. Rude, P.D.; Aller, R.C. Fluorine uptake by Amazon continental shelf sediment and its impact on the global fluorine cycle. *Cont. Shelf Res.* **1994**, *14*, 883–907. [CrossRef]
5. Wang, Q.; Makishima, A.; Nakamura, E. Determination of Fluorine and Chlorine by Pyrohydrolysis and Ion Chroma-tography: Comparison with Alkaline Fusion Digestion and Ion Chromatography. *Geostand. Geoanal. Res.* **2010**, *34*, 175–183. [CrossRef]
6. Kendrick, M.A. Halogens in seawater, marine sediments and the altered oceanic crust and lithosphere. In *The Role of Halogens in Terrestrial and Extraterrestrial Geochemical Processes*; Harlov, D., Aranovich, L., Eds.; Springer Geochemistry: Cham, Switzerland, 2018.
7. Yardley, B.W.D. 100th Anniversary Special Paper: Metal concentrations in crustal fluids and their relationship to ore formation. *Econ. Geol.* **2005**, *100*, 613–632. [CrossRef]
8. He, T.; Wang, Z.; Hu, Z. Advances in analysis for halogens in geological materials. *Earth Sci.* **2021**. Available online: https://kns.cnki.net/kcms/detail/42.1874.P.20210831.1143.004.html (accessed on 31 August 2021). (In Chinese).
9. Li, Y.; Jiang, S.; Yang, T. Br/Cl, I/Cl and chlorine isotopic compositions of pore water in shallow sediments: Implications for the fluid sources in the Dongsha area, northern South China Sea. *Acta Oceanol. Sin.* **2017**, *36*, 31–36. [CrossRef]
10. Martin, J.B. Nonconservative behavior of Br-/Cl- ratios during alteration of volcaniclastic sediments. *Geochim. Cosmochim. Acta* **1999**, *63*, 383–391. [CrossRef]
11. Oosting, S.E.; Von Damm, K.L. Bromide/chloride fractionation in seafloor hydrothermal fluids from 9-10° N East Pacific Rise. *Earth Planet. Sci. Lett.* **1996**, *144*, 133–145. [CrossRef]
12. Von Damm, K.L.; Buttermore, L.G.; Oosting, S.E.; Bray, A.M.; Fornari, D.J.; Lilley, M.D.; Shanks, W.C., III. Direct observation of the evolution of a seafloor 'black smoker' from vapor to brine. *Earth Planet. Sci. Lett.* **1997**, *149*, 101–111. [CrossRef]
13. Gieskes, J.M.; Simoneit, B.R.T.; Goodfellow, W.D.; Baker, P.A.; Mahn, C. Hydrothermal geochemistry of sediments and pore waters in Escanaba Trough-ODP Leg 169. *Appl. Geochem.* **2002**, *17*, 1435–1456. [CrossRef]
14. Wu, S.F.; You, C.F.; Valsami-Jones, E.; Baltatzis, E.; Shen, M.L. Br/Cl and I/Cl systematics in the shallow-water hydrothermal system at Milos Island, Hellenic Arc. *Mar. Chem.* **2012**, *140*, 33–43. [CrossRef]
15. Li, H.; Hermann, J. The effect of fluorine and chlorine on trace element partitioning between apatite and sediment melt at subduction zone conditions. *Chem. Geol.* **2017**, *473*, 55–73. [CrossRef]
16. Langenauer, M.; Krähenbühl, U.; Filler, V.; Wyttenbach, A. Determination of fluorine, chlorine, bromine and iodine in seven geochemical reference samples. *Geostand. Newsl.* **1992**, *16*, 41–44. [CrossRef]
17. Zhang, B.; Guo, Y.; Chen, J.; Jiang, X.; Wang, S.; Zheng, S. Occurrence characteristics and release potential of fluoride in sediment of DaiHai Lak. *China Environ. Sci.* **2020**, *40*, 1748–1756. (In Chinese)
18. Liu, Q.; Zhang, B.; Zhai, D.; Han, Z.; Chi, Q.; Nie, L.; Zhou, J.; Xu, S.; Liu, H.; Wang, W.; et al. Continental-scale distribution and source identification of fluorine geochemical provinces in drainage catchment sediment and alluvial soil of China. *J. Geochem. Explor.* **2020**, *214*, 106537. [CrossRef]
19. Shinonaga, T.; Ebihara, M.; Nakahara, H.; Tomura, K.; Heumann, K.G. Cl, Br and I in igneous standard rocks. *Chem. Geol.* **1994**, *115*, 213–225. [CrossRef]
20. Ozaki, H.; Ebihara, M. Determination of trace halogens in rock samples by radiochemical neutron activation analysis coupled with the k0-standardization method. *Anal. Chim. Acta.* **2007**, *583*, 384–391. [CrossRef]

21. Sekimoto, S.; Ebihara, M. Accurate Determination of chlorine, bromine, and iodine in sedimentary rock reference samples by radiochemical neutron activation analysis and a detailed comparison with inductively coupled plasma mass spectrometry literature data. *Anal. Chem.* **2013**, *85*, 6336–6341. [CrossRef]

22. Sekimoto, S.; Ebihara, M. Accurate Determination of chlorine, bromine and iodine in U.S. Geological Survey geochemical reference materials by radiochemical neutron activation analysis. *Geostand. Geoanal. Res.* **2017**, *41*, 213–219. [CrossRef]

23. Sano, T.; Fukuoka, T.; Hasenaka, T. Determination of chlorine contents in Geological Survey of Japan reference materials by prompt gamma neutron activation analysis. *Geostand. Geoanal. Res.* **2004**, *28*, 443–448. [CrossRef]

24. Li, X.L.; Chen, X.; Ge, H.J.; Wu, Y.T.; Xian, X.L.; Han, W. XRFS determination of chlorine etc multi-elements in sea sediments with sample preparation by fusion. *Phys. Test. Chem. Anal.* **2011**, *47*, 1420–1423. (In Chinese)

25. Yuan, J.; Xia, C.; Liu, G.; Liu, X. Determination of F, Ca and other major elements in high fluoride concentration samples by X-Ray fluorescence spectrometry. *Uranium Geol.* **2016**, *32*, 175–179. (In Chinese)

26. Xia, C.; Jiang, Y.; Zheng, J.; Qian, H. XRFS determination of chlorine in geological samples. *Phys. Test. Chem. Anal.* **2017**, *53*, 775–779. (In Chinese)

27. A, L.; Zhang, P.; He, P.; Yang, Z.; Liang, Y.; Yang, Y. Determination of sulphur and fluorine in geological samples by X-ray fluorescence spectrometry. *Chin. J. Inorg. Anal. Chem.* **2019**, *9*, 50–53. (In Chinese)

28. Anazawa, K.; Tomiyasu, T.; Sakamoto, H. Simultaneous determination of fluorine and chlorine in rocks by ion chromatography in combination with alkali fusion and cation-exchange pretreatment. *Anal. Sci.* **2001**, *17*, 217–219. [CrossRef]

29. Michel, A.; Villemant, B. Determination of halogens (F, Cl, Br, I), sulfur and water in seventeen geological reference materials. *Geostand. Newsl. J. Geostand. Geoanal.* **2003**, *27*, 163–171. [CrossRef]

30. Shimizu, K.; Itai, T.; Kusakabe, M. Ion Chromatographic Determination of Fluorine and Chlorine in Silicate Rocks Following Alkaline Fusion. *Geostand. Geoanal. Res.* **2006**, *30*, 121–129. [CrossRef]

31. Balcone-Boissard, H.; Michel, A.; Villemant, B. Simultaneous Determination of Fluorine, Chlorine, Bromine and Iodine in Six Geochemical Reference Materials Using Pyrohydrolysis, Ion Chromatography and Inductively Coupled Plasma-Mass Spectrometry. *Geostand. Geoanal. Res.* **2009**, *33*, 477–485. [CrossRef]

32. Shimizu, K.; Suzuki, K.; Saitoh, M.; Konno, U.; Kawagucci, S.; Ueno, Y. Simultaneous determinations of fluorine, chlorine, and sulfur in rock samples by ion chromatography combined with pyrohydrolysis. *Geochem. J.* **2015**, *9*, 113. [CrossRef]

33. Bianchini, G.; Brombin, V.; Marchina, C.; Natali, C.; Godebo, T.R.; Rasini, A.; Salani, G.M. Origin of Fluoride and Arsenic in the Main Ethiopian Rift Waters. *Minerals* **2020**, *10*, 453. [CrossRef]

34. Ro, S.; Hur, S.D.; Hong, S.; Chang, C.; Moon, J.; Han, Y.; Jun, S.J.; Hwang, H.; Hong, S. An improved ion chromatography system coupled with a melter for highresolution ionic species reconstruction in Antarctic firn cores. *Microchem. J.* **2020**, *159*, 105377. [CrossRef]

35. Li, T.; Min, H.; Li, C.; Yan, C.; Zhang, L.; Liu, S. Simultaneous Determination of Trace Fluorine and Chlorine in Iron Ore by Combustion-Ion Chromatography (C-IC). *Anal. Lett.* **2021**, *54*, 2498–2508. [CrossRef]

36. Pereira, L.S.F.; Enders, M.S.P.; Iop, G.D.; Mello, P.A.; Flores, E.M.M. Determination of Cl, Br and I in soils by ICP-MS: Microwave-assisted wet partial digestion using H_2O_2 in an ultra-high pressure system. *J. Anal. At. Spectrom.* **2018**, *33*, 649–657. [CrossRef]

37. He, T.; Hu, Z.; Zhang, W.; Chen, H.; Liu, Y.; Wang, Z.; Hu, S. Determination of Cl, Br, and I in Geological Materials by Sector Field Inductively Coupled Plasma Mass Spectrometry. *Anal. Chem.* **2019**, *91*, 8109–8114. [CrossRef] [PubMed]

38. Zhang, W.; Hu, Z. Recent advances in sample preparation methods for elemental and isotopic analysis of geological samples. *Spectrochim. Acta B* **2019**, *160*, 105690. [CrossRef]

39. Yin, X.; Wang, X.; Chen, S.; Ma, Y.; Kun, G.; Zeng, Z. Trace Element Determination in Sulfur Samples Using a Novel Digestion Bomb Prior to ICP-MS Analysis. *Atom. Spectrosc.* **2018**, *39*, 137–141. [CrossRef]

40. Wang, X.; Yin, X.; Zeng, Z.; Chen, S. Multi-element Analysis of Ferromanganese Nodules and Crusts by Inductively Coupled Plasma Mass Spectrometry. *Atom. Spectrosc.* **2019**, *40*, 153–160. [CrossRef]

41. Tan, X.; Wang, Z. General High-Pressure Closed Acidic Decomposition Method of Rock Samples for Trace Element Determination Using Inductively Coupled Plasma Mass Spectrometry. *J. Anal. Chem.* **2020**, *75*, 1295–1303.

Journal of
*Marine Science
and Engineering*

Article

Near-Bottom Magnetic Anomaly Features and Detachment Fault Morphology in Tianxiu Vent Field, Carlsberg Ridge, Northwest Indian Ocean

Shuang Du [1,2], Zhaocai Wu [2,*], Xiqiu Han [1,2,3,*], Yejian Wang [1,2], Honglin Li [2] and Jialing Zhang [2,3]

[1] Institute of Sedimentary Geology, Chengdu University of Technology, Chengdu 610059, China
[2] Key Laboratory of Submarine Geosciences & Second Institute of Oceanography,
Ministry of Natural Resources, Hangzhou 310012, China
[3] Ocean College, Zhejiang University, Zhoushan 316021, China
* Correspondence: wuzc@sio.org.cn (Z.W.); xqhan@sio.org.cn (X.H.)

Abstract: As a product of hydrothermal mineralization at spreading centers, seafloor massive sulfides (SMS) have become a research hotspot in the field of prospecting and exploring deep-sea mineral resources owing to their enrichment of various strategic metals. Since hydrothermal circulation changes the magnetic properties of host rocks and can generate magnetic anomalies, near-bottom magnetic surveying is an effective method to determine magnetic anomaly features of the seafloor. This technology has been applied to the detection of SMS deposits, in addition to its use in understanding hydrothermal fluid flow conduits and associated hydrothermal alterations. The Tianxiu Vent Field (TVF) is a detachment-fault-controlled, ultramafic-associated hydrothermal system located on the Carlsberg Ridge, Northwest Indian Ocean. During China's DY57th cruise in 2019, near-bottom magnetic data were collected by an autonomous underwater vehicle. In this paper, we use bathymetric and magnetic data, as well as rock sampling information, to analyze and discuss the magnetic anomaly features of the TVF region. Then, we apply 2.5D magnetic anomaly profile forward modeling to determine the shallow magnetic structure and the pattern of detachment faults in the subsurface. Our results show that TVF is characterized by a significant positive magnetic anomaly, where stronger magnetization exists in the area with active hydrothermal vent clusters. The detachment fault has a dip of less than 30° at shallow depths, which steepens to a dip of ~70° at depths of around 300 m.

Keywords: Tianxiu Vent Field; ultramafic-associated hydrothermal systems; near-bottom magnetic anomaly; detachment fault

Citation: Du, S.; Wu, Z.; Han, X.; Wang, Y.; Li, H.; Zhang, J. Near-Bottom Magnetic Anomaly Features and Detachment Fault Morphology in Tianxiu Vent Field, Carlsberg Ridge, Northwest Indian Ocean. *J. Mar. Sci. Eng.* **2023**, *11*, 918. https://doi.org/10.3390/jmse11050918

Academic Editor:
Assimina Antonarakou

Received: 4 March 2023
Revised: 7 April 2023
Accepted: 13 April 2023
Published: 25 April 2023

1. Introduction

Since the first observation of seafloor hydrothermal activity at the East Pacific Rise in 1977 [1], modern seafloor hydrothermal activity and the accompanying seafloor massive sulfides (SMS) have attracted widespread attention from the scientific community [2]. At present, the main techniques for SMS exploration include plume and water anomaly detection, geological sampling, and geophysical exploration. Among these, geophysical methods are important for understanding the spatial distribution of sulfide deposits and hydrothermal circulation processes [3]. Magnetic surveys of seafloor hydrothermal fields have revealed that the hydrothermal alteration of rocks has led to magnetic changes under the influence of hydrothermal fluids [4]. The hydrothermal areas of different bedrock types exhibit different magnetic anomaly characteristics. For basalt-hosted hydrothermal systems, oceanic crust rocks—such as basalt, diabase, and gabbro—present reduced or weakly magnetic anomalies owing to hydrothermal alteration or thermal demagnetization [5–7]. Ultramafic-hosted hydrothermal systems present strong positive magnetic anomalies due to the abundant magnetite produced by the serpentinization of peridotite [8,9]. In addition, the high hydrogen content of high-temperature hydrothermal fluids maintains a strongly reducing environment, which protects magnetite from oxidation [10]. Therefore, the acquisition of near-bottom magnetic fields when exploring and evaluating seafloor sulfide resources is becoming increasingly prominent [11]. In addition, the investigation of magnetic anomalies is not only conducive to examining the different types of seafloor hydrothermal systems [12–14], but also reflects the geometry and distribution of this subsurface crustal structure of active hydrothermal vent systems. As a result, a new research perspective is emerging for systematically exploring the hydrothermal circulation mechanism [15,16]. For example, Tivey et al. [17] analyzed the high-resolution near-bottom magnetic profiles obtained through a near-bottom draped survey over the active Trans-Atlantic Geotraverse (TAG) mound. They found a very-short-wavelength (<100 m) magnetic low directly over the active mound, which was interpreted as a subsurface alteration pipe beneath the active mound or a thermally demagnetized upflow zone. Szitkar et al. [18] reported that both the Rainbow and Ashadze hydrothermal areas on the mid-Atlantic ridge exhibited strong magnetization. This observation reflects the presence of a wide mineralized zone beneath these sites, the stockwork, where several chemical processes concur to create and preserve strongly magnetized magnetite. Tontini et al. [19] applied 3D focused inversion for the near-bottom magnetic data of the hydrothermal system of Brothers Volcano in New Zealand, South Pacific Ocean. This result showed, in particular, how the subsurface 3D magnetization distribution correlates with different vents' field characteristics at focused and diffuse sites. Galley et al. [16] applied minimum-structure inverse modeling to the near-bottom magnetic data collected in Solwara I of the East Manus Basin in the coastal area of Papua New Guinea, imaging the entire structure of a high-temperature convection column, also identifying the top of the underlying magma chamber. However, although some progress has been made, due to the associated technical difficulties and high cost, the application of near-bottom magnetic anomaly surveying at mid-ocean ridges has still been very limited to date.

The Tianxiu Vent Field (TVF) is a detachment-fault-controlled ultramafic-associated hydrothermal system located on Carlsberg Ridge [20]. However, the morphology of the detachment fault and spatial distribution features of the hydrothermal system are still unclear. In 2019, during the China DY57th cruise, we collected near-bottom magnetic and bathymetric data in the Tianxiu Vent Field and its surrounding region by using the "Qianlong III" Autonomous Underwater Vehicle (AUV). In this paper, we analyze the features and sources of near-bottom magnetic anomalies in the Tianxiu Vent Field using near-bottom magnetic and bathymetric data, seafloor observation, and geological sampling. Moreover, the 2.5 dimensional (2.5D) forward modeling of magnetic anomaly profiles [21,22] is used, with the aim of revealing the shallow magnetic structure of the TVF and the geometry of subsurface detachment faults. Our work provides a valuable basis for the establishment of geological models for the seafloor polymetallic sulfide deposits distributed in the TVF and its surrounding regions.

2. Geological Background

The slow-spreading Carlsberg Ridge is located in the Northwest Indian Ocean within 2° S–10° N, with a total length of about 1500 km and a full spreading rate of 22–32 mm/a [23,24]. The Tianxiu Vent Field (63°50′ E, 3°41′ N) is a typical off-axis vent system, and lies on the south slope of the asymmetric expansion segment of Carlsberg Ridge, with an off-axis of about 5 km at a water depth of 3000–3600 m (Figure 1a,b). Various rock types were collected in hydrothermal sites and the surrounding region, including basalt, peridotite, and gabbro [25]. A total of 8 hydrothermal black smoker clusters were observed in the study area, which are all developed within 300 m of the termination of a detachment fault. Among them, 3 clusters (A1, A2, and A3) were active and emitting hydrothermal fluid, and 5 clusters (E1, E2, E3, E4, and E5) were extinct (Figure 1b,c) [26].

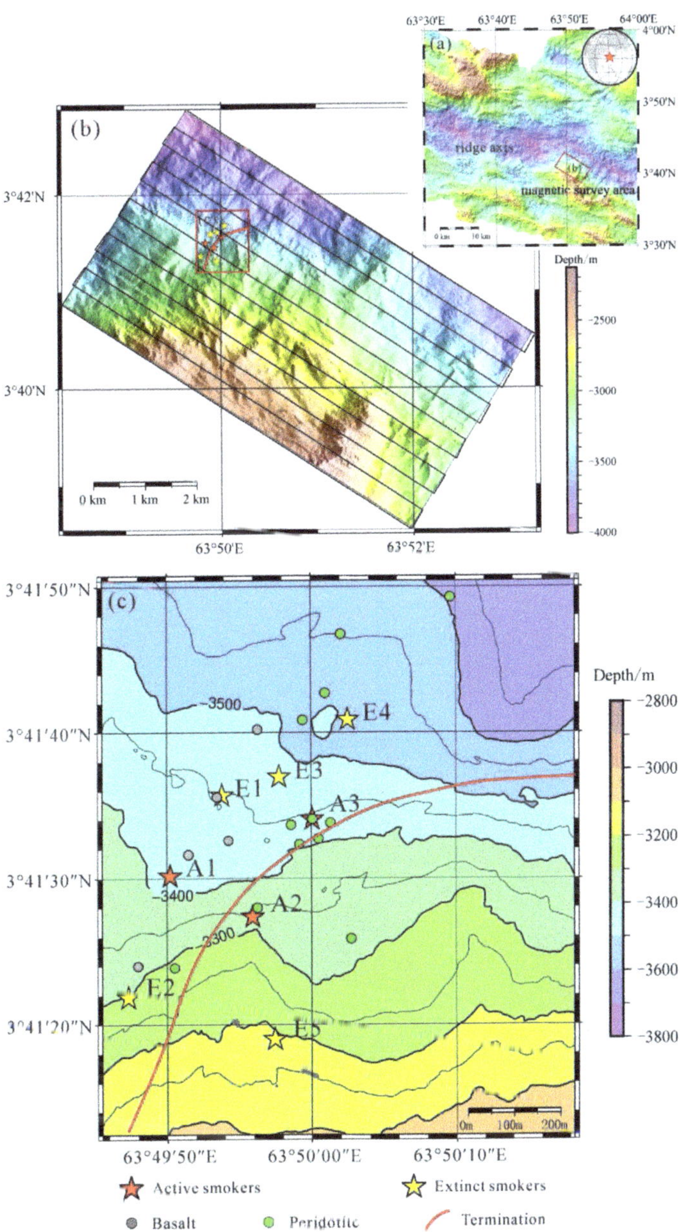

Figure 1. Bathymetry map of the survey area on the Carlsberg Ridge and location of the Tianxiu Vent Field (TVF) (bathymetry data from China DY24 cruise; resolution 50 m.). (**a**) Regional location of the Carlsberg Ridge and the survey area; (**b**) bathymetry map showing the survey area and the near-bottom magnetic detection lines (black solid lines); (**c**) bathymetry map of TVF showing the locations of sulfide chimneys (stars), sampling stations (dots), and the detachment fault (solid red line) inferred from topography and geological sampling.

3. Data Collection and Processing

High-resolution vector magnetic field data were collected using the three-axis fluxgate magnetometer system installed at the tail of the AUV. The vehicle posture (heading, roll, and pitch) data of the AUV were collected by the inertial navigation system installed in the middle of the AUV. A total of 12 near-bottom magnetic survey profiles were designed in the study area. Each line was 8000 m long, and the spacing between the profiles was 400 m. The AUV operated at the nominal altitude of 100 m above the seafloor, which varies between 50 to 150 m due to the complex terrain.

The near-bottom magnetic data processing primarily involves corrections for the vehicle-induced field and the normal magnetic field. The former mainly involved eliminating the influence of the AUV magnetic field from the magnetic data. In this study, the calibration operation and correction methods of Honsho et al. [27] and Bloomer et al. [28] were applied for the carrier magnetic disturbance correction. The latter was performed on the total magnetic field by using the international geomagnetic reference field (IGRF). Considering the short duration of the near-bottom magnetic measurements and the small effect of diurnal geomagnetic variation on the near-bottom magnetism relative to the thousands of nT amplitude variations in the near-bottom magnetism, no such correction was conducted for the data.

To eliminate the high-frequency short-wavelength noise of the magnetic anomaly caused by inconsistent terrain clearance, the data were filtered using a Butterworth low-pass filter to eliminate the high mutational sites. Next, the magnetic anomaly was converted to the new observation height of a terrain constant altitude difference (100 m) after filtering using the COMPU-DRAPETM technique [29], realizing height correction. Subsequently, low-pass filtering of the magnetic anomaly was performed to smooth the data. Figure 2 shows the processing result of the representative profile L7. After all of the 12 profiles were processed, the resulting magnetic anomaly data were interpolated to form grid data with 90 m spacing.

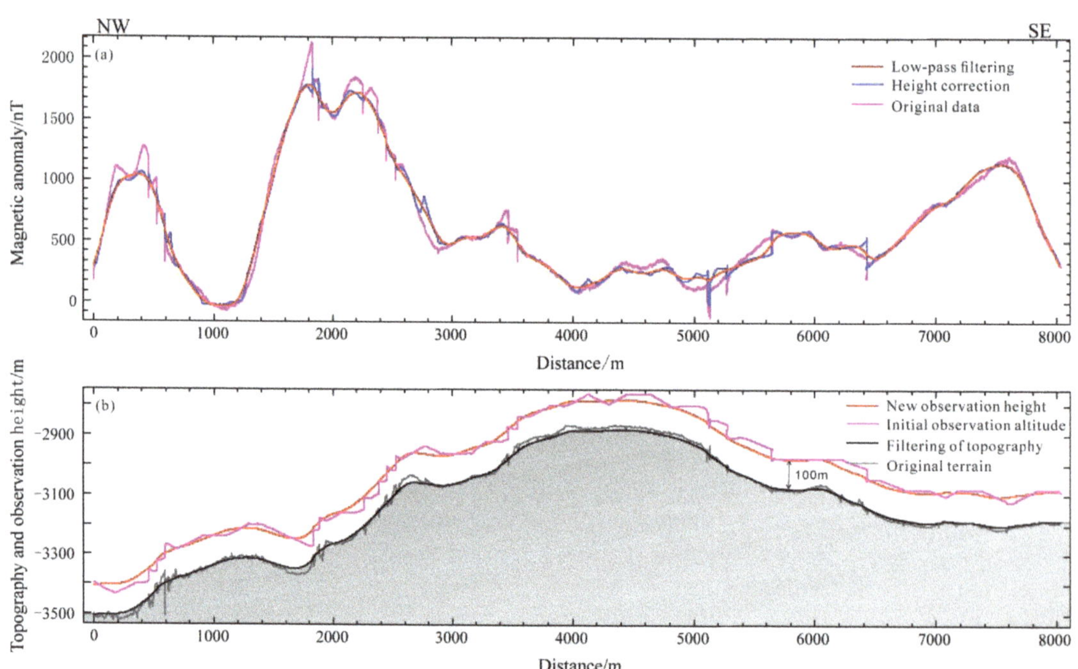

Figure 2. Example of a typical measurement line (L7) after magnetic anomaly height correction and low-pass filtering. (**a**) Magnetic anomaly data and (**b**) bathymetric profile and observation height.

The geological structures and hydrothermal fluid channels beneath the hydrothermal vents are of great research interest. Consequently, two survey lines (L7 and L8), which run through Tianxiu active hydrothermal smoker clusters, were chosen for the 2.5D forward modeling of the magnetic anomaly profile (the profile position is the red straight line in Figure 3a). To accomplish this, a base geological model was first constructed to calculate the magnetic responses based on a standard oceanic crust model as well as a mid-ocean ridge detachment fault model. Then, the semi-automatic trial-and-error method was used to adjust the distribution of seafloor magnetic structures, which were limited by lithology and rock sampling. The method's accuracy was tested by comparing the error between the model magnetic response and the observed data to ensure the best fit [21,22,30]. The forward modeling in this paper was carried out using the GM-SYS module of Geosoft's Oasis montaj software. GM-SYS Profile is a program that calculates the gravitational and magnetic response from a geological profile model. It provides an easy-to-use interface for interactively creating and debugging models to create rapid geological models. The following hypotheses were also made before forward modeling: (a) The geomagnetic inclination of the model field was considered to be consistent with that of the Earth's magnetic field. In other words, the induced and residual magnetizations were considered comprehensively as induced magnetization; therefore, the set magnetic susceptibility values differ from the actual measured magnetic susceptibility of rock, and were only used for the interpretation of the correspondence between the geological body and the magnetic susceptibility. (b) The magnetic susceptibility within lithologic blocks is uniform. According to IGRF, the magnetic anomaly calculation parameters were set as follows: total geomagnetic field = 38312 nT, declination = $-2.7865°$, inclination = $-7.5445°$, and azimuth angle of the survey line = 122.6°.

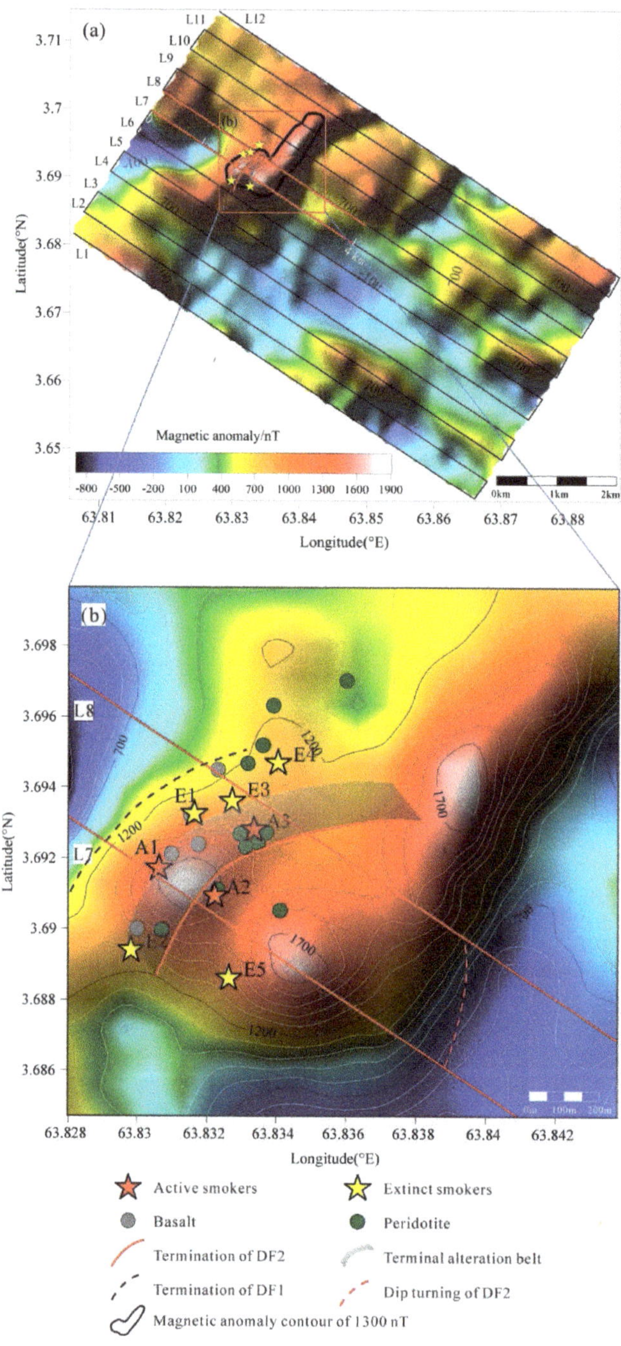

Figure 3. Distribution of magnetic anomalies in the TVF and its surrounding regions. (**a**) The survey area and (**b**) the TVF area. The red straight lines show the segments of profiles L7 and L8 used for forward modeling.

4. Results and Discussion

4.1. Magnetic Anomaly Features and Sources

The magnetic anomalies in the survey area vary from −810 to 2010 nT (Figure 3a). In the central and southern parts of the survey area, they are relatively low and vary from −200 to 200 nT. In the western and northern parts, they are characterized by positive magnetic anomalies, varying from 1000 to 2000 nT. The area with magnetic anomalies exceeding 1300 nT is approximately 0.9 km^2. The surface of the footwall of the detachment fault evidenced by the presence of serpentinized peridotites generally presents a strong positive magnetic anomaly. There are three peaks located in the south, southwest, and north of the area reaching 1800 nT. Conversely, the intensity of the magnetic anomaly of the hanging wall of the detachment fault evidenced by the presence of basalts is significantly reduced (Figure 4b). The positive magnetic anomaly intensity of the hanging wall of the detachment fault decreased significantly. The line of the E4-E3-E1-A1-E2 hydrothermal vents is close to the 1300 nT contour. The intensity of the magnetic anomaly gradually decreases to the northwest from this contour until it crosses a NE–SW magnetic anomaly gradient boundary at the 200 m position on the northwest side of A1 and quickly reduces to less than 1000 nT.

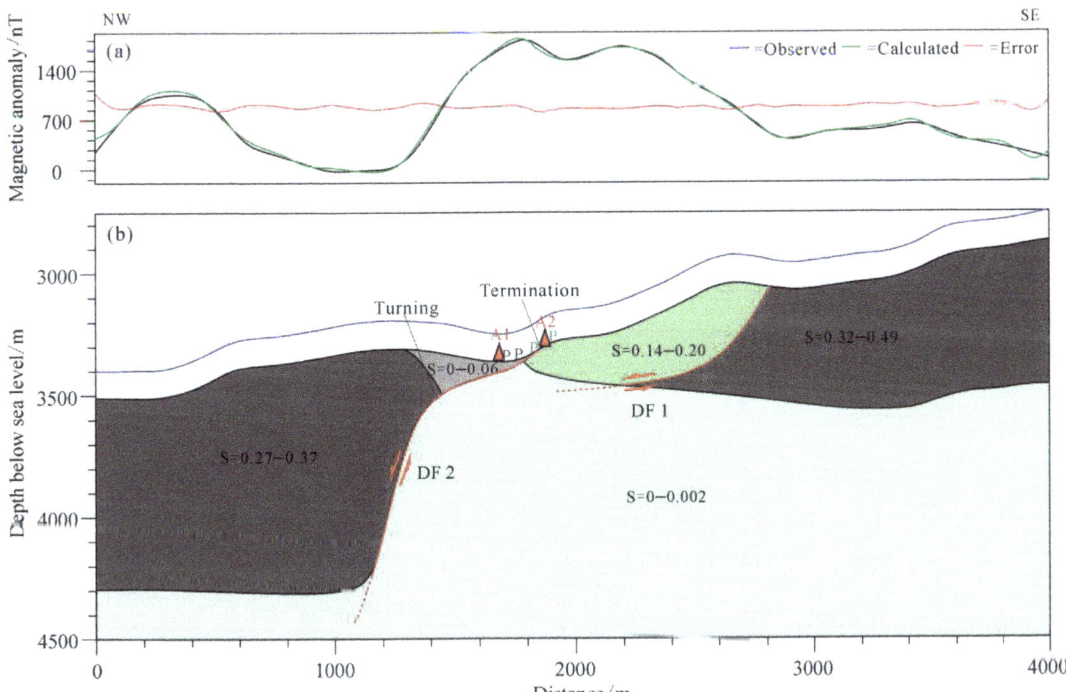

Figure 4. Forward model of the magnetic anomaly of the L7 profile in the TVF (the profile position is the red straight line in Figure 3a). (**a**) Observed and calculated magnetic anomaly and (**b**) the subsurface magnetic structure and detachment fault morphology. In Figure (**b**), the blue line denotes the observation surface, which was 100 m away from the bottom and the red triangle represents the projection positions of black smokers A1 and A2 on the profile. The gray "P" and the green "P" refer to the projection positions of basalt and peridotite sampling stations on the profile. The red solid line corresponds to the modeled detachment fault distribution, while the red dotted line represents the deduced part of the detachment fault. S denotes the magnetic susceptibility.

The observation of thin sections of rock samples under a microscope shows that the peridotite samples have been subject to severe serpentinization, with abundant magnetite grains present in serpentine veins [25]. Therefore, it is considered that the high magnetic anomalies observed in the footwall of the detachment fault in the TVF are caused by the serpentinization of peridotite. The peridotite originally in the upper mantle is uplifted to the seafloor during the detaching process, and interacts with seawater to create strongly magnetized magnetite accumulated in the mineralized zones below the hydrothermal sites. Meanwhile, the high temperature environment of hydrothermal eruption will promote the increase in magnetite proportion in serpentinization products [31]. This is consistent with the positive magnetic anomaly of other ultramafic-associated hydrothermal systems [9,31,32]. However, in the northern part of the area near E4, serpentinized peridotite was sampled, but the positive magnetic anomaly was not very strong. Therefore, this region could belong to the hanging wall of the detachment fault, with the presence of peridotite talus originating from the footwall. This is consistent with the topography of the region, which is apron-like. According to the distribution of the magnetic anomaly and rock sampling information, it is determined that the termination of the detachment fault is a 120 m wide alteration belt extending from the southwest to the northeast in the area (Figure 3b). Among the three locations with extremely high magnetic anomalies, hydrothermal black smokers have been observed in the vicinity of both the southwestern and southern ones, while the northeastern region has not been explored yet. We infer that an ultramafic-hosted hydrothermal deposit is likely to exist in this region. This is to be confirmed by future investigations.

4.2. Subsurface Magnetic Structure and Detachment Fault Morphology

Figure 4 shows the forward model of the magnetic anomaly for profile L7. The dark gray areas (magnetic susceptibility: 0.27–0.37, international system of units, denoted as S below) are dominated by mafic rocks, while the light green region (S: 0–0.002) is dominated by peridotite. These two zones are dissected by detachment faults. The green region (S: 0.14–0.20) at the footwall of the detachment fault may represent the oceanic core complex (OCC) dominated by serpentinized mantle peridotite and gabbro, with a thickness of ~300 m and a width of ~1000 m. The wedge-shaped light gray region at the hanging wall near the termination of the detachment fault is ~350 m wide and ~100 m thick, with a magnetic susceptibility reduced to 0–0.06. Based on geological sampling, this region is dominated by hydrothermally altered basalt, which explains the decrease in magnetism. Figure 5 illustrates the forward model of the magnetic anomaly for profile L8, which is similar to the model for L7. The dark-gray region (S: 0.18–0.30) corresponds to the layer dominated by mafic rocks, while the light-green region (S: 0–0.002) denotes the layer dominated by peridotite. This may represent the green area of the OCC dominated by serpentinized mantle peridotite and gabbro (S: 0.14–0.20), with an average thickness of ~280 m and a width of ~800 m. The light-gray area dominated by altered basalt is ~500 m wide and ~150 m deep, with a magnetic susceptibility reduced to 0–0.07. In both models, the peridotite alteration zone on the superficial layer extends about 100 m to the northwest beyond the termination, which supports the previous inference that the termination is an alteration zone.

Based on the obtained results of forward modeling of a magnetic anomaly in the model, it can be argued that the detachment fault (DF2) slipped from 2800 m to 1850 m along the NW in the horizontal direction on the L7 profile, and the detaching distance is about 950 m. The superficial dip angle of the fault is ~15°, changing suddenly to 70° at a depth of 300 m. The fault has extended towards deep areas for more than 1000 m. On the L8 profile, the detachment fault (DF2) has slipped from 2660 m to 2050 m along the NW in the horizontal direction, with a detaching distance of about 600 m. The superficial dip angle is ~30°, changing to 70° at a depth of 220 m. The fault extended towards deep areas for more than 800 m. The now-deactivated fault (DF1) may be located on the magnetization boundary between the OCC and the basaltic host-rock in the footwall of the two profiles. This fault is hidden below the detachment surface of DF2, and the dip angle is very small (~8°). The location where the dip of DF2 turns in the L7 and L8 profile models is located 350 m northwest of the DF2's termination in the planar projection, and the termination of DF1 is about 700 m southeast of the DF2's termination in the planar projection (marked as a black dashed line and red dashed line in Figure 3b, respectively). They both roughly coincide with the magnetic anomaly gradient boundary, indicating that the magnetic distribution of the hydrothermal sites can reflect the subsurface morphology of the detachment fault. The range of high magnetic anomalies on the L7 profile is wider than that on the L8 profile, with two adjacent maximum points of magnetic anomalies. Additionally, the detaching range of the detachment fault is wider, the extension depth is deeper, and the alteration range of the footwall peridotite and basalt is wider, indicating that the fracture structure below L7 is more developed than that below L8.

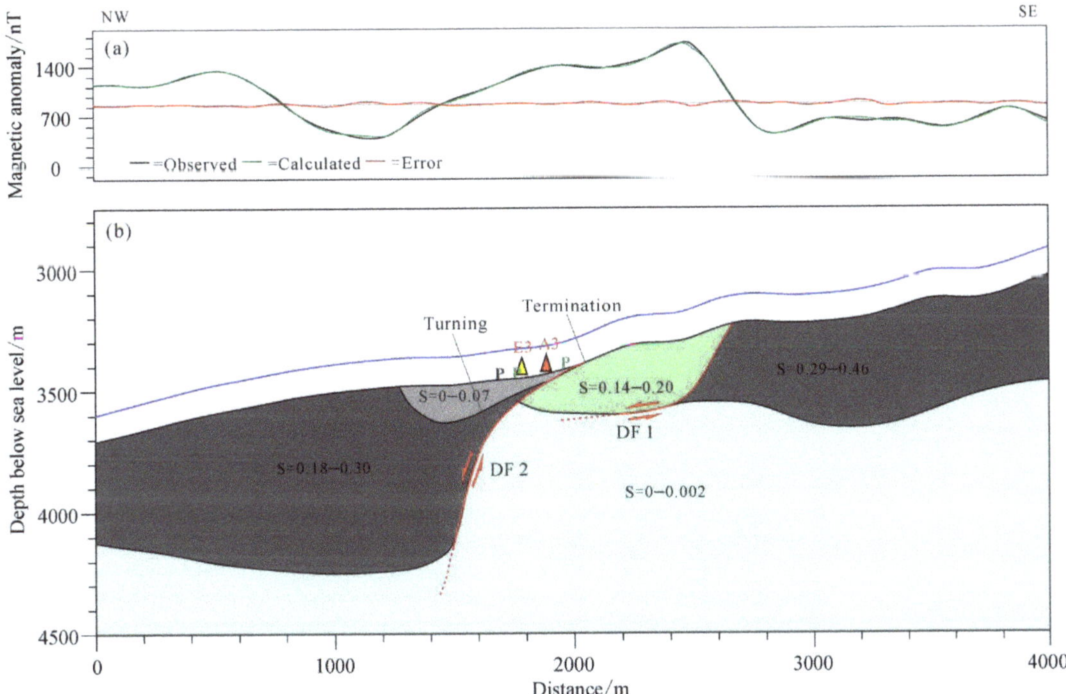

Figure 5. Forward model of the magnetic anomaly of the L8 profile in the TVF (the profile position is the red straight line in Figure 3a). (**a**) Observed and calculated magnetic anomaly and (**b**) the subsurface magnetic structure and detachment fault morphology. In Figure (**b**), the blue line denotes the observation surface, which was 100 m away from the bottom. The red and yellow triangles represent the projection positions of black smokers A3 and E3 on the profile. The gray "P" and the green "P" denote the projection positions of basalt and peridotite sampling stations on the profile, respectively. The red solid line corresponds to the modeled detachment fault distribution, while the red dotted line represents the deduced part of the detachment fault. S denotes the magnetic susceptibility.

In comparison to general detachment faults with a low angle and large displacement [33], the detachment fault (DF2) in TVF has a short detachment surface and a steep dip angle, with a morphology similar to that of the detachment fault in Rainbow hydrothermal areas [34]. According to Mccaig et al. [35], different types of hydrothermal systems are related to the evolution stages of the detachment fault: (1) In the early detaching stage, ultra-mafic rocks are not exposed to the seabed, which might produce hydrothermal vents similar to the TAG type in the Atlantic. Most of these vents are located in basalt at the hanging wall of the fault, and fluids hardly penetrate the footwall of the fault. (2) With continuous detaching, the deep gabbro and peridotite are exposed, inducing the development of Rainbow-like high-temperature hydrothermal vents at the ultrabasic basement of the footwall. (3) With further detaching of the footwall and increasing off-axis distances, the heat source might gradually cool, causing high-temperature hydrothermal vents to shift to ultra-mafic low-temperature hydrothermal vents similar to those in the Lost City. Thus, the detachment fault of TVF is relatively young and still in the second stage, which is the developing stage, conforming to the rolling hinge model of detachment faults [36]. In other words, normal faults with high dip angles begin to detach at deep and shallow regions into low angles through the rotation of the footwall. The development of the detachment fault and secondary fissure structure (e.g., DF1 and DF2) provides channels for hydrothermal circulation, facilitating the upward and lateral transport of hydrothermal fluids to bring the surrounding rocks into a full reaction with hydrothermal fluids. As a result, focused hydrothermal vents have formed in the terminal fracture zone of the detachment fault, which manifest as active and extinct black smokers. This area has high permeability, and the hydrothermal fluid is fully mixed with cold seawater. This suggests that the evolution of hydrothermal vents and sulfide deposits is greatly controlled by the evolution of the detachment fault over the long term.

5. Conclusions

Based on the first near-bottom magnetic surveys of the ultramafic-associated Tianxiu Vent Field on the slow-spreading Carlsberg Ridge, positive magnetic anomalies are detected at the hydrothermal sites. Through the processing and analysis of magnetic anomaly data, combining with the rock samples, two survey lines are selected for 2.5D magnetic anomaly profile forward modeling, leading to the following conclusions:

(1) The area where active hydrothermal vent clusters are located exhibited stronger magnetization due to the presence of magnetite produced by peridotite serpentinization. Based on the correspondence between magnetic anomalies and rock samples, it is presumed that the termination of the detachment fault is a terminal alteration belt approximately 120 m wide.

(2) The major detachment fault (DF2) running through the hydrothermal sites has a relatively short detachment surface, extending about 900 m from southeast to northwest. Its shallow dip angle is less than 30° and increases to approximately 70° at a depth of ~300 m. The fault has extended downward for more than 1000 m. There might be an inactive detachment fault with a low angle (about 8°) beneath the DF2's detachment surface. The turning point of the dip angle in the deep areas of the DF2 corresponds to the gradient boundary of the near-bottom magnetic anomaly, indicating that the magnetic anomalies can reflect the deep morphology of the detachment fault.

(3) There may be undiscovered hydrothermal vents near sites with high magnetic anomalies about 600 m east of Tianxiu hydrothermal black smokers, where hydrothermal activity may have been happening. This area is recommended for further investigation.

Author Contributions: Conceptualization, Z.W. and X.H.; methodology, Z.W. and X.H.; validation, Z.W., X.H., Y.W. and H.L.; formal analysis, Z.W. and X.H.; investigation, Z.W., X.H. and Y.W.; resources, Z.W. and X.H.; data curation, Z.W. and X.H.; writing—original draft preparation, S.D.; writing—review and editing, S.D., Z.W., X.H., H.L. and J.Z.; supervision, Z.W. and X.H.; project administration, Z.W. and X.H.; funding acquisition, Z.W. and X.H. All authors have read and agreed to the published version of the manuscript.

Funding: This research was funded by the National Key Research and Development Program of China (No. 2021YFF0501304 and No. 2021YFF0501301), the China Ocean Mineral Resources R&D Association project (DY135-S2-1), the Talent Program of Zhejiang Province (Grant No. R51003) and Natural Science Youth Fundation of Zhejiang Province (No. LQ19D060006).

Institutional Review Board Statement: Not applicable.

Informed Consent Statement: Not applicable.

Data Availability Statement: Not applicable.

Acknowledgments: We would like to thank the captain, the crew, and the scientific parties onboard during the DY57th cruise in 2019. We are grateful to all the reviewers and editors for their contributions to this paper. We would like to thank Mingcai Hou of Chengdu University of Technology for his contribution to this article.

Conflicts of Interest: The authors declare no conflict of interest.

References

1. Corliss, J.B.; Dymond, J.; Gordon, L.I.; Edmond, J.M.; Von Herzen, R.; Ballard, R.D.; Green, K.; Williams, D.; Bainbridge, A.; Crane, K. Submarine Thermal Springs on the Galapagos Rift. *Science* **1979**, *203*, 1073–1083. [CrossRef]
2. Hannington, M.D.; de Ronde, C. Sea-Floor Tectonics and Submarine Hydrothermal Systems. In *One Hundredth Anniversary Volume*; GeoScienceWorld: McLean, VA, USA, 2005; Volume 1905, pp. 111–141.
3. Tao, C.; Li, H.; Jin, X.; Zhou, J.; Wu, T.; He, Y.; Deng, X.; Gu, C.; Zhang, G.; Liu, W. Seafloor hydrothermal activity and polymetallic sulfide exploration on the southwest Indian ridge. *Chin. Sci. Bull.* **2014**, *59*, 2266–2276. [CrossRef]
4. Rona, P.A.; Hannington, M.D.; Raman, C.V.; Thompson, G.; Tivey, M.K.; Humphris, S.E.; Lalou, C.; Petersen, S. Active and Relict Sea-Floor Hydrothermal Mineralization at the Tag Hydrothermal Field, Mid-Atlantic Ridge. *Econ. Geol.* **1993**, *88*, 1989–2017. [CrossRef]
5. Kent, D.V.; Gee, J. Magnetic alteration of zero-age oceanic basalt. *Geology* **1996**, *24*, 703–706. [CrossRef]
6. De Ronde, C.E.J.; Hannington, M.D.; Stoffers, P.; Wright, I.C.; Ditchburn, R.G.; Reyes, A.G.; Baker, E.T.; Massoth, G.J.; Lupton, J.E.; Walker, S.L.; et al. Evolution of a Submarine Magmatic-Hydrothermal System: Brothers Volcano, Southern Kermadec Arc, New Zealand. *Econ. Geol.* **2005**, *100*, 1097–1133. [CrossRef]
7. Szitkar, F.; Dyment, J.; Choi, Y.; Fouquet, Y. What causes low magnetization at basalt-hosted hydrothermal sites? Insights from inactive site Krasnov (MAR 16°38′ N). *Geochem. Geophys. Geosyst.* **2014**, *15*, 1441–1451. [CrossRef]
8. Dyment, J.; Tamaki, K.; Horen, H.; Fouquet, Y.; Nakase, K.; Yamamoto, M.; Ravilly, M.; Kitazawa, M. A Positive Magnetic Anomaly at Rainbow Hydrothermal Site in Ultramafic Environment. In *AGU Fall Meeting Abstracts*; American Geophysical Union: Washington, DC, USA, 2005; Volume 2005, p. OS21C-08.
9. Tivey, M.A.; Dyment, J. The Magnetic Signature of Hydrothermal Systems in Slow Spreading Environments. In *Diversity Of Hydrothermal Systems On Slow Spreading Ocean Ridges*; Rona, P.A., Devey, C.W., Dyment, J., Murton, B.J., Eds.; American Geophysical Union: Washington, DC, USA, 2010; pp. 43–66.
10. Charlou, J.L.; Donval, J.P.; Konn, C.; OndréAs, H.; Fouquet, Y.; Jean-Baptiste, P.; Fourré, E. High Production and Fluxes of H_2 and CH_4 and Evidence of Abiotic Hydrocarbon Synthesis by Serpentinization in Ultramafic-Hosted Hydrothermal Systems on the Mid-Atlantic Ridge. *Geophys. Union Geophys. Monogr. Ser.* **2010**, *188*, 265–296.
11. Kowalczyk, P. Geophysical Exploration for Submarine Massive Sulfide Deposits. In Proceedings of the OCEANS'11 MTS/IEEE KONA, Waikoloa, HI, USA, 19–22 September 2011.
12. Wu, T.; Tao, C.; Liu, C.; Li, H.; Wu, Z.; Wang, S.; Chen, Q. Geomagnetic Models and Edge Recognition of Hydrothermal Sulfide Deposits at Mid-ocean Ridges. *Mar. Georesour. Geotechnol.* **2015**, *34*, 630–637. [CrossRef]
13. Tivey, M.A. High-resolution magnetic surveys over the Middle Valley mounds, northern Juan de Fuca Ridge. *Proc. Ocean. Drill. Program Sci. Results* **1994**, *139*, 29–35.
14. Johnson, H.P.; Karsten, J.L.; Vine, F.J.; Smith, G.C.; Schonharting, G. A low-level magnetic survey over a massive sulfide ore body in the Troodos ophiolite complex, Cyprus. *Mar. Technol. Soc. J.* **1982**, *16*, 76–80.
15. Tao, C.; Wu, T.; Liu, C.; Li, H.; Zhang, J. Fault inference and boundary recognition based on near-bottom magnetic data in the Longqi hydrothermal field. *Mar. Geophys. Res.* **2017**, *38*, 17–25. [CrossRef]
16. Galley, C.G.; Jamieson, J.W.; Lelièvre, P.G.; Farquharson, C.G.; Parianos, J.M. Magnetic imaging of subseafloor hydrothermal fluid circulation pathways. *Sci. Adv.* **2020**, *6*, c6844. [CrossRef] [PubMed]
17. Tivey, M.A.; Rona, P.A.; Schouten, H. Reduced crustal magnetization beneath the active sulfide mound, TAG hydrothermal field, Mid-Atlantic Ridge at 26° N. *Earth Planet. Sci. Lett.* **1993**, *115*, 101–115. [CrossRef]
18. Szitkar, F.; Dyment, J.; Fouquet, Y.; Honsho, C.; Horen, H. The magnetic signature of ultramafic-hosted hydrothermal sites. *Geology* **2014**, *42*, 715–718. [CrossRef]
19. Caratori Tontini, F.; de Ronde, C.E.J.; Yoerger, D.; Kinsey, J.; Tivey, M. 3-D focused inversion of near-seafloor magnetic data with application to the Brothers volcano hydrothermal system, Southern Pacific Ocean, New Zealand. *J. Geophys. Res. Solid Earth* **2012**, *117*, B10102. [CrossRef]
20. Han, X.; Wang, Y.; Li, X. First Ultramafic Hosted Hydrothermal Sulfide Deposit Discovered on the Carlsberg Ridge, Northwest Indian Ocean. In Proceedings of the Third InterRidge Theoretical Insitute, Hangzhou, China, 25–27 September 2015.
21. Rasmussen, R.; Pedersen, L.B. End corrections in potential field modeling. *Geophys. Prospect.* **1979**, *27*, 749–760. [CrossRef]
22. Talwani, M.; Heirtzler, J.R. Computation of Magnetic Anomalies Caused by Two Dimensional Structures of Arbitary Shape. In *Computers in the Mineral Industries*; Stanford University Press: Redwood City, CA, USA, 1964; pp. 464–480.
23. Kamesh Raju, K.A.; Chaubey, A.K.; Amarnath, D.; Mudholkar, A. Morphotectonics of the Carlsberg Ridge between 62°20′ and 66°20′ E, northwest Indian Ocean. *Mar. Geol.* **2008**, *252*, 120–128. [CrossRef]

24. Han, X.; Wu, Z.; Qiu, B. Morphotectonic Characteristics of the Northern Part of the Carlsberg Ridge near the Owen Fracture Zone and the Occurrence of Oceanic Core Complex Formation. In Proceedings of the AGU Fall Meeting Abstracts, Washington, DC, USA, 25–27 December 2012; American Geophysical Union: Washington, DC, USA, 2012; Volume 2012, p. OS13B-1722.

25. Chen, Y.; Han, X.; Wang, Y.; Lu, J. Precipitation of Calcite Veins in Serpentinized Harzburgite at Tianxiu Hydrothermal Field on Carlsberg Ridge (3.67° N),Northwest Indian Ocean: Implications for Fluid Circulation. *J. Earth Sci.* **2020**, *31*, 91–101. [CrossRef]

26. Zhou, P.; Han, X.; Wang, Y.; Wu, X.; Zong, T.; Yu, X.; Liu, J.; Li, H.; Qiu, Z. Hydrothermal Alteration of Basalts in the Ultramafic-Associated Tianxiu Vent Field, Carlsberg Ridge. *Mar. Geol.* **2023**, *under review*. [CrossRef]

27. Honsho, C.; Ura, T.; Kim, K. Deep-sea magnetic vector anomalies over the Hakurei hydrothermal field and the Bayonnaise knoll caldera, Izu-Ogasawara arc, Japan. *J. Geophys. Res. Solid Earth* **2013**, *118*, 5147–5164. [CrossRef]

28. Bloomer, S.; Kowalczyk, P.; Williams, J.; Wass, T.; Enmoto, K. Compensation of Magnetic data for Autonomous Underwater Vehicle Mapping Surveys. In Proceedings of the 2014 IEEE/OES Autonomous Underwater Vehicles (AUV), Oxford, MS, USA, 6–9 October 2014.

29. Paterson Norman, R.; Reford Stephen, W.; Kwan Karl, C.H. Continuation of Magnetic Data between Arbitrary Surfaces: Advances and Applications. In *SEG Technical Program Expanded Abstracts 1990*; Society of Exploration Geophysicists: Houston, TX, USA, 1990; Volume 60, pp. 666–669.

30. Lim, A.; Brönner, M.; Johansen, S.E.; Dumais, M.-A. Hydrothermal Activity at the Ultraslow-Spreading Mohns Ridge: New Insights From Near-Seafloor Magnetics. *Geochem. Geophys. Geosyst.* **2019**, *20*, 5691–5709. [CrossRef]

31. Szitkar, F.; Tivey, M.A.; Kelley, D.S.; Karson, J.A.; Früh-Green, G.L.; Denny, A.R. Magnetic exploration of a low-temperature ultramafic-hosted hydrothermal site (Lost City, 30° N, MAR). *Earth Planet. Sci. Lett.* **2017**, *461*, 40–45. [CrossRef]

32. Fujii, M.; Okino, K.; Sato, T.; Sato, H.; Nakamura, K. Origin of magnetic highs at ultramafic hosted hydrothermal systems: Insights from the Yokoniwa site of Central Indian Ridge. *Earth Planet. Sci. Lett.* **2016**, *441*, 26–37. [CrossRef]

33. Escartín, J.; Canales, J.P. Detachments in Oceanic Lithosphere: Deformation, Magmatism, Fluid Flow, and Ecosystems. *Eos Trans. Am. Geophys. Union* **2011**, *92*, 31. [CrossRef]

34. Gràcia, E.; Charlou, J.L.; Radford-Knoery, J.; Parson, L.M. Non-transform offsets along the Mid-Atlantic Ridge south of the Azores (38° N–34° N): Ultramafic exposures and hosting of hydrothermal vents. *Earth Planet. Sci. Lett.* **2000**, *177*, 89–103. [CrossRef]

35. Mccaig, A.M.; Cliff, R.A.; Escartin, J.; Fallick, A.E.; Macleod, C.J. Oceanic detachment faults focus very large volumes of black smoker fluids. *Geology* **2007**, *35*, 935–938. [CrossRef]

36. Buck, W.R. Flexural rotation of normal faults. *Tectonics* **1988**, *5*, 959–973. [CrossRef]

MDPI

St. Alban-Anlage 66

4052 Basel

Switzerland

www.mdpi.com

Journal of Marine Science and Engineering Editorial Office

E-mail: jmse@mdpi.com

www.mdpi.com/journal/jmse